Major Evolutionary Transitions
in Flowering Plant Reproduction

MAJOR EVOLUTIONARY TRANSITIONS IN FLOWERING PLANT REPRODUCTION

EDITED BY

Spencer C. H. Barrett

THE UNIVERSITY OF CHICAGO PRESS
CHICAGO AND LONDON

Spencer C. H. Barrett is University Professor and Canada Research Chair
at the University of Toronto.

The University of Chicago Press, Chicago 60637
The University of Chicago Press, Ltd., London
© 2008 by The University of Chicago
Originally published as *International Journal of Plant Sciences* Volume 169,
Number 1, Special Issue
All rights reserved. Published 2008
Printed in the United States of America

17 16 15 14 13 12 11 10 09 08   1 2 3 4 5

ISBN-13: 978-0-226-03816-2 (paper)
ISBN-10: 0-226-03816-5 (paper)

Library of Congress Cataloging-in-Publication Data
Major evolutionary transitions in flowering plant reproduction / edited by Spencer C. H. Barrett.
    p. cm.
    Includes index.
    ISBN-13: 978-0-226-03816-2 (pbk. : alk.paper)
    ISBN-10: 0-226-03816-5 (pbk. : alk.paper) 1. Angiosperms—Reproduction.
  2. Angiosperms—Phylogeny. I. Barrett, Spencer Charles Hilton
    QK495.A1M35 2008
    571.8'2—dc22
                                                              2008024588

∞ The paper used in this publication meets the minimum requirements of the
American National Standard for Information Sciences—Permanence of Paper for
Printed Library Materials, ANSI Z39.48-1992.

Contents

ASEXUAL REPRODUCTION AND POLYPLOIDY

Int. J. Plant Sci. 169(1):1–5. 2008.
1058-5893/2008/16901-0001$15.00   DOI: 10.1086/522511

MAJOR EVOLUTIONARY TRANSITIONS IN FLOWERING PLANT REPRODUCTION: AN OVERVIEW

Spencer C. H. Barrett

Department of Ecology and Evolutionary Biology, University of Toronto, 25 Willcocks Street, Toronto, Ontario M5S 3B2, Canada

Evolutionary transitions are functionally significant changes in organismal traits that largely result from the action of natural selection. They first appear within populations when novel traits replace the ancestral state because of increased fitness. Evolutionary transitions take on broader significance and can be considered major when newly established traits persist, are maintained through multiple speciation events, and ultimately become well-established features of lineages. The shifts in character state that constitute an evolutionary transition are key elements of biological diversification, and the identification and study of major transitions during the history of life now represent an important research program in evolutionary biology (Maynard Smith and Szathmáry 1995).

Similar character state transitions often occur repeatedly among unrelated lineages, and these situations are of particular interest because they usually indicate equivalent selective mechanisms and convergence in function. Estimating the number of transitions using phylogenetic methods and character reconstructions can provide important information for more in-depth studies of the ecological basis of selection. Multiple transitions can also provide outstanding opportunities to investigate whether similar molecular and developmental mechanisms are responsible, although this area of study is still in its infancy. Another important issue concerns the polarity of transitions and to what extent particular changes are irreversible or evolutionarily labile. The developmental complexity of traits can play an important role in determining the degree of asymmetry, although this is not always the case. Finally, when transitions involve similar suites of characters, determining the order of acquisition of component parts can provide insight into adaptation and the ecological drivers of change. Research on the causes and consequences of character transitions are at the heart of modern evolutionary biology and, to be successful, require the integration of both microevolutionary and macroevolutionary approaches.

Among plant life-history traits, reproductive characters are particularly important in affecting microevolutionary processes and macroevolutionary patterns. These fundamental roles arise because reproductive characters influence genetic transmission, population genetic structure, selection response, and patterns of evolutionary diversification. Major reproductive transitions are often associated with changes to other components of life history and also with modifications in the genetic system. In flowering plants, resource allocation, pollination, and mating influence reproductive success in an integrated manner, and, as a result, functional correlations between these components of reproduction are a pervasive feature of phenotypic evolution.

Studies of reproductive trait transitions therefore should not be conducted in isolation from other changes to phenotype involving nonreproductive traits. In addition, phyletic history, development, and genetic architecture can all influence pathways of change and need to be considered.

Flowering plants exhibit exceptional diversity in floral traits and reproductive mechanisms, with closely related species often possessing different modes of sexual and asexual reproduction, contrasting pollination and mating systems, and diverse gender strategies. This variation was first studied in detail by Charles Darwin (1877) and was later exploited during the 1950s and 1960s by G. Ledyard Stebbins (e.g., 1974), Verne Grant (e.g., 1965), and Herbert Baker (e.g., 1959) during the golden era of plant biosystematics. These workers recognized that important insights into variation and evolution of reproductive traits within families and genera could often be obtained by careful studies of intraspecific variation, especially in wide-ranging species adapted to different ecological circumstances. The development of population biology in the 1970s led to field studies of natural selection (reviewed in Endler 1986) and the development of theoretical models investigating key parameters involved in the selective mechanisms responsible for particular reproductive transitions (e.g., Lloyd and Webb 1992). More recently, comparative and phylogenetic approaches have been used to investigate reproductive transitions in an effort to link macroevolutionary patterns of diversification with the ecology and genetics of species (reviewed in Weller and Sakai 1999).

Studies of major reproductive transitions in flowering plants are today the focus of considerable research in plant evolutionary biology. This volume highlights a selection of current work by leading authors in the field. The contributions feature new research findings, reviews, and synthesis and include diverse approaches for understanding the pathways of reproductive trait evolution in flowering plants. These include comparative and phylogenetic methods, theoretical models, investigations of structure and development, molecular genetics, and experimental studies of the ecology and genetics of wild populations. Given the exceptional reproductive diversity of flowering plants, it has not been possible to be comprehensive, and important major transitions involving flower morphology and development (e.g., evolution of zygomorphy), pollen biology (e.g., evolution of trinucleate pollen), life history (e.g., evolution of monocarpy), and fruit and seed dispersal (e.g., evolution of fleshy fruits) await future treatment. Instead this volume focuses on a selected group of topics, emphasizing how they can be tackled using complementary approaches. The volume is divided into three sections that in turn deal with flowers and pollination, mating patterns and gender strategies, and asexual reproduction and polyploidy.

Flowers and Pollination

Flowering plants exhibit spectacular variation in flower design and display, and much of the functional basis of this diversity is associated with the evolution of pollination systems. Mark Rausher begins the section on flowers and pollination by providing a comprehensive review of transitions in flower color in animal-pollinated lineages, focusing in particular on the change from blue to red flowers. He takes a critical approach to existing evidence about the mechanisms responsible for shifts in flower color and challenges the widespread assumption that these changes are inevitably related to pollinator-mediated selection and adaptation to novel pollinators. Although not disputing that pollinators may play an important role in flower color transitions, his review of the evidence indicates that a convincing case is still to be made. He proposes that some transitions in flower color could result from selection on the pleiotropic effects of flower color alleles by nonpollinating agents. Rausher also points out that transition rates from blue to red flowers are usually asymmetrical and associated with loss-of-function mutations and inactivation of branches of the anthocyanin pathway. His review provides a valuable lesson in the complexities of studying the evolution and adaptive significance of a seemingly simple trait such as flower color.

The next article, by James Thomson and Paul Wilson, also features changes in flower color from blue to red associated with pollinator transitions from hymenoptera, especially bees (melittophily), to pollination by hummingbirds (ornithophily). These transitions are commonplace, especially in western North America, where they have occurred at least 100 times in diverse lineages of herbs. Thomson and Wilson's approach is to consider the ecological and genetic mechanisms that might account for the destabilization of pollination syndromes, using *Penstemon*, *Mimulus*, *Ipomoea*, *Costus*, *Aquilegia*, *Silene*, and *Salvia* as examples. They focus specifically on three main topics: differences in pollen transfer efficiency among bees and birds, the role of mutations with large effects on floral phenotypes, and the ecological conditions that change visitation rates of pollinators and hence the nature of pollen dispersal. Among the various factors reviewed, Thomson and Wilson consider ecological change as the most likely initial driver of pollinator shifts but point out that little concrete information is currently available on precisely what these changes involve.

Risa Sargent and Jana Vamosi take up this topic further in the next article by investigating the extent to which ecological context influences evolutionary transitions in the degree of pollinator specialization. They examine the hypothesis that ecological shifts to environments with different light conditions are accompanied by transitions in pollinator guild. Using phylogenetically independent contrasts and data collected largely from tropical forest environments, they examine the degree to which transitions in canopy position are associated with particular pollinators and pollinator guild size. Their analysis demonstrates that species that tend to occupy the same position in the forest canopy are more closely related than would

be expected by chance, as are species with particular pollinator syndromes (e.g., bee or bird). Transitions to generalist pollination are strongly associated with beetle and fly pollination and with position in the canopy above the forest floor. Their results suggest that evolutionary transitions between specialized and generalized interactions are unlikely to be subject to phylogenetic constraint or specific requirements for particular light environments.

In contrast to several of the major angiosperm reproductive transitions featured in this special volume (e.g., the evolution of selfing and dioecy), remarkably little is known about the evolution of wind pollination from animal pollination. This is surprising because this shift in pollination mode, with at least 65 independent transitions, represents one of the major transformations in the reproductive biology of flowering plants. Jannice Friedman and Spencer Barrett use comparative approaches to investigate the correlated evolution and order of trait acquisition between pollination mode and a range of ecological and reproductive characters. One of their most interesting findings is that wind pollination evolves more frequently in lineages that already possess unisexual flowers, and they propose a novel hypothesis to account for this association. Populations with unisexual flowers may evolve wind pollination as a mechanism of reproductive assurance ensuring more effective pollen dispersal between plants and relieving pollen limitation. Reproductive assurance is usually invoked to explain the evolution of selfing; however, the presence of unisexual flowers would prevent selfing by autonomous self-pollination in most groups. Pollen limitation may therefore promote strikingly different evolutionary transitions in pollination systems, depending on the sexual traits of ancestral populations.

In the next article, Lawrence Harder and Steven Johnson tackle the intriguing problem of why some flowering plants disperse their pollen in groups. The evolution from individual monads to pollen aggregation, including tetrads, polyads, pollen threads, and pollinia, has at least 39 independent origins and therefore represents a significant functional transition in angiosperm pollination. However, the adaptive benefits of pollen aggregation have not been explored either theoretically or empirically. It seems likely that transitions to pollen aggregation require special conditions because diminishing returns through male function during pollination, as well as the genetic benefits of multiple paternity, should favor the dispersal of individual pollen grains. Harder and Johnson explore the reproductive circumstances that are likely to favor the different forms of pollen aggregation and propose that this variation comprises alternative strategies for relieving contrasting limitations on siring ability. They also consider several aspects of plant reproduction that are consequences of the evolution of pollen aggregation, focusing in particular on the evolution of pollinia in orchids and proposing that the ability of orchid pollinia to reduce diminishing returns during pollination may explain the widespread occurrence of deceit pollination in this clade and its exceptional floral diversity.

In the last article in this section, William Friedman, Eric Madrid, and Joseph Williams provide a novel evolutionary and developmental perspective on the structural diversity of female gametophytes in angiosperms. They argue that female gametophytes are iteratively expressed modular entities and that structural diversity results from variation in the relative timing of the establishment of modules as well as ontogenetic events that determine their number and degree of modification from the plesiomorphic condition. By linking structural diversity in the female gametophyte to endosperm biology, they demonstrate that variation in developmental patterns determines variation in endosperm genetics. Friedman and colleagues also review potential selective mechanisms that may drive changes in endosperm genetics. They show that hypotheses based on heterozygosity, ploidy level, and sexual conflict make similar predictions concerning the evolution of female gametophyte development. This article demonstrates that future investigation of evolutionary transitions in female gametophyte development and endosperm genetics cannot be examined independently.

Mating Patterns and Gender Strategies

The evolution of predominant selfing from obligate outcrossing has received more attention than any other reproductive transition in flowering plants. Indeed, Stebbins (1974) suggested that this transition has occurred more often than any other. This section begins with three contrasting articles concerned with various facets of this frequent change in the mating system of populations.

The self-incompatibility polymorphism is the principal and most effective mechanism preventing self-fertilization in hermaphroditic flowering plants. In the first article, Boris Igic, Russell Lande, and Joshua Kohn examine the breakdown of self-incompatibility and its evolutionary consequences. They begin by reviewing the available literature on the frequency distribution of self-incompatibility and find that it is reported from 100 taxonomically diverse families and ca. 39% of angiosperm species. They then consider why self-incompatibility is often lost but rarely, if ever, regained during angiosperm diversification. They suggest that loss of self-incompatibility occurs because transitions to self-compatibility are generally accompanied by losses in allelic diversity at the S-locus, because the variation becomes selectively neutral in self-compatible populations, and because of the accumulation of loss-of-function mutations, for which there is considerable evidence in *Arabidopsis* and *Solanum*. Assuming that the loss of self-incompatibility is irreversible, they develop a theoretical model that examines the evolutionary processes required to maintain self-incompatibility. They show that stable maintenance of self-incompatibility can only occur if it is associated with increased diversification relative to self-compatible lineages, with the balance of transition and diversification rates determining the frequency distribution of mating systems they consider. Lower diversification of self-compatible lineages may occur because mutations causing self-compatibility are commonly associated with increased selfing rates, and this can lead to lower genetic diversity and the possibility of an increased risk of extinction.

In the next article, Stephen Wright, Rob Ness, John Paul Foxe, and Spencer Barrett review the genomic consequences of selfing and outcrossing, picking up on the theme discussed in the preceding article regarding the influence of predominant self-fertilization on the evolutionary fate of populations. They review available genomic data contrasting selfing and outcrossing populations and discuss opportunities for selfing populations to avoid an irreversible decline in fitness and extinction. Transitions to selfing are expected to cause a reduction in effective population size, an increase in fixation rates of slightly deleterious mutations, and a decrease in fixation of advantageous mutations; however, the existing evidence does not suggest a significant reduction in the efficacy of selection associated with high selfing rates. Although the available data are sparse, Wright and colleagues also examine evidence that recombination rates may evolve in response to changes in mating patterns, thus limiting the deleterious effects of inbreeding. The abundance and activity of selfish genetic elements may also be reduced in selfing lineages. A reduction in genomic conflict can increase mean fitness, reduce deleterious mutation rates, and reduce genome size. Using comparative data, Wright and colleagues show that highly selfing species have smaller genomes in comparison with outcrossing relatives, consistent with reduced activity and spread of repetitive elements in inbreeders. One of the main messages of this article is that as genomic data rapidly accumulate over the coming years, there will be exciting new opportunities to test evolutionary theory within a phylogenetic framework using comparative analyses of closely related selfing and outcrossing species.

Early theories on mating systems proposed that group selection favored some optimum level of recombination within species. Today, these ideas are no longer widely accepted, and models based on individual selection are used to explore the evolution of selfing from outcrossing. In the next article, Daniel Schoen and Jeremiah Busch reconsider the importance of group-level selection of mating systems. Their investigation is opportune in light of evidence, discussed earlier in this section, for differences in diversification and extinction rates of predominantly selfing versus outcrossing plants and the general observation that transitions to predominant self-fertilization may be unidirectional. Schoen and Busch develop models of mating system evolution using a metapopulation framework to investigate factors that may operate when group-level selection occurs. A particular focus of these models is to examine situations in which individual and group-level selection oppose one another, as well as those that result in stable mixed mating. They find that if group-level selection in a metapopulation is sufficiently strong, it may limit transition rates from outcrossing to selfing and therefore counteract individual selection for selfing through reproductive assurance. The models generally show that selection among populations can maintain outcrossing through higher extinction rates of selfing groups and through reduced transition rates from outcrossing to selfing. Further studies of the role of multilevel selection in the evolution of self-fertilization will require detailed information on the extent to which transitions to selfing influence population viability and longevity and also how genetic architecture and the details of floral development influence the tempo by which selfing variants can spread.

The remaining two articles in this section consider transitions in gender strategies, particularly the evolution of gender dimorphism from monomorphism. This represents a prominent transition in the sexual systems of plants that, in common with the evolution of selfing from outcrossing, has received considerable theoretical and empirical attention. John Pannell, Marcel Dorken, Benoit Pujol, and Regina Berjano employ microevolutionary approaches to investigate transitions between sexual systems in the annual herb *Mercurialis annua*, in which dioecious, monoecious, and androdioecious populations occur in different parts of the European and North African range. This species complex has provided outstanding opportunities for investigating the ecological and genetic mechanisms driving sexual system evolution. Pannell and colleagues point out that hybridization and polyploidy have played an important role in initiating transitions between sexual systems. However, they also demonstrate that some transitions are not confounded with changes to the genetic system, and these can provide valuable insights into the ecological and demographic mechanisms involved. Investigations of geographical transitions between monoecious and androdioecious populations support a metapopulation model in which differential selection for reproductive assurance during colonization at the regional level plays a key role. Pannell and colleagues also present new experimental data that illustrate the importance of phenotypic plasticity in hermaphrodite sex allocation in regulating male frequencies in androdioecious populations. This contribution illustrates how the resource status of plants and their local mating environment can play critical roles in regulating gender strategies and sex ratios.

A complementary approach to understanding evolutionary transitions in gender strategies is to investigate genera or families that contain sexual system diversity using phylogenetic analysis and the reconstruction of character evolution. In their article, Andrea Case, Sean Graham, Terence Macfarlane, and Spencer Barrett address some of the difficulties associated with inferences about historical transitions in sexual systems using *Wurmbea*, a small genus of monocotyledons from the Southern Hemisphere. *Wurmbea* is divided into two well-supported clades, each defined by geography and variation in sexual system. Case and colleagues explore the influence of tree uncertainty, taxon sampling and extinction, the evolutionary lability of characters, and several other sources of ambiguity for maximum likelihood (ML)–based inferences of sexual system evolution. They find that the interspersion of species across trees that vary in sexual system is the main cause of ambiguity in their historical reconstructions. Another source of uncertainty that they identify concerns the nonmonophyly of two sexually polymorphic species, *Wurmbea dioica* and *Wurmbea biglandulosa*. These geographically widespread taxa have been the subject of detailed ecological and genetic studies over the past two decades aimed at understanding the selective mechanisms driving the evolution of gender dimorphism from monomorphism. Clearly, the finding that these taxa are not monophyletic will have important implications for future interpretation of character transitions. This article provides examples of some of the challenges in linking macroevolutionary pattern with microevolutionary processes for evolutionarily labile traits in plant groups that possess diverse sexual systems. It should caution workers interested in reconstructing character evolution that using these approaches is not always as straightforward as is sometimes assumed.

Asexual Reproduction and Polyploidy

Asexual reproduction and polyploidy are both very widespread among flowering plant families and, in some groups,

are commonly associated. In the final section of this issue, evolutionary transitions to asexual reproduction and genome duplication through polyploidy are examined. In the first article, Jonathan Silvertown evaluates the costs and benefits of asexual and sexual reproduction in plant species that have both reproductive modes. He is intrigued by the fact that the evolutionary transition to clonal reproduction has rarely, if ever, replaced sexual reproduction entirely. Silvertown compares the genotypic diversity of populations, a proxy for the relative success of recruitment through sexual and asexual means, across a range of ecological conditions in 218 species from 74 families to determine why one mode of reproduction may be favored over the other. After controlling for bias resulting from marker type, scale of sampling, sampling design, and the number of populations sampled, Silvertown is able to make several generalizations from his data set. First, most populations in the survey are multiclonal, with populations of aquatic plants, especially those recently founded through vegetative dispersal, often the exception. Second, clonality occurs more commonly than sexual reproduction in older populations maintained by a lack of disturbance, in geographically marginal environments, and also in rare or alien species. Silvertown concludes by proposing that the shift to exclusive clonal reproduction is rare because clonality is not a substitute for sex but, rather, prolongs the time to extinction when restrictive ecological conditions prevent sex from occurring.

In contrast to clonal propagation, the evolution of asexual reproduction through apomixis has occurred frequently in sexual lineages, and in some polyploidy groups this can be obligate, although mixed reproductive modes are more common. Apomixis is reported from many angiosperm families, with particular concentrations of species in the Asteraceae, Poaceae, and Rosaceae. In the next article, Jeannette Whitton, Christopher Sears, Eric Baack, and Sarah Otto provide a comprehensive review of the evolution of apomixis, focusing on its genetic basis and on the population genetic and ecological factors that affect the spread of asexual lineages. They emphasize the importance of occasional bouts of sexual reproduction to the establishment and proliferation of asexuality and, specifically, the role that pollen plays in this context. Evidence indicates that many apomicts retain residual pollen function, providing opportunities for the spread of apomixis through male gametes. Whitton and colleagues also propose several hypotheses to explain the association between gametophytic apomixis and polyploidy and argue that determining whether polyploidy involves autopolyploidy or allopolyploidy may provide critical insights into the causal mechanism(s) responsible for this association. Review of their own work on two North American apomictic complexes in the Asteraceae illustrates some of the challenges in interpreting the origins and spread of apomixis in systems with complex phylogenetic histories. The article concludes by addressing general issues concerning the evolution of sex and the long-standing issue of why it is so prevalent. Future studies of apomictic plants should provide exciting opportunities for investigating these problems through experimental studies of the benefits of sex and the costs of asexuality.

Populations at range limits often reproduce primarily through asexual reproduction, and an important question is the extent to which this results in the dissolution of sex function. This problem has been considered for clonal species, but in the next article, Stacey Lee Thompson, Gina Choe, Kermit Ritland, and Jeannette Whitton investigate the presence and extent of asexuality and recombination within populations of the polyploid apomict *Townsendia hookeri* (Easter daisy) at the extreme northern limit of the species range in Yukon Territory, Canada. Their study nicely illustrates some of the complexities that can arise as a result of the evolutionary transition to asexuality in apomictic lineages. Using genetic markers, surveys of genome size using flow cytometry, estimates of pollen viability, and a novel procedure for estimating long-term recombination, they demonstrate considerable variation within and among populations in ploidal level, male fertility, and the reproductive modes of populations. Of particular significance is their finding that in apparently male sterile polyploid populations, there is evidence of a low level of sexuality with an estimated equilibrium rate of approximately three sexual events every two generations. This study by Thompson and colleagues adds to a growing literature on apomictic polyploid groups that were initially believed to be strictly asexual but instead display cryptic sexuality.

The final article in this volume, by Brian Husband, Barbara Ozimec, Sara Martin, and Lisa Pollock, brings together two recurrent themes featured in earlier articles—polyploidy and mating systems. Because polyploidy affects the entire genome, it is perhaps not surprising that it influences many aspects of the phenotype, including the mating system. However, although it has long been recognized that the evolutionary transition from diploidy to polyploidy may result in correlated changes in mating patterns, the theoretical and empirical evidence is limited and often contradictory. Several theoretical models predict an increase in the rate of self-fertilization in polyploids, and a survey of mating patterns in related diploid and polyploid congeners in this article provides support for this prediction. However, increased selfing was a feature only of allopolyploids and not autopolyploids, raising the important question of what factors limit selfing in autopolyploid species. To address this problem, Husband and colleagues use *Chamerion angustifolium* to compare the magnitude of inbreeding depression in established autopolyploids and neo-

polyploids synthesized using colchicine. They find that the cost of selfing in neopolyploids is negligible compared with extant polyploids but that there is some evidence that inbreeding depression increases with the history of inbreeding. Their results suggest that any initial increase in selfing may be transient in autopolyploids and that selection may ultimately favor mixed or outcrossed mating. An important conclusion from this study is that the costs of selfing in polyploids are likely to be dynamic, changing with the age of the polyploid and history of mating.

Acknowledgments

To conclude this volume, I would like to acknowledge individuals who have assisted in this project. First, I thank the editors of the journal, particularly Larry Hufford and Manfred Ruddat, for encouragement and the opportunity to produce a volume about plant reproduction on a topic of my choice. Second, I thank Dennis Keppeler, managing editor of *IJPS*, for his continuous support and prompt advice throughout the birth and development of this issue. Third, I thank the authors for their essential contributions and efforts to produce a fully integrated and up-to-date volume of high standards. Synergy among authors was greatly facilitated by the use of a dedicated Web site set up and maintained by Bill Cole at the University of Toronto. Finally, on behalf of the authors, I greatly appreciate the efforts of the many individuals who reviewed articles for this volume and by doing so helped to improve the quality of the contributions, including Scott Armbruster, Thomas Bataillon, Christian Brochmann, Deborah Charlesworth, Mark Chase, Timothy Dickinson, Marcel Dorken, Christopher Eckert, Norman Ellstrand, Charles Fenster, Mark Fishbein, David Haig, Lawrence Harder, Donald Levin, Peter Linder, Barbara Mable, Jill Miller, David Moeller, Michael Mogie, John Pannell, Richard Ree, Ophélie Ronce, Paula Rudall, Ann Sakai, Douglas Schemske, Pamela Soltis, James Thomson, Marcy Uyenoyama, Peter van Dijk, Mario Vallejo-Marín, Jana Vamosi, Stephen Weller, Justen Whittall, and Sarah Yakimowski.

Literature Cited

Baker HG 1959 Reproductive methods as factors in speciation. Cold Spring Harbor Symp Quant Biol 24:177–191.

Darwin C 1877 The different forms of flowers on plants of the same species. J Murray, London.

Endler JA 1986 Natural selection in the wild. Princeton University Press, Princeton, NJ.

Grant V, KA Grant 1965 Flower pollination in the Phlox family. Columbia University Press, New York.

Lloyd DG, CJ Webb 1992 The selection of heterostyly. Pages

179–207 *in* SCH Barrett, ed. Evolution and function of heterostyly. Springer, Berlin.

Maynard Smith J, E Szathmáry 1995 The major transitions in evolution. Oxford University Press, Oxford.

Stebbins GL 1974 Flowering plants: evolution above the species level. Belknap, Harvard University Press, Cambridge, MA.

Weller SG, AK Sakai 1999 Using phylogenetic approaches for the analysis of plant breeding system evolution. Annu Rev Ecol Syst 30: 167–199.

Int. J. Plant Sci. 169(1):7–21. 2008.
1058-5893/2008/16901-0002$15.00   DOI: 10.1086/523358

EVOLUTIONARY TRANSITIONS IN FLORAL COLOR

Mark D. Rausher

Department of Biology, Duke University, Durham, North Carolina 27708, U.S.A.

The tremendous diversity in flower color among angiosperms implies that there have been numerous evolutionary transitions in this character. The conventional wisdom is that a large proportion of these transitions reflect adaptation to novel pollinator regimes. By contrast, recent research suggests that many of these transitions may instead have been driven by selection imposed by nonpollinator agents of selection acting on pleiotropic effects of flower color genes. I evaluate the evidence for these alternative hypotheses and find that while there is circumstantial evidence consistent with each hypothesis, there are no definitive examples of flower color evolution conforming to either hypothesis. I also document four macroevolutionary trends in flower color evolution: color transitions rates are often asymmetrical; biases favoring loss of pigmentation or favoring gain of pigmentation are both observed, but bias favoring transition from blue to red flowers seems more common than the reverse bias; transitions from blue to red often involve inactivation of branches of the anthocyanin pathway; and color transitions often involve loss-of-function mutations. Finally, I discuss how these trends may be related to one another.

Keywords: floral evolution, anthocyanin, natural selection, pollination biology, pollinator choice, pleiotropy.

Introduction

Angiosperms exhibit a tremendous diversity of flower colors, with sister species often differing in the intensity, hue, or patterning of the corolla. This diversity implies that there have been numerous evolutionary transitions in flower color. The observation that floral color is often correlated with other floral traits, resulting in the common recognition of "pollination syndromes" (Faegri and van der Pijl 1966; Fenster et al. 2004), suggests that many of these transitions have been adaptive. Moreover, the apparent importance of showy flowers in attracting pollinators has led to the common interpretation that pollinators are the primary selective agents influencing flower color and that transitions to different colors represent adaptation to different suites of pollinators, a proposition I call the "conventional wisdom" (Faegri and van der Pijl 1966; Grant 1993; Fenster et al. 2004).

In the first part of this article, I inquire into the causes of evolutionary transitions in flower color. I first evaluate the evidence supporting the conventional wisdom. I then consider evidence supporting alternative interpretations, including the possibilities that many flower color transitions are nonadaptive or that many reflect natural selection on pleiotropic effects of genetic variants that affect flower color.

The astounding variety of floral hues, intensity, and patterns of pigmentation makes it appear as if there are few constraints on the evolution of flower color. However, constraints become apparent when one examines macroevolutionary trends in floral color transitions. In the second part of this article, I discuss how properties of flower color genes

and the biochemical pathways they encode may contribute to establishing these trends.

Pollinators as Selective Agents on Flower Color

The idea that evolutionary change in flower color reflects adaptation to novel pollinators can be traced back at least as far as Darwin (for historical review, see Fenster et al. 2004). The primary evidence supporting this contention is the existence of "pollination syndromes," groups of floral traits that occur together typically in plants pollinated by a particular agent. Examples include (1) bird-pollinated flowers, which are typically red or orange and have elongated floral tubes, reduced floral limbs, exserted stigmas, and copious dilute nectar; (2) bee-pollinated flowers, which are typically blue or purple and have short, wide tubes, wide limbs, inserted stigmas, and small amounts of concentrated nectar; and (3) moth-pollinated flowers, which are typically white and fragrant, have long floral tubes, and open at night.

Although the generality of floral syndromes has been questioned (Robertson 1928; Ollerton 1996, 1998; Waser et al. 1996) and, clearly, not all flowers exhibit standard syndromes, there is substantial evidence to indicate that many species conforming to a particular syndrome are in fact pollinated most effectively by the agent associated with that syndrome (Fenster et al. 2004). This evidence indicates that we should take seriously the hypothesis that interactions with pollinators have driven the evolution of flower color in many, if not all, species.

While the existence of pollination syndromes is consistent with this hypothesis, it is also consistent with others. For example, like any evolutionary change, a particular flower color transition may have occurred by genetic drift. Alternatively, it may have been driven by selection on pleiotropic effects of flower color alleles, even if the change is deleterious with respect to pollinator attraction. In either case, if the flower color change subsequently attracts novel pollinators, selection imposed by these pollinators may mold other floral characteristics to produce a standard floral color syndrome. It is therefore possible for transitions between pollination syndromes with different flower colors to occur without direct selection on flower color by pollinators. Consequently, no matter how common pollination syndromes may be, their existence cannot be taken as definitive evidence that pollinator-mediated selection drives the evolution of floral color.

What type of evidence would constitute a definitive demonstration that a floral color change reflects adaptation to novel pollinators? Ideally, it must be shown that (1) the change was caused by selection and (2) the agent causing that selection was pollinators. As will be seen below, there are no species for which both of these conditions have been demonstrated. This lack of compelling evidence does not necessarily indicate, however, that pollinators are unimportant effectors of flower color evolution. Rather, it just as likely reflects the difficulty of assessing both of these criteria simultaneously for any given plant species. Because of this, it is worthwhile considering how much evidence supports the generality of each condition separately. If, for example, investigations seldom detect selection on flower color variants, we may be led to think that genetic drift is more commonly responsible for flower color evolution than is currently believed.

For information about selection on and pollinator responses to flower color variation, I conducted a literature search on the Web of Science (keywords used were "flower color variation," "flower color evolution," "selection on flower color," and "flower color polymorphism"). Studies were included only if they attempted to determine either whether flower color variants differed in some component of fitness or whether pollinators respond differently to different color variants (table 1). Information was obtained on 24 different species that exhibited variation in floral color. For two species (*Antirrhinum majus* and *Ipomoea purpurea*), information was available for two different color polymorphisms, and these are listed separately in table 1. Nine of the examples involve color divergence between populations or interspecific hybrid zones in which selection was examined on potentially introgressing color phenotypes (table 1A). The remainder involved within-population color polymorphisms (table 1B). Although this search cannot claim to be exhaustive, the investigations it turned up are likely to be representative of studies that have examined selection on and pollinator discrimination among flower color variants.

Selection on Floral Color Variation

If the results from these investigations are taken at face value, selection on flower color variants is ubiquitous. Of 21 species that have been examined, 18 exhibit evidence of selection on flower color phenotype (table 1). However, a number of biases and limitations associated with many of these studies restrict the degree of confidence that can be placed on this conclusion. The first is possible reporting bias: it is likely that evidence for selection is more likely to be reported than lack of evidence for selection.

A second limitation is that in most of the investigations listed in table 1, it is impossible to differentiate between selection on flower color itself and selection on genetically linked traits. In only two of the investigations listed in table 1 was any attempt made to randomize or otherwise account for the effects of the genetic background (the two different polymorphisms in *I. purpurea*), and even in these investigations, associations between the focal polymorphism and moderately linked loci were probably not disrupted. The inability to identify unambiguously the target of selection may mean that direct selection on flower color variation is much less common than the sample seems to indicate.

A third possible bias in many of these studies arises from the fact that in most investigations that attempt to quantify fitness experimentally, selection is measured for only some components of fitness. In cases in which fitness differences are found in one or more fitness components, it is unlikely that fitness differences in unmeasured components will just compensate the measured components to produce net neutrality. By contrast, in cases in which no fitness differences were detected, it is very possible that differences would be exhibited in other fitness components and, hence, in net fitness. This type of bias will lead to underestimating the prevalence of selection. Because lack of selection was reported for only three of the species examined, however, the extent of this bias will be minor in the sample reported here. It should be noted that the four studies that infer fitness differences by comparing distributions of color morphs along a transect with distribution of neutral markers ("cline" in column 3 of table 1) do not suffer from this bias because the approach implicitly considers all components of fitness.

Given these limitations, what is to be concluded about the prevalence of selection on floral color variation? The most legitimate conclusion is that the evidence suggests that color variants are usually not selectively neutral but that this has been shown definitively for only a handful of species (those exhibiting flower color clines). Thus, the current data are consistent with the hypothesis that selection on floral color variation is ubiquitous and therefore likely responsible for many, if not most, evolutionary transitions in flower color, but they also do not exclude alternative hypotheses. This issue will likely be settled only by future investigations that distinguish between direct selection on floral phenotypes and selection on linked traits. This will be most easily achieved by introgressing different flower color alleles into the same genetic background (e.g., Bradshaw and Schemske 2003) and measuring fitness on the resulting isogenic lines. To be definitive, however, this approach will require genetic documentation that the lines are truly isogenic, i.e., that they differ at only the flower color locus.

Selection Driven by Pollinators

As described above, the conventional explanation for floral color transitions is that they represent adaptations to exploiting different types of pollinators. While I argued that available evidence suggests that color transitions are often adaptive, there is little definitive evidence that pollinators are driving that adaptation. It is certainly true that in most cases that have been investigated (13 of 15 species in table 1),

Table 1

Investigations Examining either Selection on or Pollinator Discrimination among Floral Color Morphs in Nature

Species	Colors compared	Fitness differences	Divergent selection	Differential pollinator visitation	Pollinators impose selection?	Pleiotropy suggested	References
A. Cases involving population divergence or divergence between closely related species in flower color:							
Antirrhinum majus	Red/yellow	Cline	Yes	...	...	...	Whibley et al. 2006
Antirrhinum majus	Yellow/white	sp, ss	...	D[a]	...	...	Jones and Reithel 2001
Aquilegia caerulea	Blue/white	...	...	...	D	...	Miller 1981
Aquilegia formosa, Aquilegia pubescens[b]	Dark/pale red	Cline	Yes	...	...	...	Hodges and Arnold 1994
Ipomopsis aggregata, Ipomopsis tenuituba[b]	Red/white	fp, sp, ss	...	D	Yes	...	Campbell et al. 1997; Melendez-Ackerman et al. 1997; Melendez-Ackerman and Campbell 1998; Anderson and Paige 2003
Iris fulva, Iris brevicaulis[b]	Blue/red	...	...	D	...	...	Wesselingh and Arnold 2000
I. fulva, Iris hexagona[b]	Blue/red	...	...	D	...	...	Emms and Arnold 2000
Linanthus parryae	Blue/white	Cline, sp	Yes	No	...	Yes	Schemske and Bierzychudek 2001, forthcoming
Mimulus aurantiacus	Red/yellow	Cline	Yes	D	...	...	Streisfeld and Kohn 2005
Mimulus lewisii, Mimulus cardinalis[b]	Pink/orange	...	...	D	...	...	Schemske and Bradshaw 1999; Bradshaw and Schemske 2003
B. Cases involving within-population flower color polymorphisms:							
Phlox pilosa	Pink/white	sp	...	I	...	...	Levin and Kerster 1967
Phlox drummondii	Red/pink	sp	...	I	...	...	Levin 1969, 1972, 1985
Platystemon californicus	Yellow/white	spf	...	D	...	...	Hannan 1981
Clarkia gracilis	Purple/white	sp, ss	...	...	...	...	Jones 1996
Claytonia virginica	Four discrete colors	None (fp)	...	...	...	...	Frey 2004
Dactylorhiza sambucina	Yellow/red	pr, sp, ss	...	...	...	...	Gigord et al. 2001; Pellegrina et al. 2005
Delphinium nelsonii	Blue/white	sp	...	D	Yes	...	Waser and Price 1981, 1993
Hydrophyllum appendiculatum	Blue/white	sp	...	No	...	...	Wolfe 1993
Ipomoea purpurea	Blue/white (ww)	ss, sm, fln	...	D	...	Yes	Rausher and Fry 1993; Rausher et al. 1993; Fry and Rausher 1997
I. purpurea	Blue/white (aa)	v, sp	...	...	...	Yes	Coberly 2003; Coberly and Rausher 2003
Linaria canadensis	Purple/blue	fls, frs	...	...	...	...	Wolfe and Sellers 1997
Linum pubescens	Purple/yellow	None (fn, sp, sm)	...	...	...	...	Wolfe 2001
Lobelia maritima	Purple/white	None (sp)	...	...	...	...	Gomez 2000
Phlox drummondii	Red/white	v, sp	...	...	...	Yes	Levin and Brack 1995
Raphanus raphanistrum	Yellow/white	ss	...	D	...	...	Kay 1976; Stanton et al. 1989
Raphanus sativus	Yellow/others	ss	...	I	Yes	...	Irwin and Strauss 2005

Note. Ellipsis in table indicates relevant data not reported. For fitness differences, cline = selection deduced from comparison of color and neutral marker distributions; fln = flower number; fls = flower size; fn = fruit number; fp = fruit production; frs = fruit size; pr = pollen removal; sm = seed mass; sp = seed production; spf = seeds per fruit; ss = siring success; v = viability; none (x, y) = fitness components x and y examined but no differences detected between morphs. For divergent selection, yes = experimental confirmation that different morphs favored in different areas. For differential pollinator visitation, D = direct observation indicates differential pollinator visitation; I = indirect evidence indicates differential pollinator visitation. For whether pollinators impose selection, yes = experimental confirmation that pollinators impose selection on color. For pleiotropy suggested, yes = evidence provided that fitness effects not due to interactions with pollinators.

[a] Experiments conducted in nonnative habitat.

[b] Experiments conducted in or across hybrid zone.

pollinators discriminate among color phenotypes and usually visit some morphs more frequently than others. However, while such discrimination could cause fitness differences among color morphs, one can imagine situations in which no fitness differences result from discrimination. For example, consider a situation in which individual pollinators specialize on one flower color morph or the other, so that there is little pollinator movement between morphs. Even though there may be an inherent tendency to specialize on one morph, so that more individual pollinators visit that morph, this differential visitation will not cause differences in either male or female fitness of the two morphs, as long as seed production is not pollen limited. Therefore, simply demonstrating that pollinators discriminate among morphs is insufficient for concluding that pollinators impose selection on flower color.

Instead, some sort of experimental demonstration that pollinators cause fitness differences is required, either by manipulating pollinator access or by manipulating floral characters. Of the 15 species in table 1 that show evidence of pollinator discrimination, only three have been investigated in this way. Irwin and Strauss (2005) examined a two-locus color polymorphism in a naturalized population of *Raphanus sativus*. Using progeny analysis, they estimated the proportions of the four paternal haploid haplotypes in each of two treatments: one in which pollination occurred naturally and one in which maternal plants were pollinated with a mixture of pollen in which the haplotype frequencies were equal to those produced by flowers in the population. Proportions were different between the two treatments, indicating that some aspect of pollinator behavior generated differences in the male component of fitness among the flower color genotypes.

Using a similar approach, Waser and Price (1981) found that rare white-flowered individuals of *Delphinium nelsonii*, compared to blue-flowered individuals, had reduced seed production when pollination was by natural pollinators. By contrast, when plants were hand pollinated, there was no difference in seed production. Moreover, the observation that pollinators visited white-flowered plants at a lower rate suggests that the difference in seed production is due to greater pollinator limitation in the whites, caused by pollinator discrimination. This study convincingly demonstrates that pollinator preferences cause differences in the female component of fitness between the floral color morphs.

Finally, Melendez-Ackerman and Campbell (1998) examined a hybrid zone between *Ipomopsis aggregata* and *Ipomopsis tenuituba*, in which red-flowered *I. aggregata* had higher seed production and outcross male success than did hybrid genotypes or the white-flowered *I. tenuituba*. In addition, pollinators (primarily hummingbirds) preferentially visited *I. aggregata*. The causal link between differential visitation and fitness differences was established by experimentally manipulating floral color to disassociate effects of flower color from other floral characteristics. When *I. aggregata* flowers were painted either red or white, red flowers received more visits and sired more seeds than did white flowers. In another experiment, when flowers of the two pure species and the hybrids were painted red, differences in both visitation rates and fitness were almost completely eliminated. It should be noted, though, that selection by pollinators has not been demonstrated outside the hybrid zone; it is therefore unclear that the selection documented in this study is responsible for flower color divergence between the two species.

Because evidence for pollinator-mediated selection has been found when sought, it is tempting to infer that, in general, differential pollinator visitation among color morphs will tend to cause fitness differences. With only three species examined, however, one cannot, at this point, place too much confidence in this inference.

Selection for Divergence

More than half of the species listed in table 1 (i.e., those in part B of the table) report on intrapopulation color polymorphisms rather than color divergence between populations. Moreover, for most of these species, it is believed (admittedly, often without much evidence) that some sort of balancing selection maintains the polymorphism. Information from these species may be misleading regarding the validity of the conventional wisdom that flower color transitions reflect adaptation to novel pollinators because it is not clear that the selective agents responsible for balancing selection are necessarily similar to those responsible for the divergent selection between populations or species that are presumed to generate evolutionary transitions in flower color. Instead, it may be most meaningful to focus only on species that exhibit geographic divergence in flower color (table 1A).

For such cases, we would like to know (1) whether the divergence is adaptive and (2) whether pollinators are the selective agents driving divergence. One way of demonstrating that divergence is adaptive is to show that spatial divergence in color is much greater than expected on the basis of divergence in the frequencies of neutral genetic markers (e.g., Spitze 1993; Storz 2002). Evidence of this type has been reported for four species (*A. majus*, *Linanthus parryae*, *Mimulus aurantiacus*, and a hybrid zone of *Aquilegia formosa* and *Aquilegia pubescens*; see references in table 1A). Unfortunately, only for *M. aurantiacus* is there evidence of pollinator discrimination between morphs, and in this investigation, there is no experimental evidence showing that pollinators cause fitness differences.

For many of the remaining species exhibiting flower color divergence, there is evidence for both differential pollinator visitation and differential fitness between morphs when measured in a single population or hybrid zone (table 1A). However, none of the studies reporting fitness differences conducted reciprocal transplants, so it is not known whether selection is divergent, as opposed to, say, purifying. Moreover, in only one species (*I. aggregata* and *I. tenuituba* hybrid zone) is there convincing evidence that pollinators impose selection on color morphs, and in this case, there is no evidence that flower color divergence is adaptive.

Remarkably, then, we lack any example in which flower color divergence can be reliably attributed to adaptation to different types of pollinators. This, of course, does not mean that the conventional wisdom is incorrect: arguably, for no species have investigators conducted the entire set of experiments required to verify or falsify this hypothesis. Nevertheless, the absence of firm support for this hypothesis indicates that alternative explanations for flower color transitions should be seriously entertained.

Color Transitions Not due to Selection by Pollinators

Genetic Drift

One alternative to the conventional wisdom is that color divergence is caused by genetic drift. The only case for which the neutrality hypothesis has been seriously argued is the blue-white polymorphism in *Linanthus parryae*, which is controlled by variation at a single Mendelian locus. Using data on spatial distribution of color morphs, Wright (1943) argued that this variation is selectively neutral and that even extreme divergence in gene frequencies between populations was caused by genetic drift. By contrast, subsequent observations and experiments led Epling et al. (1960) to conclude that this variation was subject to selection, but Wright (1978) continued to disagree. Only recently has this debate apparently been resolved in Epling's favor by Schemske and Bierzychudek (2001, forthcoming), who have demonstrated that neutral markers do not exhibit the same clines as seen in flower color frequencies and in the higher fitness of the local morph in reciprocal transplant experiments.

One reason that neutrality of color variation has seldom been championed is undoubtedly the apparent ubiquity of selection on flower color variants. Nevertheless, it is premature to conclude that neutrality is rare. As discussed above, for only a few species can fitness differences among morphs be unambiguously ascribed to variation in floral color genes rather than to variation at linked genes. It is thus possible that much of the flower color variation we see in nature is actually selectively neutral.

Selection on Pleiotropic Effects

Many of the enzymes involved in anthocyanin synthesis are also required for the synthesis of other flavonoid compounds (fig. 1). Noting that these compounds in turn influence ecological and physiological traits in addition to flower color in plants (Shirley 1996), a number of authors have suggested that flower color evolution may often be influenced by selection on these pleiotropic effects rather than, or in addition to, selection imposed by pollinators (Rausher and Fry 1993; Simms and Bucher 1996; Fineblum and Rausher 1997; Armbruster 2002; Irwin et al. 2003). Indirect selection on flower color through genetic correlations thus provides an alternative hypothesis for explaining floral color transitions.

Because evidence for pleiotropic fitness effects of color variants has been reviewed recently by Strauss and Whittall (2006), I will simply summarize and comment on their conclusions. These authors (see their table 7.2) report evidence of fitness differences among morphs in survival, flower production, seed production, or biomass for 10 species, with flower color polymorphisms and differences in damage by natural

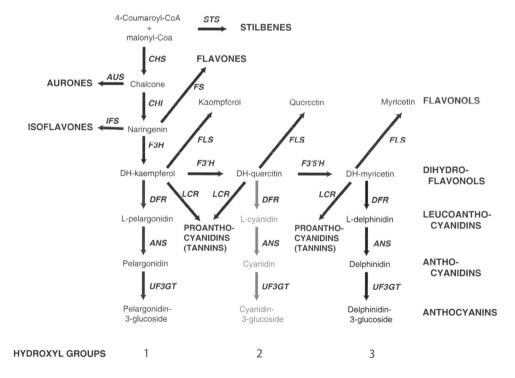

Fig. 1 Schematic diagram of anthocyanin pathway. Each of the three branches of the pathway leading to different classes of anthocyanins is colored differently. Branches leading to other types of flavonoids are not colored. Specific compounds are listed in lowercase. Classes of compounds are listed in bold uppercase. Enzymes are listed in uppercase. *STS* = stilbene synthase. *CHS* = chalcone synthase. *CHI* = chalcone-flavanone isomerase. *F3H* = flavanone-3-hydroxylase. *DFR* = dihydroflavonol 4-reductase. *ANS* = anthocyanidin synthase. *UF3GT* = UDP flavonoid glucosyltransferase. *F3′H* = flavonoid 3′hydroxylase. *F3′5′H* = flavonoid 3′5′hydroxylase. *LCR* = leucoanthocyanidins reductase. *FS* = flavone synthase. *FLS* = flavonol synthase. *IFS* = isoflavone synthase. *AUS* = aureusidin synthase. *DH* = dihydro. *L* = leuco. Hydroxylation state is the number of hydroxyl groups on the beta ring of the anthocyanidin.

enemies for three additional species. Unfortunately, because they do not report investigations in which no fitness differences were reported (e.g., table 1), their data do not allow even a crude assessment of how frequently such pleiotropic effects occur. Nevertheless, it is clear from their compilation that pleiotropic fitness effects of color variants may be common. At least two caveats must be attached to this conclusion, however. First, as described above, evidence that these effects are due to true pleiotropy of flower color genes rather than to effects of linked genes is nonexistent. Second, it cannot be ruled out that fitness differences were not caused by pollinators in many of the cases reported by Strauss and Whittall.

The most convincing example of selection on pleiotropic effects from Strauss and Whittall (2006) is represented by Levin and Brack (1995), who showed that color morphs of *Phlox drummondii* differed in viability. Because it is difficult to imagine pollinators affecting survival before flowering begins, the fitness differences in this species can be ascribed to an agent other than pollinators. Also, in *L. parryae*, extensive pollinator observations (Schemske and Bierzychudek 2001) indicated no pollinator discrimination among morphs, which makes it difficult to believe that the selection on flower color documented by Schemske and Bierzychudek (2001, forthcoming) was imposed by pollinators, although pollinator-imposed selection was not directly examined.

In the remaining studies listed by Strauss and Whittall, fitness traits for which differences among morphs were reported were flower number, seed number (or fruit number), or biomass. Biomass by itself is not a fitness component but rather a trait that contributes to fitness components such as viability, seed production, and siring success. While biomass is often highly correlated with true fitness components (e.g., seed production) in plants, this correlation needs to be demonstrated in each investigation in order for biomass by itself to be used as evidence of differential fitness. One of the studies reported by Strauss and Whittall (on *Echium plantagineum*; Burdon et al. 1983) should be discounted for this reason.

For six of the remaining species, the primary evidence for the importance of pleiotropic effects is differential seed production among morphs. While differential seed production may be due to pleiotropic effects, it may also be caused by differential pollinator visitation (e.g., Waser and Price 1981), and it is not clear that, for any of these six species, differences in visitation can be ruled out as causing the observed differences in morph fitness. Five of the species were examined by Warren and Mackenzie (2001), who allowed open pollination by natural pollinators in their experiment. Although in their experiment the color morph producing the most seeds switched between drought and watered treatments, the authors do not rule out the possibility that pollinator behavior also switches in these treatments (e.g., morphs may switch rankings in nectar production, causing a change in pollinator preference). Because pollinator effects were not decoupled experimentally from other sources of selection, we cannot be certain that pollinators did not impose the fitness differences observed in Warren and Mackenzie's investigation, albeit on a possible pleiotropic effect of flower color variation.

Finally, the three intriguing studies demonstrating differential susceptibility to natural enemies reported by Strauss and Whittall (2006) demonstrate the potential for these kinds of

organisms to act as agents of selection on flower color. However, differential damage or infection by itself does not imply differential fitness because many plants are tolerant of damage and infection (Kniskern and Rausher 2006; Weis and Franks 2006). In this context, while Frey (2004) reported both differential herbivory and differential fitness among color morphs in *Claytonia virginica*, selection on flower color was not significant when examined in a multivariate context that included floral size and leaf area, so it is not clear that either pollinators or natural enemies caused fitness differences among flower color genotypes.

What, then, can one conclude about the hypothesis that floral color transitions are often driven by selective agents other than pollinators? About the same that was concluded for the conventional pollinator-driven hypothesis, that the evidence is suggestive but that there is not a single definitive example to support the hypothesis. The issue is not whether loci affecting flower color also have pleiotropic effects; in a variety of species, flower color variants also influence characters such as trichome density, root hair formation, cell morphology, and vacuolar size (Walker et al. 1999; Spelt et al. 2002; Morita et al. 2006), any of which could, in turn, affect fitness. Rather, the issue is that there is little evidence indicating that these or other pleiotropic effects are subject to selection. What will be needed to evaluate this hypothesis conclusively for any particular situation? Again, as for the conventional hypothesis, two things will be required: (1) population differentiation in flower color must be shown to be due to natural selection on flower color genes and not linked genes and (2) the agent of selection must be shown unequivocally to be something other than pollinators, which will require some sort of experimental manipulation of purported selective agents. Until both of these criteria have been addressed for a number of species, we will be unable to say anything conclusive about whether pollinators or other agents are the primary cause of evolutionary transitions in flower color or about the importance of genetic drift.

Macroevolutionary Trends

In many angiosperm taxa, bee pollination and blue-purple flowers are ancestral characters, while hummingbird pollination and the typical red/orange flowers associated with bird pollination are derived characters (see Ackermann and Weigend 2006; also see Thomson and Wilson 2008). By contrast, in these taxa, transitions from bird to bee pollination, with an accompanying color transition, are generally perceived as rare. These observations imply an asymmetry in evolutionary transition rates between these two colors. In this section, I examine the evidence for this type of asymmetry and consider why it might exist. In doing so, I first identify four general macroevolutionary trends in flower color evolution that the literature suggests are relatively common: (1) floral color transition rates are often unequal; (2) in general, transitions from blue to red flowers, and from pigmented to white flowers, are more common than the reverse transitions, although some taxa exhibit the reverse pattern; (3) transitions from blue to red flowers usually are caused by inactivating branches of the anthocyanin pathway, resulting in the production of different, less hydroxylated anthocyanins; and (4) transitions from blue to red flowers, and from pigmented to white flowers, typically involve loss-of-function (LOF) mutations.

I then describe how these four trends may be causally connected. I note here that this analysis is based primarily on patterns of anthocyanin pigment evolution. Information on macroevolutionary trends in flower color changes involving carotenoids, the other major floral pigment, is largely lacking.

Asymmetry of Floral Color Transitions

Demonstrating asymmetry in color transitions requires a statistical assessment of whether transition rates in one direction of change differ from rates in the other direction. The ideal approach would take into account uncertainty due to both phylogeny and ancestral character state by averaging estimated transition rates over a large number of plausible phylogenies (Pagel and Lutzoni 2002). Because diversification and extinction differences for taxa with different character states can confound estimates of transition rates (Nosil and Mooers 2005; Maddison 2006), such analyses should ideally jointly estimate both diversification/extinction rates and transition rates. Techniques for doing this are still in their infancy, with the first method being published only now (Maddison et al., forthcoming). In the absence of this approach, character state reconstruction can be used to assess the reasonableness of patterns deduced from analysis of only transition rates (Nosil and Mooers 2005).

Despite potential pitfalls of transition state analysis, one study has used this approach successfully. Using 160,000 trees generated from a Bayesian analysis, Kay et al. (2005) found the rate of transitions from bee to bird pollination in *Costus* to be significantly larger than the reverse rate, which was estimated to be zero. In this taxon, the shift from bee to bird pollination typically involved a transition from white flowers with pigmented stripes to fully pigmented red or yellow corollas.

Two other investigations strongly suggest more frequent flower color transitions in one direction but fail either to estimate actual transition rates or to sample multiple trees. Using a single tree, Whittall et al. (2006) demonstrated that in *Aquilegia*, the rate of transition from pigmented to unpigmented (white) flowers was substantially and significantly greater than the rate of gain of pigmentation. In fact, no reversals were inferred from ancestral state reconstruction. And for *Penstemon*, P. Wilson, A. D. Wolfe, W. S. Armbruster, and J. D. Thomson (unpublished manuscript) examined origins and reversals of hummingbird pollination, which involved shifts between blue and red pigments, in a sample of 1000 bootstrapped phylogenetic trees. Although they did not statistically assess transition rates, they found numerous (minimum 10, maximum 21) blue/purple to red transitions and no convincing reverse transitions (see also Thomson and Wilson 2008).

When we use less rigorous criteria, several other taxa appear to exhibit biased transition rates. In *Ipomoea*, Rausher (2006) identified, without any statistical analysis but based on a phylogeny and ancestral state reconstruction of Miller et al. (1999, 2004) and Zufall (2003), eight unambiguous,

evolutionarily independent transitions from pigmented to unpigmented flowers and four unambiguous, evolutionarily independent transitions from blue to red flowers (we have subsequently identified a fifth, *Ipomoea horsfalliae*). By contrast, no reverse transitions of either type were inferred. In the Sinningieae, white hawkmoth- or bat-pollinated flowers have evolved independently five times, with no instances of regain of pigmentation (Perret et al. 2003). And in *Iochroma*, floral anthocyanins have been lost five times, with only one regain (Smith and Baum, forthcoming). Moreover, an analysis using 100 Markov chain Monte Carlo–generated trees shows that transition rates to white are substantially higher than the reverse transition rate (S. D. Smith and M. D. Rausher, unpublished data).

While there are doubtless exceptions to this pattern of asymmetry in floral color transitions, it seems clear that in many angiosperm taxa, marked asymmetry is the rule. One other pattern that may be associated with this asymmetry is "tipness." Rausher (2006) noted that in *Ipomoea*, white-flowered species tend to arise from pigmented clades and have short branch lengths. Very few deep nodes are reconstructed as having white flowers. A similar pattern is evident in the phylogenies of *Penstemon* presented by Wolfe et al. (2006), in which red-flowered, hummingbird-pollinated lineages tend to be confined to the tips, and in the phylogenies of *Sinningia* reported by Perret et al. (2003), in which white-flowered, hawkmoth- and hummingbird-pollinated lineages appear to be confined to the tips. Unfortunately, although this pattern may have implications for evolutionary potential, it has not been verified statistically by any analysis, largely because the appropriate methodology is only now appearing (e.g., Maddison et al., forthcoming).

Directions of Asymmetric Transition

I have documented several taxa in which color transitions between pigmented and white were asymmetric. Although the sample size is small, four taxa (*Ipomoea*, *Sinningia*, *Iochroma*, and *Aquilegia*) appear to show greater rates of transition from pigmented to white flowers, while only one taxon (*Costus*) exhibits the reverse pattern. In addition, in both *Dalechampia* and *Acer*, there have been three independent gains of pigmentation and no losses (Armbruster 2002), although transition rates have not been compared statistically for these species. It thus appears that in different taxa, the favored direction of transition between pigmented and unpigmented may differ.

While I am unaware of any taxa in which the transition rate from red to blue/purple is greater than the reverse transition rate, a number of taxa exhibit the opposite pattern. As described above, in *Penstemon* and *Ipomoea*, there are numerous transitions from blue to red but no reversals. The same appears to be true in the Antirrhineae (three transitions from blue to red and no reversals; Ghebrehiwet et al. 2000). By contrast, in *Sinningia*, transitions from red to blue appear to be as common as the reverse transitions. Again, although the sample size is small, there appears to be a trend for the transition from blue to red to be favored, although reverse transitions do occur.

Pigment Changes in Blue-Red Transitions

The most common and widely distributed type of floral pigment is the anthocyanins. Within angiosperms, there are three major classes of anthocyanin pigments: those based on the anthocyanidins pelargonidin, cyanidin, and delphinidin (fig. 1). These differ primarily in the number of hydroxyl groups on the beta ring of the molecule, with pelargonidins having the fewest, cyanidin having one more, and delphinidin having yet one more. As hydroxyl groups are added, the peak of the absorbance spectrum shifts from the red end of the visible spectrum toward the blue end (Tanaka et al. 1998). Consequently, delphinidin-based anthocyanins tend to be purple, violet, or dark blue, while cyanidin-based anthocyanins tend to be blue, magenta, or sometimes red and pelargonidin-based anthocyanins are almost always red or orange.

Although floral color changes can occur without alteration of the pigment that is produced (e.g., most notably by altering vacuolar pH; Griesbach 1996), the general color differences among classes of anthocyanins suggest that most evolutionary transitions from blue/purple to red/orange involve a switch from producing more hydroxylated anthocyanins to producing less hydroxylated anthocyanins. Several examples support this suggestion. In the genus *Ipomoea*, the red-flowered, hummingbird-pollinated species in the *Mina* clade produce only pelargonidin-based anthocyanins. By contrast, the ancestral state, represented by closely related species, consisted of blue-flowered, bee-pollinated species that produce cyanidin-based anthocyanins (Zufall and Rausher 2004). Moreover, in three out of four evolutionary transitions from blue to red flowers in *Ipomoea* for which data exist, the pigments shift from entirely cyanidin to entirely pelargonidin anthocyanins (Zufall 2003). In a fifth species (*Ipomoea purga*), the transition has been from producing solely cyanidin-based anthocyanins to producing a mixture of pelargonidin- and cyanidin-based compounds. The flowers of this species are intermediate in color (magenta) but exhibit several traits characteristic of hummingbird pollination (e.g., a reduced width : length ratio, increased nectar volume, reduced nectar concentration, and exserted anthers and stig-

mas; Zufall 2003). There are no known cases of shifts from cyanidin to pelargonidin production without a change in flower color from blue to red.

Scogin and Freeman (1987) noted a similar correlation between flower color and anthocyanidin class in *Penstemon*. Because they did not have available a reliable phylogeny of this genus, however, they were not able to convincingly demonstrate that color transitions were caused by shifts in the class of anthocyanidin produced. With the publication of a phylogeny of this group by Wolfe et al. (2006), though, a phylogenetically based analysis of this correlation is possible.

A crude analysis of the data now available for the genus *Penstemon* corroborates Scogin and Freeman's (1987) conclusion. The *trnC-D/T-L* tree of Wolfe et al. (2006) was pruned to include only those species for which information is known on both flower color and floral pigments (see appendix). This pruned tree is presented as figure A1 in the appendix. Wilson et al. (2004; P. Wilson, A. D. Wolfe, W. S. Armbruster, J. D. Thomson, unpublished manuscript) demonstrate that each of the red-flowered lineages portrayed in this figure represents an independent transition from blue to red flowers. In order to determine whether these transitions are correlated with transition to a lower hydroxylation state, a series of independent contrasts of sister groups were performed (table 2; for details, see appendix). Each contrast included one red-flowered species (all hummingbird pollinated) and one or more blue-flowered species (all bee pollinated). In some cases, more than one blue-flowered species was used because of unresolved polytomies. However, in these cases, the results were similar, regardless of which blue-flowered species was compared to the red-flowered species.

In seven of nine contrasts, the shift from blue to red was coupled with a reduction in the hydroxylation state of the anthocyanins (table 2). In the two other cases, whether there was a reduction depended on which neighbor species was used as a comparison.

For *Penstemon kunthii*, if *Penstemon incertus*, *Penstemon caespitosus*, or *Penstemon confusus* was used as the paired blue-flowered species, then there was also a reduction in hydroxylation state. Only if *Penstemon clevelandii* was used as

Table 2

Correlation between Transition from Hummingbird-Pollinated Red and Bee-Pollinated Blue *Penstemon* Lineages and Corresponding Transition in Anthocyanidins Produced

Bee-pollinated species	Hummingbird-pollinated species	Anthocyanidin transition
Antirrhinum majus	*Keckiella ternata*	1 → 3, 2 → 3, or 3 → 3
P. barrettiae	*P. newberryi*	1 → 3
P. incertus group	*P. rostriflorus*	1 → 4, 2 → 4, or 3 → 3
P. incertus group	*P. lanceolatus*	1 → 5, 2 → 5, or 3 → 5
P. linarioides and *P. incertus* groups	*P. hartwegii/P. isophyllus*	1 → 5, 2 → 5, or 3 → 5
P. incertus group	*P. kunthii*	1 → 3, 2 → 3, or 3 → 3
P. speciosus	*P. labrosus*	2 → 5
P. thompsonii	*P. centranthifolius*	1 → 5
P. perpulcher group	*P. barbatus*	1 → 5

Note. All transitions are from bee to hummingbird pollination. See appendix for additional information. For anthocyanidin transition, 1 = delphinidin; 2 = delphinidin + cyanidin; 3 = cyanidin; 4 = cyanidin + pelargonidin; 5 = pelargonidin.

the paired blue-flowered species was there no change in hydroxylation level. For *Keckiella ternata*, comparison to any of eight blue-flowered species in the *Penstemon albertinus* group indicated a reduction in hydroxylation state; only when compared to the ninth species, *Penstemon caesius*, was there no change in hydroxylation state. Effectively, these results mean that in eight or nine of the nine comparisons, there was a decrease in hydroxylation state, and in none of the nine comparisons was there a change to a higher hydroxylation state. The probability of this correlation occurring by chance is <0.01.

It thus seems that in at least two major angiosperm taxa, evolutionary transitions from blue to red flowers, usually associated with transitions from bee to bird pollination, are highly correlated with changes in the class of pigment produced, although for both species, more detailed analyses that incorporate uncertainty in tree topology are warranted. Our understanding of the anthocyanin biosynthetic pathway in turn indicates that this change in pigment type is most likely due to inactivation of one or more of the major branches of that pathway.

The anthocyanin pathway has three major branches (fig. 1), leading to pelargonidin-, cyanidin-, and delphinidin-based anthocyanins. The branching enzymes F3′H and F3′5′H are responsible for adding the additional hydroxyl groups characteristic of the longer branches. Which class(es) of anthocyanins is (are) produced depends on the amount of flux down each of these branches. The ancestral state of many angiosperm groups, and perhaps of angiosperms as a whole, is the production of blue-purple flowers that produce primarily or exclusively delphinidin-based anthocyanins (Rausher 2006). Mutational studies of model organisms such as *Antirrhinum*, *Petunia*, and *Ipomoea* demonstrate, however, that when branching enzymes are inactivated, flux is often redirected along one of the other branches, indicating that the enzymes in these branches are often fully functional. For example, wild-type *Petunia* produce deep purple flowers with delphinidin anthocyanins. Mutational inactivation of F3′5′H, however, results in the production of redder flowers that produce primarily cyanidin anthocyanins (Griesbach 1996). In *Ipomoea*, the delphinidin-producing branch of the pathway was inactivated early in the diversification of the genus, so that most blue species produce only cyanidin (Zufall 2003). In three of these species, mutations in the branching enzyme F3′H inactivate the cyanidin branch, which directs flux down the pelargonidin branch to produce red/pink flowers (Hoshino et al. 2003; Zufall and Rausher 2004). A similar situation obtains in *Antirrhinum majus* (Stickland and Harrison 1974). These observations suggest that, in general, nonutilized pathway branches are potentially functional and all that is needed to produce redder flowers is inactivating the branch that is currently associated with maximum flux.

The genetics of blue-red flower color transition has been examined in *Ipomoea quamoclit* and supports the hypothesis that this transition is commonly accomplished by inactivating a branch of the anthocyanin pathway. In this species, *F3′H* is almost completely downregulated, and the enzyme DFR has become a substrate specialist, unable to process the precursor to cyanidin (Zufall and Rausher 2004). These two changes completely block flux down the cyanidin branch and redirect it down the pelargonidin branch. The correlation between

red flowers and the reduced hydroxylation state described above for *Penstemon* also indicates that branches of the pathway have repeatedly been inactivated in the transition to red flowers, although the genes involved have not been described.

Transitions Often Involve LOF Mutations

Available evidence indicates that many, if not most, flower color transitions result from LOF mutations in pigment pathway genes. LOF mutations can be of two types: mutations in coding regions that abolish enzyme function or mutations in regulatory regions (including coding regions of transcription factors) that reduce or eliminate protein expression. Extensive work on spontaneous mutations in model organisms such as *Petunia*, *Antirrhinum*, and *Ipomoea* has demonstrated that LOF mutations in both structural and regulatory genes of the anthocyanin pathway typically result in either loss of pigmentation or change in color of the corolla (Holton and Cornish 1995; Mol et al. 1998).

The genetic changes associated with floral color change in nature have been documented for eight taxa (table 3; note that for *Ipomoea purpurea*, there are three different color change polymorphisms). Four of the studies involve natural flower color polymorphisms, and seven involve fixed changes within a species. In every case, except *Mimulus aurantiacus*, one or sometimes two LOF mutations have been identified as causing the flower color change. (In *M. aurantiacus*, the direction of the color change has not been determined, so it is not known whether the change was a gain or a loss of function.) These mutations are roughly equally divided among structural genes and transcription factors and between coding region and cis-regulatory region mutations. One caveat is that in four taxa (*I. quamoclit*, *Ipomoea alba*, *Ipomoea ochracea*, and *Aquilegia* spp.), the genetic changes identified have not been shown by crosses to be the only changes influencing floral color. Consequently, they may not be the original changes that produced the flower color transition. Instead, they may reflect subsequent degeneration of one or more branches of the pathway after the original mutation abolished function (Zufall and Rausher 2004). Even if this is the case, however, the documented changes provide indirect evidence that LOF mutations were responsible for flower color change because they would not have been fixed if the pathway had not been previously inactivated.

LOF mutations cause floral color change either by blocking branches of the pathway, forcing flux down other branches (e.g., blue-red transitions), or by blocking the entire pathway so that no anthocyanins are produced (pigmented-nonpigmented transitions). Once the branch or pathway has been inactivated, there is expected to be no natural selection to maintain function in other elements of that branch or pathway. Consequently, additional LOF mutations are expected to accumulate by genetic drift, leading to degeneration of the branch or pathway. With the accumulation of more than one LOF mutation, restoration of branch or pathway function is expected to be essentially impossible because it would require multiple simultaneous mutations restoring function (Rausher 2006).

This type of degeneration has been reported for the blue-red transition in *I. quamoclit* (Zufall and Rausher 2004). Another possible example may be the transition to white flowers

Table 3

Species for Which Genes Responsible for Naturally Occurring Color Morph Differences Have Been Identified

Species	Locus	Gene	Phenotype	LOF mutation	References
A. Within-population polymorphisms:					
Antirrhinum	*Rosea*	*myb* (tf)	Yellow[a]	?	Whibley et al. 2006
Ipomoea purpurea	*W*	*Ipmyb1* (tf)	White	Coding region deletions	Chang et al. 2005
I. purpurea	*A*	*CHS* (e)	White	Coding region insertion	Habu et al. 1998
I. purpurea	*P*	*F3'H* (e)	Pink	Coding region deletion	Zufall and Rausher 2004
B. Fixed change within a species:					
Antirrhinum (five species)	*Rosea*	*myb* (tf)	White	?	Schwinn et al. 2006
Aquilegia (*chrysantha* clade)	...	*DFR* (e)	Yellow	Downregulation	Whittall et al. 2006
Ipomoea alba	...	*CHS* (e), *DFR* (e)	White	Downregulation	Durbin et al. 2003
Ipomoea ochracea	...	*DFR* (e), *CHI* (e)	White	Downregulation	Durbin et al. 2003
Ipomoea quamoclit	...	*F3'H* (e), *DFR* (e)	Red	Downregulation	Zufall and Rausher 2004
Mimulus aurantiacus	...	*DFR* (e), *myb* (tf)	Yellow[a]	Downregulation	M. A. Streisfeld and M. D. Rausher, unpublished manuscript
Petunia axillaris	*An2*	*myb* (tf)	White	Coding region frameshift	Quattrocchio et al. 1999

Note.   Ellipsis in table indicates relevant data not reported. LOF = loss of function. For gene, e = codes for enzyme; tf = codes for transcription factor.

[a] Anthocyanins absent.

in *I. alba*. In this species, both *CHS* and *DFR* are markedly downregulated, while *CHI* is not (Durbin et al. 2003). Because at least *CHS* and *CHI* are coordinately regulated by transcription factors (Mol et al. 1998) in all species that have been examined, it seems unlikely that a single mutation in a transcription factor is responsible for downregulaton of *CHS* and *DFR*. Instead, it seems much more probable that either independent LOF mutations in the promoter region of these two genes or independent mutations in different transcription factors have led to the downregulation of these two genes.

Causal Relationship among Trends

It is very likely that the four macroevolutionary trends identified above are causally interrelated. In particular, the asymmetry of transitions from blue to red and from pigmented to white seen in many taxa can be explained by the observation that these transitions tend to involve LOF mutations. LOF changes are expected to be difficult to reverse for two reasons. First, they often involve coding region insertions or deletions of more than one base pair that cause frame shifts or premature termination, often as a result of imprecise transposon exsertion. Simple point mutations will be unable to reverse such changes. Moreover, once these knockout mutations have arisen, subsequent mutations that would be deleterious in a functional gene product are free to accumulate by genetic drift, rendering reversal even less likely. Second, inactivation of one gene in the pathway facilitates pathway degeneration, which again would presumably make reversal virtually impossible because it would require multiple simultaneous mutations. Examples of the former type include the *I. purpurea* mutations at all three loci and the *Petunia axillaris* mutant described in table 3—all four of the cases for which we know the nucleotide changes responsible for LOF. Examples of pathway degeneration have been given previously.

Although loss of floral pigments seems to occur at rates higher than those for regain of color in many taxa, it is clear that reversals occur, even frequently in some taxa (e.g., *Cos-*

tus, *Dalechampia*). Although no cases of the regain of pigmentation have been examined at the molecular level, these reversals most likely occur when the original LOF mutation is easily reversible. Cis-regulatory mutations that greatly reduce the expression of an anthocyanin structural gene or transcription factor are the most probable candidates for reversal because lost promoter binding sites can be recovered by a variety of mutational processes (Wray et al. 2003). Moreover, it seems probable that gene and pathway degeneration are less likely for LOF cis-regulatory mutants than for coding sequence mutants. Anthocyanin genes often exhibit differential expression patterns in different tissues, implying that expression in different tissues is controlled by different cis-regulatory elements. LOF mutations that abolish floral pigments will thus frequently not abolish anthocyanin production in vegetative tissues. Because anthocyanins and other flavonoids in vegetative tissues serve many ecologically and physiologically important functions (Shirley 1996), selection will tend to prevent degeneration of anthocyanin structural genes. This argument thus suggests that although there are no documented cases of cis-regulatory mutants causing changes in floral pigment color, we may have been looking in the wrong taxa: instead of taxa showing transition rates that are biased in favor of LOF, we should perhaps be examining taxa in which floral pigments have been regained multiple times.

While the predominance of LOF mutations can explain asymmetry in color transition rates, it cannot by itself explain a second apparent feature of floral color evolution, the tipness of transitions, whereby transitions tend to appear late in a phylogeny. There are at least two explanations for this pattern. The first is that ecological conditions favoring transitions became more common in recent times. If this explanation is correct, then tipness reflects a temporal change in environments and selection regimes and a real change in the rate at which transitions occur. The second possible explanation is that lineages with nonancestral flower color have a shorter time to extinction or lower speciation rate (Maddison 2006), perhaps because inactivation of major parts of the anthocyanin pathway allows less evolutionary flexibility. For example, shifts from blue to red associated with shifts to hummingbird pollination, as well as shifts from pigmented to white associated with moth or bat pollination, may involve a transition to increased pollinator specialization, which could render the lineage more susceptible to extinction. Alternatively, pleiotropic effects, such as increased susceptibility to herbivory, that have been ascribed to floral pigment change could also increase the probability of extinction of white-flowered lineages.

Differential rates of speciation and extinction among lineages with different characteristics have also been suggested for the concentration of self-compatible plant lineages at the tips of phylogenies (Igic et al. 2003; see Igic et al. 2008). If this explanation is correct, then the absence of deep nodes with a derived flower color results not from change in the rate of transitions but from differential extinction of lineages in which transitions occur early. Although no information exists to decide between these two explanations, both involve interesting macroevolutionary processes that deserve further study.

Acknowledgments

I thank three anonymous reviewers for constructive comments on the manuscript. This article was written while I was a Triangle Sabbatical Fellow at the National Evolutionary Synthesis Center. Support was provided by National Science Foundation grant DEB-0448889.

Appendix

Correlation between Flower Color Change and Pigment Change in *Penstemon*

To examine the correlation between change in flower color and change in anthocyanidin pigment class in *Penstemon*, I began with the *trnC-D/T-L* consensus tree for *Penstemon* and *Keckiella* (with *Antirrhinum majus* as outgroup) reported by Wolfe et al. (2006). This tree is the consensus of 2000 equally parsimonious trees. Floral color states were taken from work by Wilson et al. (2004, 2006) and personal observation. Blue, violet, and purple flowers were considered "blue" in this analysis. All species of these colors are visited primarily by bees (Wilson et al. 2004). Floral anthocyanidin content was taken from work by Scogin and Freeman (1987) and unpublished data from our laboratory.

Nine independent transitions from blue to red flowers were identified based on this phylogeny and the information presented by Wilson et al. (2004; P. Wilson, A. D. Wolfe, W. S. Armbruster, and J. D. Thomson, unpublished manuscript). The red-flowered species involved in this transition was then paired with a blue-flowered sister species or species group, and the anthocyanidin contents of the two species were compared. In some cases (e.g., *Penstemon kunthii*), there were multiple blue-flowered sister species that could have been used for the comparison because of lack of phylogenetic reso-

lution. In these cases, all possible comparisons were examined. In all cases of this type, the multiple comparisons gave consistent results. *Penstemon kunthii*, *Penstemon rostriflorus*, and *Penstemon lanceolatus* were compared to the same blue species (the four species in the clade with *Penstemon anguineus*). Although technically this makes these three comparisons nonindependent, they most likely represent three independent transitions to red (P. Wilson, A. D. Wolfe, W. S. Armbruster, and J. D. Thomson, unpublished manuscript), as evidenced by the presence of the blue-flowered *Penstemon lentus* separating these two red-flowered species (broken blue line in fig. A1).

Anthocyanidin content was coded 1 through 5, with increasing numbers corresponding to decreasing average hydroxylation state. States 1, 3, and 5 correspond to the production of one major anthocyanidin (delphinidin, cyanidin, and pelargonidin, respectively), while states 2 and 4 correspond to two major anthocyanidin components (delphinidin + cyanidin and cyanidin + pelargonidin, respectively). Transitions between states for the paired red and blue species are reported in table 2 as transitions from the anthocyanidin state of the blue-flowered species to the anthocyanidin state of the red-flowered species.

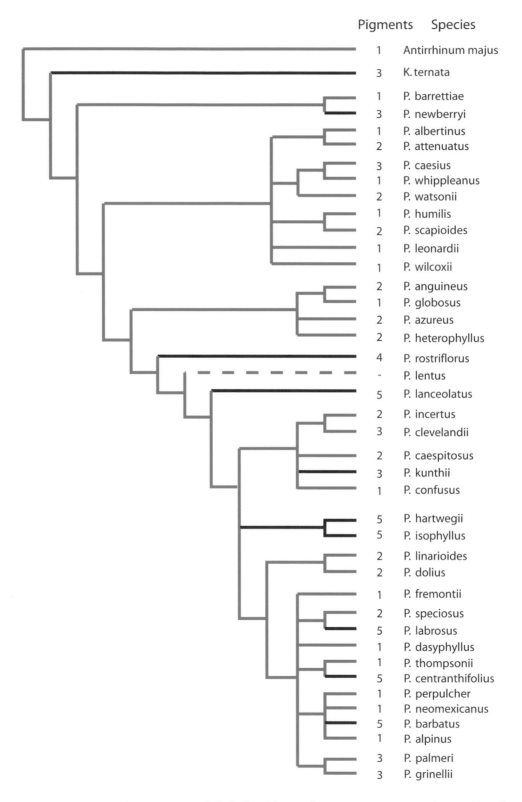

	Pigments	Species
	1	Antirrhinum majus
	3	K. ternata
	1	P. barrettiae
	3	P. newberryi
	1	P. albertinus
	2	P. attenuatus
	3	P. caesius
	1	P. whippleanus
	2	P. watsonii
	1	P. humilis
	2	P. scapioides
	1	P. leonardii
	1	P. wilcoxii
	2	P. anguineus
	1	P. globosus
	2	P. azureus
	2	P. heterophyllus
	4	P. rostriflorus
	-	P. lentus
	5	P. lanceolatus
	2	P. incertus
	3	P. clevelandii
	2	P. caespitosus
	3	P. kunthii
	1	P. confusus
	5	P. hartwegii
	5	P. isophyllus
	2	P. linarioides
	2	P. dolius
	1	P. fremontii
	2	P. speciosus
	5	P. labrosus
	1	P. dasyphyllus
	1	P. thompsonii
	5	P. centranthifolius
	1	P. perpulcher
	1	P. neomexicanus
	5	P. barbatus
	1	P. alpinus
	3	P. palmeri
	3	P. grinellii

Fig. A1 The *trnC-D/T-L* consensus tree for *Penstemon* and *Keckiella* (with *Antirrhinum majus* as outgroup) from Wolfe et al. (2006), pruned to include only species (except *Penstemon lentus*) for which both floral color and floral anthocyanidins are known. Branch colors: blue = blue/purple flowers that are bee pollinated; red = red/orange flowers that are hummingbird pollinated. Pigments: *1* = delphinidin; *2* = delphinidin + cyanidin; *3* = cyanidin; *4* = cyanidin + pelargonidin; *5* = pelargonidin. Broken line for *P. lentus* indicates no anthocyanidin data available for this species.

Literature Cited

Ackermann M, M Weigend 2006 Nectar, floral morphology and pollination syndrome in Loasaceae subfam. Loasoideae (Cornales). Ann Bot 98:503–514.

Anderson LL, KN Paige 2003 Multiple herbivores and coevolutionary interactions in an *Ipomopsis* hybrid swarm. Evol Ecol 17:139–156.

Armbruster WS 2002 Can indirect selection and genetic context contribute to trait diversification? a transition-probability study of blossom-colour evolution in two genera. J Evol Biol 15:468–486.

Bradshaw HD, DW Schemske 2003 Allele substitution at a flower colour locus produces a pollinator shift in monkeyflowers. Nature 426:176–178.

Burdon JJ, DR Marshall, AHD Brown 1983 Demographic and genetic changes in populations of *Echium plantagineum*. J Ecol 71:667–679.

Campbell DR, NM Waser, EJ Melendez-Ackerman 1997 Analyzing pollinator-mediated selection in a plant hybrid zone: hummingbird visitation patterns on three spatial scales. Am Nat 149:295–315.

Chang S-M, Y Lu, MD Rausher 2005 Neutral evolution of the nonbinding region of the anthocyanin regulatory gene *Ipmyb1* in *Ipomoea*. Genetics 170:1967–1978.

Coberly LC 2003 The cost of white flowers: pleiotropy and the evolution of floral color. PhD diss. Duke University, Durham, NC.

Coberly LC, MD Rausher 2003 Analysis of a chalcones synthase mutant in *Ipomoea purpurea* reveals a novel function for flavonoids: amelioration of heat stress. Mol Ecol 12:1113–1124.

Durbin ML, KE Lundy, PL Morrell, CL Torres-Martinez, MT Clegg 2003 Genes that determine flower color: the role of regulatory changes in the evolution of phenotypic adaptations. Mol Phylogenet Evol 29:507–518.

Emms SK, ML Arnold 2000 Site-to-site differences in pollinator visitation patterns in a Louisiana iris hybrid zone. Oikos 91:568–578.

Epling C, H Lewis, FM Ball 1960 The breeding group and seed storage: a study in population dynamics. Evolution 14:238–255.

Faegri K, L van der Pijl 1966 The principles of pollination ecology. Pergamon, Oxford. 248 pp.

Fenster CB, WS Armbruster, P Wilson, MR Dudash, JD Thomson 2004 Pollination syndromes and floral specialization. Annu Rev Ecol Evol Syst 35:375–403.

Fineblum WL, MD Rausher 1997 Do genes influencing floral pigmentation also influence resistance to herbivores and pathogens? the *W* locus in *Ipomoea purpurea*. Ecology 78:1646–1654.

Frey FM 2004 Opposing natural selection from herbivores and pathogens may maintain floral-color variation in *Claytonia virginica* (Portulacaceae). Evolution 58:2426–2437.

Fry JD, MD Rausher 1997 Selection on a floral color polymorphism in the tall morning glory (*Ipomoea purpurea* L.): transmission success of the alleles through pollen. Evolution 51:66–78.

Ghebrehiwet M, B Bremer, M Thulin 2000 Phylogeny of the tribe Antirrhinae (Scrophulariaceae) based on morphological and *ndh*F sequence data. Plant Syst Evol 220:223–239.

Gigord LDB, MR Macnair, A Smithson 2001 Negative frequency-dependent selection maintains a dramatic flower color polymorphism in the rewardless orchid *Dactylorhiza sambucina* (L.) Soo. Proc Natl Acad Sci USA 98:6253–6255.

Gomez JM 2000 Phenotypic selection and response to selection in *Lobelia maritima*: importance of direct and correlational components of natural selection. J Evol Biol 13:689–699.

Grant V 1993 Origin of floral isolation between ornithophilous and sphingophilous plant species. Proc Natl Acad Sci USA 90:7729–7733.

Griesbach RJ 1996 The inheritance of flower color in *Petunia hybrida* Vilm. J Hered 87:241–244.

Habu Y, Y Hisatomi, S Iida 1998 Molecular characterization of the mutable *flaked* allele for flower variegation in the common morning glory. Plant J 16:371–376.

Hannan GL 1981 Flower color polymorphism and pollination biology of *Platystemon californicus* Benth. (Papaveraceae). Am J Bot 68:233–243.

Hodges SA, ML Arnold 1994 Floral and ecological isolation between *Aquilegia formosa* and *Aquilegia pubescens*. Proc Natl Acad Sci USA 91:2493–2496.

Holton TA, EC Cornish 1995 Genetics and biochemistry of anthocyanin biosynthesis. Plant Cell 7:1071–1083.

Hoshino A, Y Morita, J-D Choi, N Saito, K Toki, Y Tanaka, S Iida 2003 Spontaneous mutations of the flavonoid 3'-hydroxylase gene conferring reddish flowers in three morning glory species. Plant Cell Physiol 44:990–1001.

Igic B, R Lande, JR Kohn 2008 Loss of self-incompatibility and its evolutionary consequences. Int J Plant Sci 169:93–104.

Igic G, L Bohs, JR Kohn 2003 Historical inferences from the self-incompatibility locus. New Phytol 161:97–105.

Irwin RE, SY Strauss 2005 Flower color microevolution in wild radish: evolutionary response to pollinator-mediated selection. Am Nat 165:225–237.

Irwin RE, SY Strauss, S Storz, A Emerson, G Gibert 2003 The role of herbivores in the maintenance of a flower color polymorphism in wild radish. Ecology 84:1733–1743.

Jones KN 1996 Fertility selection on a discrete floral polymorphism in *Clarkia* (Onagraceae). Evolution 50:71–79.

Jones KN, JS Reithel 2001 Pollinator-mediated selection on a flower color polymorphism in experimental populations of *Antirrhinum* (Scrophulariaceae). Am J Bot 88:447–454.

Kay KM, PA Reeves, RG Olmstead, DW Schemske 2005 Rapid speciation and the evolution of hummingbird pollination in Neotropical *Costus* subgenus *Costus* (Costaceae): evidence from nrDNA ITS and ETS sequences. Am J Bot 92:1899–1910.

Kay QON 1976 Preferential pollination of yellow-flowered morphs of *Raphanus raphanistrum* by *Pieris* and *Eristalis* spp. Nature 261:230–232.

Kniskern JM, MD Rausher 2006 Environmental variation mediates the deleterious effects of *Coleosporium ipomoeae* on *Ipomoea purpurea*. Ecology 87:675–685.

Levin DA 1969 The effect of corolla color and outline on interspecific pollen flow in *Phlox*. Evolution 23:444–455.

——— 1972 The adaptedness of corolla-color variants in experimental and natural populations of *Phlox drummondii*. Am Nat 106:57–70.

——— 1985 Reproductive character displacement in *Phlox*. Evolution 39:1275–1281.

Levin DA, ET Brack 1995 Natural selection against white petals in phlox. Evolution 49:1017–1022.

Levin DA, HW Kerster 1967 Natural selection for reproductive isolation in *Phlox*. Evolution 21:679–687.

Maddison WP 2006 Confounding asymmetries in evolutionary diversification and character change. Evolution 60:1743–1746.

Maddison WP, PE Midford, SP Otto 2007 Estimating a binary character's effect on speciation and extinction. Evolution (forthcoming).

Melendez-Ackerman EJ, DR Campbell 1998 Adaptive significance of flower color and inter-trait correlations in an *Ipomopsis* hybrid zone. Evolution 52:1293–1303.

Melendez-Ackerman EJ, DR Campbell, NM Waser 1997 Hummingbird behavior and mechanisms of selection on flower color in *Ipomopsis*. Ecology 78:2532–2541.

Miller RB 1981 Hawkmoths and the geographic patterns of floral variation in *Aquilegia caerulea*. Evolution 35:763–774.

Miller RE, JA McDonald, PS Manos 2004 Systematics of *Ipomoea* subgenus *Quamoclit* (Convolvulaceae) based on ITS sequence data and a Bayesian phylogenetic analysis. Am J Bot 91:1208–1218.

Miller RE, MD Rausher, PS Manos 1999 Phylogenetic systematics of *Ipomoea* (Convolvulaceae) based on ITS and *waxy* sequences. Syst Bot 24:209–227.

Mol J, E Grotewold, R Koes 1998 How genes paint flowers and seeds. Trends Plant Sci 3:212–217.

Morita Y, M Saitoh, A Hoshino, E Nitasaka, S Iida 2006 Isolation of cDNAs for R2R3-MYB, bHLH and WDR transcriptional regulators and identification of *c* and *ca* mutations conferring white flowers in the Japanese morning glory. Plant Cell Physiol 47:457–470.

Nosil P, AO Mooers 2005 Testing hypotheses about ecological specialization using phylogenetic trees. Evolution 59:2256–2263.

Ollerton J 1996 Reconciling ecological processes with phylogenetic patterns: the apparent paradox of plant-pollinator systems. J Ecol 84:767–769.

——— 1998 Sunbird surprise for syndromes. Nature 394:726–727.

Pagel M, F Lutzoni 2002 Accounting for phylogenetic uncertainty in comparative studies of evolution and adaptation. Pages 148–161 *in* M Lässig, A Valleriani, eds. Biological evolution and statistical physics. Springer, Berlin.

Pellegrina G, F Bellusci, A Musacchio 2005 Evidence of post-pollination barriers among three colour morphs of the deceptive orchid *Dactylorhiza sambucina* (L.) Soo. Sex Plant Reprod 18:179–185.

Perret M, A Chautems, R Spichiger, G Kite, V Savolainen 2003 Systematics and evolution of tribe Sinningieae (Gesneriaceae): evidence from phylogenetic analysis of six plastid DNA regions and nuclear ncpGS. Am J Bot 90:445–460.

Quattrocchio F, J wing, K van der Woude, E Souer, N de Vetten, J Mol, R Koes 1999 Molecular analysis of the *anthocyanin2* gene of *Petunia* and its role in the evolution of flower color. Plant Cell 11:1433–1444.

Rausher MD 2006 The evolution of flavonoids and their genes. Pages 175–211 *in* E Grotewold, ed. The science of flavonoids. Springer, Berlin.

Rausher MD, D Augustine, A Vanderkooi 1993 Absence of pollen discounting in genotypes of *Ipomoea purpurea* exhibiting increased selfing. Evolution 47:1688–1695.

Rausher MD, JD Fry 1993 Effects of a locus affecting floral pigmentation in *Ipomoea purpurea* on female fitness components. Genetics 134:1237–1247.

Robertson C 1928 Flowers and insects: lists of visitors of four hundred and fifty-three flowers. Published by the author, Carlinville, IL.

Schemske DW, P Bierzychudek 2001 Perspective: evolution of flower color in the desert annual *Linanthus parryae*: Wright revisited. Evolution 55:1269–1282.

——— Forthcoming Spatial differentiation for flower color in the desert annual *Linanthus parryae*: was Wright right? Evolution, doi: 10.1111/j.1558-5646.2007.00219.x.

Schemske DW, HD Bradshaw 1999 Pollinator preference and the evolution of floral traits in monkeyflowers (*Mimulus*). Proc Natl Acad Sci USA 96:11910–11915.

Schwinn K, J Venail, Y Shang, S Mackay, V Alm, E Butelli, R Oyama, P Bailey, K Davies, C Martin 2006 A small family of MYB-regulatory genes controls floral pigmentation intensity and patterning in the genus *Antirrhinum*. Plant Cell 18:831–851.

Scogin R, CE Freeman 1987 Floral anthocyanins of the genus *Penstemon*: correlations with taxonomy and pollination. Biochem Syst Ecol 15:355–360.

Shirley BW 1996 Flavonoid biosynthesis: "new" functions for an "old" pathway. Trends Plant Sci 1:377–382.

Simms EL, MA Bucher 1996 Pleiotropic effects of flower-color intensity on herbivore performance on *Ipomoea purpurea*. Evolution 50:957–963.

Smith SD, DA Baum Forthcoming Systematics of Iochrominae (Solanaceae): patterns in floral diversity and interspecific crossability. Acta Horticult.

Spelt C, F Quattrocchio, J Mol, R Koes 2002 ANTHOCYANIN1 of petunia controls pigment synthesis, vacuolar pH, and seed coat development by genetically distinct mechanisms. Plant Cell 14:2121–2135.

Spitze K 1993 Population structure in *Daphnia obtusa*: quantitative genetic and allozyme variation. Genetics 135:367–374.

Stanton ML, AA Snow, SN Handel, J Bereczky 1989 The impact of a flower-color polymorphism on mating patterns in experimental populations of wild radish (*Raphanus raphanistrum* L.). Evolution 43:335–346.

Stickland G, BJ Harrison 1974 Precursors and genetic control of pigmentation. I. Induced biosynthesis of pelargonidin, cyanidin and delphinidin in *Antirrhinum majus*. Heredity 33:108–112.

Storz JF 2002 Contrasting patterns of divergence in quantitative traits and neutral DNA markers: analysis of clinal variation. Mol Ecol 11:2537–2551.

Strauss SY, JB Whittall 2006 Non-pollinator agents of selection on floral traits. Pages 120–138 *in* LD Harder, SCH Barrett, eds. Ecology and evolution of flowers. Oxford University Press, Oxford.

Streisfeld MA, JR Kohn 2005 Contrasting patterns of floral and molecular variation across a cline in *Mimulus aurantiacus*. Evolution 59:2548–2559.

Tanaka Y, S Tsuda, T Kusumi 1998 Metabolic engineering to modify flower color. Plant Cell Physiol 39:1119–1126.

Thomson JD, P Wilson 2008 Explaining evolutionary shifts between bee and hummingbird pollination: convergence, divergence, and directionality. Int J Plant Sci 169:23–38.

Walker AR, PA Davison, AC Bolognesi-Winfield, CM James, N Srinivasan, TL Blundell, JJ Esch, MD Marks, JC Gray 1999 The TRANSPARENT TESTA GLABRA1 locus, which regulates trichome differentiation and anthocyanin biosynthesis in *Arabidopsis*, encodes a WD40 repeat protein. Plant Cell 11:1337–1349.

Warren J, S Mackenzie 2001 Why are all colour combinations not equally represented as flower-color polymorphisms? New Phytol 151:237–241.

Waser NM, L Chittka, MV Price, NM Williams, J Ollerton 1996 Generalization in pollination systems, and why it matters. Ecology 77:1043–1060.

Waser NM, MV Price 1981 Pollinator choice and stabilizing selection for flower color in *Delphinium nelsonii*. Evolution 35:376–390.

——— 1993 Pollination behaviour and natural selection for flower colour in *Delphinium nelsonii*. Nature 302:422–424.

Weis AE, SJ Franks 2006 Herbivory tolerance and coevolution: an alternative to the arms race? New Phytol 170:423–425.

Wesselingh RA, ML Arnold 2000 Pollinator behaviour and the evolution of Louisiana iris hybrid zones. J Evol Biol 13:171–180.

Whibley AC, NB Langlade, C Andalo, AI Hanna, A Bangham, C Thebaud, E Coen 2006 Evolutionary paths underlying flower color variation in *Antirrhinum*. Science 313:963–965.

Whittall JB, C Voelckel, DJ Kliebenstein, SA Hodges 2006 Convergence, constraint and the role of gene expression during adaptive radiation: floral anthocyanins in *Aquilegia*. Mol Ecol 15:4645–4657.

Wilson P, MC Castellanos, JN Hogue, JD Thomson, WS Armbruster 2004 A multivariate search for pollination syndromes among penstemons. Oikos 104:345–361.

Wilson P, MC Castellanos, A Wolfe, JD Thomson 2006 Shifts between bee and bird-pollination among penstemons. Pages 47–68 *in* NM Waser, J Ollerton, eds. Plant-pollinator interactions: from specialization to generalization. University of Chicago Press, Chicago.

Wolfe AD, CP Randle, SL Datwyler, JJ Morawetz, N Arguedas, J Diaz 2006 Phylogeny, taxonomic affinities, and biogeography of *Penstemon* (Plantaginaceae) based on ITS and cpDNA sequence data. Am J Bot 93:1699–1713.

Wolfe LM 1993 Reproductive consequences of a flower color polymorphism in *Hydrophyllum appendiculatum*. Am Midl Nat 129:405–408.

—— 2001 Associations among multiple floral polymorphisms in *Linum pubescens* (Linaceae), a heterostylous plant. Int J Plant Sci 162:335–342.

Wolfe LM, SE Sellers 1997 Polymorphic floral traits in *Linaria canadensis* (Scrophulariaceae). Am Midl Nat 138:134–139.

Wray GA, MW Hahn, E Abouheif, JP Balhoff, M Pizer, MV Rockman, L Romano 2003 The evolution of transcriptional regulation in eukaryotes. Mol Biol Evol 20:1377–1419.

Wright S 1943 An analysis of local variability of flower color in *Linanthus parryae*. Genetics 28:139–156.

—— 1978 Evolution and the genetics of populations. IV. Variability within and among natural populations. University of Chicago Press, Chicago.

Zufall RA 2003 Evolution of red flowers in *Ipomoea*. PhD diss. Duke University, Durham, NC.

Zufall RA, MD Rausher 2004 Genetic changes associated with floral adaptation restrict future evolutionary potential. Nature 428:847–850.

Int. J. Plant Sci. 169(1):23–38. 2008.
1058-5893/2008/16901-0003$15.00   DOI: 10.1086/523361

EXPLAINING EVOLUTIONARY SHIFTS BETWEEN BEE AND HUMMINGBIRD POLLINATION: CONVERGENCE, DIVERGENCE, AND DIRECTIONALITY

James D. Thomson* and Paul Wilson[†]

*Department of Ecology and Evolutionary Biology, University of Toronto, Toronto, Ontario M5S 3G5, Canada, and Rocky Mountain Biological Laboratory, Crested Butte, Colorado 81224-0519, U.S.A.; and †Department of Biology, California State University, Northridge, California 91330-8303, U.S.A.

In certain angiosperm genera, closely related species have diverged from one another to converge on different pollination syndromes, whereas species with intermediate phenotypes are rare or absent. Convergent conformity to syndromes implies the existence of "evolutionary attractors" toward which phenotypes are drawn; divergent breaks from conformity show that populations can escape one attractor and be drawn to another. We discuss how these two opposed processes can be reconciled for the special case of evolutionary transitions between bee pollination and hummingbird pollination. In this case, a third phenomenon, the directional bias in favor of transitions from melittophily to ornithophily, also needs explanation. Older treatments chiefly ascribed convergence to cognitive and morphological properties of pollinators and ascribed transitions to geographical differences in pollinator availability. Those treatments did not specifically address what factors would overcome and disrupt the stabilizing selection that would be expected to preserve the pollination syndrome of a plant species. Here, we focus on possible contributors to destabilization, especially considering the possible roles of (1) differences among pollinators in pollen-transfer efficiency, (2) mutations with large effects on floral phenotypes, and (3) losses of function in the biochemical pathways that produce floral pigments. We conclude that all of these can influence the evolution of pollinator transitions but that the process usually needs to be initiated by external ecological factors that change the visitation rates of pollinators. We discuss the roles of particular floral characters in several plant genera that have undergone transitions. We expect that transitions reach completion through a "centripetal" process of selection that incrementally recruits changes in multiple characters.

Keywords: evolutionary transitions, *Mimulus*, *Penstemon*, pollination syndromes.

Introduction

Building on the natural history of pollination pioneered by Köhlreuter (1761–1766) and Sprengel (1793), Darwin (1862) illustrated his theory of evolution by showing that many floral characters can be interpreted as specific adaptations for service by various kinds of pollinating animals. Static views of function take on a dynamic dimension if one considers closely related plant species whose flowers differ in multiple characters because they are adapted to different pollinators. In such cases, we say that an evolutionary pollinator transition, or pollinator shift, has occurred. Such shifts can exemplify several evolutionary processes: adaptation, divergence from an ancestral mode of life, convergence toward a common functional phenotype, and the maintenance of reproductive isolation. Here, we focus on the circumstances that promote one type of evolutionary transition between "pollination syndromes," namely, shifts between adaptation to pollination by Hymenoptera, especially bees (the melittophily syndrome), and adaptation to pollination by hummingbirds (one type of ornithophily).

Such shifts may be a special case among pollinator transitions because the applicability of the pollination syndrome concept is greater here than in much of the rest of the angiosperms

(Fenster et al. 2004). Still, we believe that our narrow focus is warranted for several reasons. (1) Such shifts have occurred repeatedly, more than 100 times in western North America alone (Stebbins 1989; Grant 1994; Wilson et al. 2007). Importantly, species that conform to distinctly different pollination syndromes can be close relatives, even sister species. (2) There are active research programs on several genera displaying such shifts, including *Penstemon*, *Mimulus*, *Ipomoea*, *Costus*, *Aquilegia*, *Silene*, and *Salvia* (fig. 1). As these research programs mature, we expect generalities to emerge. In this review, we relate findings concerning these genera and synthesize them into a common explanation. Research on other types of shifts is patchily distributed, and a search for generality is probably premature. (3) Adaptation to bees or hummingbirds seems to produce multicharacter phenotypic discontinuities that correspond to cleanly separated pollination syndromes. For example, ordinations of *Penstemon* species based on floral characters yield clear clusters of melittophiles and ornithophiles (Wilson et al. 2004).

Pollination syndromes—defined as suites of floral characters that are adaptations to one kind of pollinator or another—are important in our account of transitions. Probably for as long as the syndrome concept has existed, its architects and defenders have warned against oversimplification (Müller and Delpino 1871; Baker 1963; Stebbins 1970; Faegri and van der Pijl 1979). Nevertheless, the telegraphic condensations in textbooks have left students with the simplistic impression that most flowers

Fig. 1 Flowers seemingly adapted to bee and bird pollination in six genera. *A*, Melittophilous *Penstemon strictus* (photo: J. Thomson). *B*, Ornithophilous *Penstemon barbatus* (photo: J. Thomson). *C*, Melittophilous *Mimulus lewisii* (photo: D. Schemske). *D*, Ornithophilous *Mimulus cardinalis* (photo: D. Schemske). *E*, Melittophilous *Ipomoea purpurea* (photo: M. Rausher). *F*, Ornithophilous *Ipomoea quamoclit* (photo: M. Rausher). *G*, Melittophilous *Costus malortieanus* (photo: D. Schemske). *H*, Ornithophilous *Costus pulverulentus* (photo: D. Schemske). *I*, Melittophilous *Aquilegia saximontana* (photo: J. Whittall). *J*, *Aquilegia formosa* (photo: S. Hodges). *K*, Entomophilous *Silene caroliniana* (photo: J. Antonovics and M. Hood). *L*, Ornithophilous *Silene virginica* (photo: J. Antonovics and M. Hood).

can be classified nonhierarchically into categories pollinated by beetles, flies, carrion flies, bees, moths, butterflies, birds, and bats (the text by Simpson [2006] is a recent example). In fact, many animal-pollinated flowers present rewards openly and advertise them with cues that are accessible to many animals. Such flowers tend to attract insects from several orders and can be more or less successfully pollinated by most of them (Herrera 1996; Waser et al. 1996). Such generalized flowers present a variety of characters, and they resist functional categorization into pollination syndromes. In contrast, other taxa conform well to syndrome characterizations in that their morphologies appear to be adapted to pollination by animals of narrower taxonomic groups (e.g., single orders or families) or functional groups (e.g., long-tongued and large-bodied nectar feeders; Fenster et al. 2004). Such plants may be in the minority, and even when a species' conformity to a syndrome indicates a history of adaptation to a particular type of pollinator, it may be visited and successfully pollinated by other types (Faegri and van der Pijl 1979; Ollerton 1996; Mayfield et al. 2001). Syndromes are most convincingly revealed by convergent evolution, but even here there is a problem: differently endowed lineages may adapt to the same pollinators in ways that do not produce convergence, although those changes may be comprehensible given the initial conditions. Some authors find syndromes generally inapplicable to the systems they know best (e.g., Corbet 2006), but we find them useful in the more restricted context considered here (see Armbruster et al. 2000; Thomson et al. 2000; Fenster et al. 2004).

In this article, we hope to extend and enrich classical evolutionary explanations for clear-cut, qualitative, evolved shifts out of melittophily (the bee-pollination syndrome) and into ornithophily (the bird-pollination syndrome). In considering such transitions, Grant (1994) emphasized the biogeographical history of range extensions that brought hummingbirds into contact with bee-pollinated progenitors; he also invoked a positive feedback process at the community level, whereby the presence of some ornithophilous taxa would pave the way for subsequent shifts by melittophiles, simply by ensuring that hummingbirds would be regular members of such communities. Stebbins (1989) concentrated on changes in progenitor lineages that would serve as preadaptations for bird pollination, including the colonization of wetter habitats and the adoption of perennial life histories. He inferred that these characteristics were important preconditions for hummingbird pollination because they are frequently associated with it. These older explanations, therefore, invoked factors that could bring about frequent contacts between hummingbirds and melittophilous plants that had the potential to shift. Those conditions probably encourage shifts, but we find them incomplete: They do not explain what could overcome the strong niche conservatism that we would expect in plants that are well adapted to one sort of pollinator. In our view, conformity to a syndrome is maintained by stabilizing selection, and transitions occur only when something happens to destabilize the interdependence of plant and pollinator. Grant (1994) and Stebbins (1989) did not address this requirement of destabilization. It is our main focus.

We begin by proposing some metaphorical language for discussing the convergence of floral phenotypes toward pollination syndrome attractors. Then, we review models that show how differences in the efficiency of pollen transfer by bees and

birds could establish a switch point that could divert a lineage from melittophily to ornithophily, and we discuss ecological circumstances that could trigger such switches. Next, we consider the sources and kinds of genetically based variation in floral traits that are involved in shifts from bee to bird pollination. Recognizing that new mutations of large effect are implicated in some bee-to-bird transitions, we ask whether such mutations are necessary in general. We then consider why bee-to-bird transitions strongly outnumber transitions in the opposite direction. We follow this by some reflections on the multicharacter nature of the difference between melittophiles and ornithophiles. Finally, we briefly consider the applicability of our approach to broader questions: transitions among pollination systems other than Hymenoptera and hummingbirds, and evolutionary transitions in general.

Metaphors for Transitional and Nontransitional Changes: Evolutionary Vortices and Adaptive Wandering

Recognizing that the contrast between melittophily versus ornithophily (examples in fig. 1) may represent an extraordinarily clear-cut distinction between syndromes, it is worth mentioning as a point of contrast how we think generalist flowers might diversify without becoming specialized. Unlike the dynamics that we want to focus on here, it may be more common for floral phenotypes to respond to pollinator-driven selection by what we have called "adaptive wandering" (Wilson and Thomson 1996).

When flowers diversify by adaptive wandering, plants in geographically separate populations diverge phenotypically in response to different selection regimes imposed by local differences in the pollinator communities they experience (as in the geographic mosaic of Thompson 1994). They adapt to their local pollinators; this adaptation results in wandering because the local differences are too slight or too brief to cause the flowers to evolve any mechanisms that exclude other pollinators upon secondary contact. Lilies in the genus *Calochortus* section *Mariposa* are varied in color patterns, petal hairs, and the shape of nectar glands. Indeed, various species of pollinators respond differently to the different species of mariposa lilies, but the plants have not permanently specialized on different pollinators, and we cannot sensibly explain the diversity among mariposa lilies by seeking pollinator shifts (Dilley et al. 2000). Adaptive wandering results in taxonomically noteworthy variation among species that is driven by local adaptation, not by genetic drift, but it does not involve a shift to a new functional group of pollinators (sensu Fenster et al. 2004). It is not a coherent change in strategy. The adaptive adjustments that do take place are likely to be transient, reversible, and idiosyncratic; they do not converge toward a different syndrome. Like floating logs pushed around on the surface of the ocean by shifting winds and currents, the phenotypes of these populations are nudged about in multivariate character space, but the journey lacks structure and predictability. One population might be influenced by a bee that has an arbitrary preference for distinctive brown spots, another by a beetle that prefers hairy petals. In time, however, those pollinators give way to others; although the flowers have diverged, the divergence has not resulted in the exclusion or discouragement of any subsequent type of pollinator. The pattern reflects little more than local adaptation plus geographic heterogeneity in the selective environment.

In contrast, bee-to-bird shifts are more deterministic, and the multicharacter syndromes seem more like specialized strategies. The lineages have gravitated toward predefined attractors. Rather than resembling logs buffeted by unstructured Brownian currents, these phenotypes are like logs that have drifted into fixed whirlpools or vortices (Gilpin and Soulé 1986), where they tend to become entrained by a kind of stabilizing selection. In multivariate character space, these phenotypes tend to be drawn toward conformity with particular pollination syndromes by a coherent set of changes in a predictable set of characters. Importantly, the approach toward a syndrome involves two sorts of phenotypic change, which Faegri and van der Pijl (1979, p. 126) term "positive" and "negative." Positive adaptations are directed to attracting and using the primary pollinator type, whereas negative adaptations lead toward repelling or excluding secondary, less desirable pollinators (Castellanos et al. 2004).

We like the metaphor of the vortex, which Gilpin and Soulé (1986) used to describe the interacting factors leading to population extinction. Other metaphorical language is available for progressive change toward alternative attractors. Community ecologists sometimes speak of "alternative stable states" or "basins of attraction" (Beisner et al. 2003), and many comparative biologists have freely borrowed Sewall Wright's metaphor of "adaptive peaks" (Armbruster 1990; Losos 1992), which is the gravitational inversion of basins and vortices. The vortex metaphor is a good catchall that incorporates both genetic processes and ecological circumstances and shows how they interact to produce a particular outcome. Because it is not burdened by as much past usage as other metaphors and because it is unassociated with any particular model of genetic control, we hope that it will encourage fresh thought. Whether a population gets drawn into a syndrome vortex, we believe, depends not only on genetics and floral phenotypes but also on complicated interactions and feedbacks that ramify through the entire plant-pollinator community, as well as on stochasticity. In this sense, the factors that draw a population toward a syndrome are similar in complexity to those that can lead a population toward extinction. Finally, the vortex metaphor lends itself to considering why one vortex may be weaker than another, i.e., easier for a species to escape from. This notion is useful in considering directional biases in pollinator shifts.

Within a single pollination vortex, there may be plenty of multivariate character space for bounded adaptive wandering. A melittophilous lineage of flowers may diversify in response to selection exerted by different communities of bees on different populations. In such a case, we might expect to see more variation in less essential characters, such as nectar guides, scents, and fringing of petals, and less variation in the less decorative characters needed for retaining and exploiting bees. These would include the reward characters most important to securing visits (Waser et al. 1996), such as nectar quantity, concentration, and composition, and the floral geometry needed to ensure anther and stigma contacts. Thus, selection would keep the various species of the lineage in the vortex while permitting them to wander adaptively. In contrast, a shift out of the melittophily vortex into the ornithophily vortex would require escaping from those stabilizing forces and coming under the influence of a qualitatively new attractor.

Differences in Pollinator Quality as Well as Quantity: Lessons from Penstemons

Our thinking about shifts between melittophily and ornithophily has been inspired by penstemons (Thomson et al. 2000; Wilson et al. 2006; fig. 1A, 1B). In this group, there have been numerous transitions in which bee-to-bird shifts have been marked by multicharacter convergence toward the hummingbird floral syndrome (Wolfe et al. 2006; Wilson et al. 2007). Our proffered evolutionary scenario for transitions from bee to hummingbird pollination rests importantly on differences between bees and birds in the quality of pollination ("efficiencies" or "efficacies") as well as the quantity of pollination ("visitation rates"). The transition we envision is outlined briefly as follows. Ancestral penstemons are adapted to bees (including pollen-collecting and therefore beelike masarid wasps), even though bees transfer relatively low proportions of the pollen that they remove from anthers. These melittophilous penstemons are not adapted to birds and are not particularly attractive to them. Therefore, bird visits are infrequent, but even without adaptation to birds, when a hummingbird does visit a patch of flowers, it delivers a high proportion of the pollen it removes. Given a strong enough difference in the pollen-delivery efficiencies of two covisiting pollinators, it can be shown that the less efficient one may become a detrimental, pollen-wasting parasite if visits by the better one are frequent enough (Thomson and Thomson 1992; Thomson 2003). The relative numbers of visits by the two pollinators establish a threshold at which a pollinator's contribution switches from beneficial to parasitic. We propose that a change in ecological circumstances can increase the frequency of bird visits enough to cross this threshold, thereby destabilizing selection and bumping the floral phenotype out of the melittophily vortex and into the ornithophily vortex. If the new condition persists for long enough, selection will favor characters that make the flowers better suited to bird pollination and less attractive to bees. Those changes will effectively lower the threshold level of bird visits, providing a positive feedback that augments the superiority of ornithophily and makes reversal unlikely.

The clade we call "penstemons" includes the large genus *Penstemon* and the smaller sister genera *Keckiella*, *Chionophila*, *Nothochelone*, and *Chelone* (Wolfe et al. 2002). Essentially all the 284 species are melittophiles or hummingbird ornithophiles, and there is little ambiguity regarding these assignments. The two syndromes form distinct clusters in an ordination of 49 species (Wilson et al. 2004); yet, even among the species that we refer to as ornithophiles, there is variation in whether bees are retained as copollinators or excluded from access to nectaries. Field observations confirm that hummingbirds are prominent visitors to plants conforming to ornithophily and that they are very low-frequency visitors to melittophiles. Of the 284 species, 41 conform more or less to the hummingbird syndrome. Phylogenetic analyses suggest as many as 21 separate transitions toward ornithophily, and certainly no fewer than 10 (Wilson et al. 2007). There is no evidence for reversals, although reversal is a fairly difficult phenomenon to rule out. In addition, there are few if any species that have evolved toward other kinds of pollinators: just one clade of two species that may use bee flies along with bees and one species that may use butterflies along with bees (Wilson et al. 2007). Both the frequency and the directionality of shifts to hummingbird pollination are striking phenomena that demand explanation.

We have explored a mechanistic hypothesis to account for a pollen-transfer efficiency difference between bees and birds (Castellanos et al. 2003), based on differences in how these animals treat pollen. Depending on the size of a penstemon flower and the size of a bee visitor, anthers deposit pollen on the dorsal surfaces of the face, head, or thorax, and stigmas receive pollen from these same regions. Female bees groom while foraging, sweeping grains from the valuable dorsal surfaces into corbiculae or scopae on the legs and undersides of the body. Surely, pollen that is moistened and packed into a corbicular pellet can no longer participate in pollination, and the chances also seem slim that a pollen grain in the ventral abdominal scopa of an *Osmia* would reach a penstemon stigma. Because the collected pollen is fed to larvae, natural selection has endowed bees with highly effective grooming and structures and behaviors for gathering up and retaining large quantities of pollen. To hummingbirds, however, pollen is useless. They have no specialized pollen-grooming structures, and they tend to preen only between foraging bouts. This difference in grooming behavior may contribute to the seeming ease and the unidirectionality of transitions from melittophily to ornithophily in penstemons. One confirmatory finding is that the anthers of ornithophilous penstemons tend to open more widely and present pollen more freely than those of closely related melittophilous species (Castellanos et al. 2006), as would be expected if birds were less wasteful of pollen.

Simple bookkeeping models of pollen grain fates (Thomson and Thomson 1992) consider how male reproductive success (i.e., pollen export to stigmas) accrues when a plant is visited by varying numbers of two pollinator types, one wasteful and one efficient. If visits by the more efficient pollinator are rare, the more wasteful pollinator will serve the plant's interests as a beneficial mutualist. If, however, visits by the more efficient pollinator become more frequent, a threshold can be reached beyond which visits by the more wasteful pollinator actually reduce successful pollen export. At that point, any genetic variants that deter visits by the wasteful pollinator would be favored, as would any variants that improve the flower's fit or attractiveness to the less wasteful pollinator. For example, red colors might spread by deterring bees that have become conditional parasites (Raven 1972); improvements such as exserting the anthers and stigma probably make birds less wasteful transporters of pollen (Castellanos et al. 2004).

In other words, we envision the following scenario for pollinator transitions in flowers like those of penstemons. An ancestral population is adapted to pollination by large bees and displays the appropriate syndrome characters. Hummingbirds may be occasional visitors—these inquisitive animals frequently investigate flowers that do not conform to the "legitimate" syndrome—but visits are inconsequentially rare because the birds find the nectar offerings too meager. Therefore, bee pollination continues to exert stabilizing selection on the bee syndrome characters. The crucial initiator is some ecological change that shifts the bee-to-bird balance: either bees begin visiting less or birds begin visiting more. The former dominance of bees is destabilized. New directional selection then acts on standing genetic variation in floral characters to gradually produce conformity to the bird syndrome. Later, bees may return

to prominence in the visitor assemblage, but there is selection to discourage and exclude them after birds have become reliable pollinators. With the increase in hummingbird pollination, visits by bees have crossed the threshold from being beneficial to being parasitic. The ornithophiles acquire antibee adaptations in addition to probird adaptations (Castellanos et al. 2004). Bees can no longer extract nectar easily, and the position of sexual organs no longer fits their bodies. The progression toward ornithophily is ratchetlike.

Our explanation has at least two weak points concerning testability. First, it is hard to demonstrate numerically whether the difference in pollen wastage between bees and birds is large enough to reach the threshold point where bee visits become parasitic. The model parameters that determine that threshold elude simple empirical measurement, partly because it is hard to track the fates of pollen grains as they move through the very messy process of pollination. Considering the melittophilous *Penstemon strictus* as a stand-in for an ancestral melittophile, Castellanos et al. (2003) confirmed that hummingbirds could remove enough pollen from anthers to be important pollinators. Total pollen deposition by birds did not exceed that by bees, and removal by birds was slightly lower, so those measurements fell slightly short of showing that birds are already better pollinators than bees. However, total deposition was recorded from short pollen carryover trials of only 15 recipient flowers; given the flatter pollen carryover curves conferred by birds, a longer series of recipients probably would have shown birds to be more efficient. Figure 2 gives a calibrated graphical extrapolation for how pollen carryover curves could translate into differences in male reproductive success. Unfortunately, repeating the experiment with longer sequences of recipients would be prohibitively tedious and probably impossible for wild-caught bees because the bees do not behave naturally in flight cages after they are well fed. The flatter carryover functions of birds may also produce higher-quality pollination because a greater proportion of grains removed from a flower will be transported to a different plant as opposed to being geitonogamously deposited. Another complication of this experiment is that *P. strictus* flowers have wide mouths that permit birds to approach laterally, rather than being constrained to probe straight in, along the principal floral axis (Castellanos et al. 2004). Had we chosen a melittophilous species with a narrower corolla, stigmatic contacts might have been more frequent. So, birds may not be better pollinators than bees on a per-visit basis to a bee-adapted penstemon, but they are close. Castellanos et al. also found that bees were completely ineffectual at transferring pollen of the extreme ornithophile *Penstemon barbatus*, whereas hummingbirds were supremely effective pollinators of this species, which is adapted to being pollinated by them.

The second sketchy point of our scenario lies in the unstudied "ecological changes" that involve a shift in the bee : bird ratio. Stebbins (1989) favored the idea that shifts to hummingbird pollination have often been preceded by a species colonizing and adapting to a novel habitat where bees are less active or birds are more active than in the ancestral habitat. He suggested that this might often happen in habitats where bees get a slow start in the morning because of fog and where hummingbirds have ample woods for nesting. Cruden (1972) also stressed poor weather as favoring dependence on birds. Otherwise,

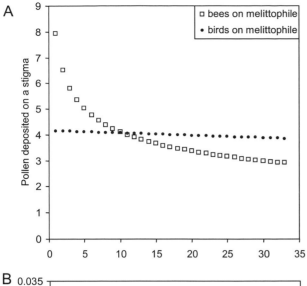

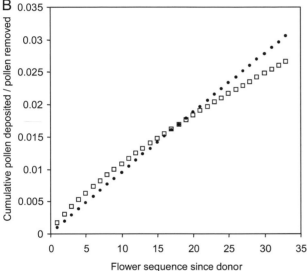

Fig. 2 Graphical depiction of why pollinators with flatter, long-tailed pollen carryover curves are favored through male function over pollinators with steeper curves. *A*, Bees export more pollen to the first recipient stigmas but less to later recipient stigmas than birds. *B*, Cumulative pollen exported divided by pollen removed in a visit shows how male fitness is accrued; taking into account pollen carryover beyond ca. 20 recipient stigmas, birds appear to be better than bees. However, these curves are extrapolated from data on only the first 15 stigmas (Castellanos et al. 2003).

the causes of ecological changes in the bee : bird ratio have not been well discussed, and they are difficult to study experimentally. Hummingbirds will investigate melittophilous penstemons, so the critical obstacle keeping birds from moving on to a bee-pollinated plant lies in reward economics. The comparatively paltry nectar volumes of melittophilous species do not induce regular visitation by birds under normal circumstances, particularly when frequent visits by bees keep the flowers nearly empty (Williams and Thomson 1998; Rodríguez-Gironés and Santamaría 2004). When Williams and Thomson continuously videotaped a plant of the melittophilous *P. strictus*

for most of a day, thousands of bee visits were recorded but no bird visits; at the same site, however, hummingbirds do sometimes visit *P. strictus* flowers at daybreak, when nectar volumes may have accumulated overnight. Using a pipette, Jordan (2004) brought nectar offerings in the melittophilous *Penstemon spectabilis* up to the 5 μL of 20% sugar found in the ornithophilous *Penstemon centranthifolius*, and hummingbirds immediately increased their visitation rate, despite the fact that the flowers were purple, had large lips, etc. This crude experiment suggests that hummingbird visitation quantity probably responds to nectar offerings directly. The study also suggests that something other than nectar offerings (flower color) may be important in determining visitation rates by bees, but unfortunately, at this study site, the bees involved were *Ceratina*, not industrious bees such as *Bombus*. After we have introduced the biology of a few other plant groups and brought up several other ideas, we will return to considerations of the ecology of shifts between pollination syndromes.

Monkeyflowers: Ecology and Genetics

The melittophilous *Mimulus lewisii* and the closely related ornithophilous *Mimulus cardinalis* (fig. 1C, 1D) have been studied by Bradshaw, Schemske, and colleagues (Bradshaw et al. 1995, 1999; Wilbert et al. 1997; Schemske and Bradshaw 1999; Bradshaw and Schemske 2003). The major articles from this team have mostly addressed the genetics of adaptation and the mechanisms of reproductive isolation, rather than the circumstances enabling the pollinator shift; nevertheless, the results have provocative implications for shifts. The team made crosses between the two species. Studying F_2 hybrids and subsequent progeny arrays, they found that a modest number of major genes control some of the characters that most influence visitation rates. In other words, the pollination syndrome difference is largely attributable to genetic loci of fairly large effect. Moreover, they used the floral variation in an F_2 progeny array to study the response of pollinators to multiple floral characters. By crossing the syndromes, they were able to recreate, in one place at one time, all the variation that must have been necessary to get from melittophily to ornithophily over evolutionary time. This allowed them to examine the different pollinators' reactions to all combinations of characters.

Hummingbird pollination has evidently arisen twice from bee pollination in *Mimulus* section *Erythranthe* (Beardsley et al. 2003). Thus, the direction is consistent with a seemingly widespread bias in favor of bee-to-bird transitions. In the species pair studied by Bradshaw and colleagues, *M. cardinalis* has evolved many ornithophilous traits compared with *M. lewisii*. Although sympatric speciation may have occurred, it is not implied by any data, nor have Bradshaw et al. pushed such a claim. For our purposes, we can envision that the two lineages might have first diverged in elevational preference, making them allopatric, with the nascent *M. cardinalis* living at lower elevations and *M. lewisii* living at higher elevations. Then later, we imagine, there was an adaptive shift in the lineage of *M. cardinalis* toward ornithophily. Eventually, the two species came to differ in nectar quantity, the extent of yellow carotenoid pigment, the amount of pinkish anthocyanin pigment, the orientation of the petal lobes, and the exsertion of anthers and stigmas. Each of these differences is partially attributable to a very few major loci (major loci being ones that account for >25% of the variance among F_2s), as well as probably some minor genes (Bradshaw et al. 1995, 1999). The extent of carotenoid pigmentation is controlled by a single genetic locus (yellow on upper petals [*yup*]; Hiesey et al. 1971). Other major quantitative trait loci (QTL) are associated with nectar offerings and petal orientation, characters that are prominent in constituting the difference in pollination syndrome.

The rates of visitation by bees and birds were studied in an array of F_2 progeny set out at intermediate elevation (Schemske and Bradshaw 1999). As displayed in figure 3, multiple regression showed that anthocyanin concentration, carotenoid concentration, nectar production, and the projected area of the corolla each affected bee or bird visitation rates when the other three variables were statistically accounted for. Therefore, each trait makes a contribution to the difference in pollination syndrome and to which kind of animal visits an individual plant. Those F_2 individuals that resemble melittophilous *M. lewisii* receive many visits by bees and few visits by birds, while those F_2 individuals that resemble ornithophilous *M. cardinalis* receive many visits by birds and few by bees. Each of the F_2 individuals was genotyped, so Schemske and Bradshaw could quantify how well any particular QTL predicted the visitation rates of bees and birds. The genotype at the *yup* locus (governing whether the flowers are pink or orange) affected bee visitation such that orange individuals received only 20% of the bee visits received by pink individuals. This genetic locus had no significant effect on bird visitation. Other loci did, however; most notably, a QTL associated with nectar production cut bird visitation by 50%. This nectar QTL had no significant effect on bee visitation.

The discovery of major genes differentiating syndromes is at least mildly unexpected. Ever since Fisher (1930), macro-mutations have often been dismissed as "hopeless monsters," so deviant that they ought to function poorly; also, genetic response to selection in a novel environment has been viewed as proceeding best when a character is affected by many additive genes of small effect. Such alleles are assumed to exist in any population and are available to be brought together to produce a phenotype more extreme than was previously present. Before Bradshaw et al. established the contrary, most evolutionists would have tended to attribute the syndrome differences in these two *Mimulus* to an indefinitely large number of loci of individually minor effect (see Orr and Coyne 1992; Coyne 1995), although perhaps Gottlieb (1984) would not have. Given Bradshaw et al.'s findings, one might imagine that a new mutation of large effect (such as *yup*) might allow the population to "jump" to a state of adaptation to hummingbirds, and that until the new mutation of large effect arose, evolution might have been limited by insufficient variation to act upon.

Bradshaw and Schemske (2003) constructed genotypes that mimicked a new mutant of this sort; i.e., they produced plants with an allele that makes the flowers orange but that otherwise had nearly the genetic background of melittophilous *M. lewisii*. (The goal was to make this flower color locus counter-correlated with other syndrome-specific characters. Unfortunately, nectar offerings were dragged along, presumably because of a closely linked nectar locus, so the artificial "mutant phenotypes" were not exactly like real mutants, as mentioned in Bradshaw and Schemske 2003; see also Wilson et al. 2006.) On the basis of visits to these phenotypes, Bradshaw and Schemske calculated that if the bee : bird abundance ratio were to fall to around one-ninth of its value in contemporary field conditions, then a mutant with the *yup* allele would receive as many visits as a wild-type *M. lewisii* and would be mostly hummingbird pollinated, whereas the wild type would be mostly bee pollinated.

This calculation raises a question about pollinator shifts. Given that a single novel mutation like *yup* can by itself produce such drastic changes in the attendances of different pollinators, ought we to view such mutations as either necessary or sufficient causes of pollinator shifts? D. W. Schemske (personal communication) does not think so. He believes that ecological changes in pollinator visitation rates are necessary before the genetics of floral characters become important, following this scenario: (1) a change in the environment favored hummingbirds over bees, perhaps a drop in the abundance of bees; (2) any mutation that increased the frequency of hummingbird pollination was favored, regardless of its genetic basis; and (3) mutations of large effect were most likely to be fixed in the early stages of adaptation when the population was far from its phenotypic optimum.

Therefore, although our comparisons of melittophilous and ornithophilous relatives in the penstemon clade focus on the efficiency of pollen transfer by different pollinators, while Bradshaw and Schemske's comparisons of similarly related *Mimulus* species focus more on the genetic control of syndrome characters and the quantitative preferences of pollinators for those characters, the two research programs do not contradict one another. Both of them invoke externally driven ecological

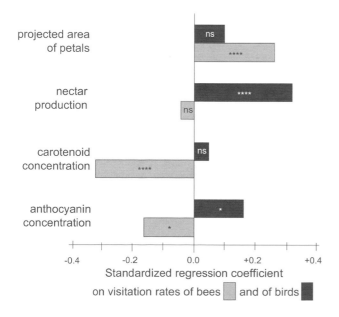

Fig. 3 Results of multiple regression of four floral traits on visitation rates by bees and by birds in an F_2 array of monkeyflowers. Significance notes: ns, $P > 0.05$; one asterisk, $P < 0.05$; four asterisks, $P < 0.0001$ (modified from Schemske and Bradshaw 1999).

changes in visitation rates as the initiators of change. Both of them suggest particular phenomena that might facilitate the transition from one syndrome to another. In penstemons, general relationships of pollinator efficiency might provide a switch point and a ratchet mechanism that would foster transition to birds. In particular, one can envision that a small increase in hummingbird visitation could nudge a plant population over a pollen-transfer efficiency threshold. In *Mimulus*, alleles with large effects on floral phenotypes and visitation rates could provide the genetic variation necessary to allow selection to shift phenotypes beyond a visitation rate threshold, also given a decrease in the bee : bird ratio.

The *Mimulus* data are tantalizing but not decisive on the possibility that a shift in pollination syndrome might be made permanent through a new mutation of large effect. When Bradshaw and Schemske (2003) put *yup* into plants that otherwise were nearly *M. lewisii*, birds found them more attractive than wild-type *M. lewisii* and bees found them less attractive, but bees were still more frequent pollinators than birds; therefore, the data are hardly in favor of a shift being "initiated" only by a new mutation. Even without referring to this result, we share Schemske's skepticism about the possibility that a new mutation, without an extrinsic change in ecology, could have been selected to fixation, thereby concomitantly transforming a lineage adapted to bees into one adapted to birds. Instead, we prefer a scenario that starts with some extrinsic ecogeographic change that increases the relative frequency of visitation by birds, then a period of response to selection using existing or new genetic variation that eventually moves the lineage out of the melittophily vortex into the ornithophily vortex. The finding of alleles of largish effect in *Mimulus* raises the possibility that a new "macromutation" might have moved the lineage past the threshold, thereby committing it to ornithophily. Did the lineage leading up to *M. cardinalis* make permanent its escape from melittophily using preexisting genetic variation affecting nectar quantity, or using a new mutation affecting color, or using some other new mutation, or was it incrementally secured by several or many new mutations during a prolonged ecological perturbation caused by a low bee : bird ratio?

We are reluctant to believe that plants such as *M. lewisii* are unable to escape the melittophily vortex for lack of one macromutation that by itself allowed the ancestors of *M. cardinalis* to escape. If a single mutation such as *yup* were globally beneficial, we suppose it would have occurred already and swept throughout the species, destroying melittophily. Because *M. lewisii* and many other melittophiles seem stable as melittophiles, we assume they are trapped in their syndrome vortex. We assume that the various characters of *M. lewisii* keep hummingbirds from being important pollinators even in an individual that has a mutation that makes the flowers orange or in an individual with a mutation that increases nectar production by 25%. Thus, we suppose that the shift between vortices was allowed because of special ecological circumstances and a series of loci being selectively substituted. In the case of *Mimulus*, the alleles involved were of fairly large effect at fairly few loci, and because of that, we assume they were new mutations. Our efficiency threshold scenario, however, is compatible with more conventional genetic assumptions. In *Penstemon*, we have not attempted QTL studies on syndrome characters, but we did study the variance among F_2 offspring

and among backcrosses of melittophilous *Penstemon spectabilis* and ornithophilous *Penstemon centranthifolius* (Jordan 2004). We studied morphological dimensions, nectar characteristics, and color spectra. In general, there was little or no deviation from additivity of parental character states, and there was no heightening of variance such as would be due to genetic segregation if a character differed because of one or two loci. The one possible exception was that color showed some suggestion of heightened variance, so perhaps major genes do play a role in pigment production.

When we envision the possible fixation of a single mutation that moves flowers past a threshold, a signaling-trait locus such as *yup* would not seem to be the most likely candidate. Signaling is important, because pollinators use signals to distinguish one phenotype from another. On that basis, one might expect that there is an especially steep selection gradient for a signaling trait, but there is a catch. Although pollinators do frequently respond to color signals, their preferences are generally developed through associative learning of colors associated with floral rewards. Preferences for the colors that are classically associated with syndromes are not always present and can be overridden by reward economics (Waser et al. 1996). For a few examples, bumblebees in northern Wisconsin prefer orange *Hieracium aurantiacum* to yellow *Hieracium florentinum* in mixed stands (Thomson 1978); hummingbirds in Colorado will seldom visit red *Penstemon barbatus* if brown *Scrophularia lanceolata* is blooming nearby (J. D. Thomson, unpublished observations); and in the lab, naive *Bombus impatiens* are more likely to prefer red artificial flowers than blue ones (Gegear and Burns 2007). In *Mimulus*, we would not expect hummingbirds to consistently restrict their visits to orange phenotypes or bees to pink, unless first there were economic advantages for them to do so (also the opinion of H. Bradshaw [personal communication]).

For monkeyflowers, we envision a scenario in which ecological circumstances interact with new mutations to produce a shift. Imagine that the incipient *M. cardinalis*, still pink, is living at a site where increased nectar production is favored, perhaps because the plants are growing at warmer low elevation where the cost of producing nectar is diminished. Both hummingbirds and bees come to be regular visitors. As in our hypotheses regarding quality, we will further suggest that hummingbirds are less wasteful at delivering a plant's pollen to stigmas. Selection would favor increases in nectar production, and hummingbird pollination would become further established by selection favoring anther/stigma exsertion, narrowing of the floral tube, reflexing the petal lobes, and other characters that increased the efficiency of placing pollen on the foreheads of birds and removing it from those foreheads. The *yup* mutation could have arisen and spread during this stage of consolidation. In short, although *yup* has potent effects on pollination and may have promoted the shift in *Mimulus* after it was under way, it is not likely to have driven the shift on its own. While we are speculating freely, we note that plants of section *Erythranthe* live in wet seeps, so that increased nectar production in lower, warmer habitats might not entail meaningful water costs. It would be interesting to have some data on the pollen-transfer efficiency of bees and birds visiting *M. cardinalis* and *M. lewisii* and on the cost of nectar at high and low elevations.

Directionality: Lessons from Three Genera

Although phylogenetic studies cannot yet support a full meta-analysis of the many genera that contain melittophiles and ornithophiles, there seem to be many more shifts from bee pollination to hummingbird pollination than the other way around (Faegri and van der Pijl 1979; Wilson et al. 2007). In our discussion of penstemons, we suggested that this is because hummingbirds have the capacity to become more efficient at transferring pollen than bees. However, there may well be other factors involved, as indicated by work on *Ipomoea*, *Costus*, and *Aquilegia*.

Morning glories. Zufall and Rausher (2004) studied the floral pigments of morning glories (fig. 1E, 1F). Comparing the red ornithophilous *Ipomoea quamoclit* to the blue-purple melittophilous *Ipomoea purpurea*, they concluded that a reversal from red to blue would be very difficult. The biochemical pathway has changed in at least two and probably three ways (fig. 4). (1) In *I. quamoclit*, F3'H, the enzyme that leads to blue cyaniden, has been downregulated almost out of existence. This prevents the pathway from moving anthocyanins toward blue cyaniden. (2) The next enzyme in the pathway, DFR, has evolved greater specificity for dihydrokaempferol, the reactant that would have been acted on by F3'H, and a lessened ability to catalyze the product of F3'H, dihydroquercetin; thus, this second enzyme now moves anthocyanins toward red pelargonidin. (3) The downregulated *f3'h* gene is apparently accumulating structural mutations, so it would not work properly even if it were turned back on. Overall, it seems very improbable that "lucky" mutations could reverse all these

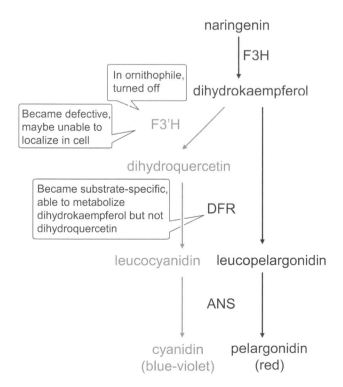

Fig. 4 Three modifications to the anthocyanin pathway in the red-flowered *Ipomoea quamoclit*. Parts of the pathway that have been deactivated are in gray (modified from Zufall and Rausher 2004).

changes simultaneously or recruit other copies of the genes while reversing other changes. Evolutionary shifts from blue-purple to red may be quite easy, but then there may be no going back. To the extent that floral colors dictate pollination, this mechanism may produce a directional bias that favors bee-to-bird shifts.

The findings of Zufall and Rausher (2004) for *Ipomoea* are not directly relevant to the characters studied in monkeyflowers. First, Bradshaw and Schemske have focused on the effects on pollinator visitation of carotenoids more than anthocyanins. Second, in monkeyflowers the crucial difference in carotenoids is whether the yellow patch is restricted to the throat of the flower or spread across the whole upper petal. Therefore, it is not a matter of turning off the pathway completely through a loss-of-function mutation. Third, there is a difference in anthocyanins, but it depends as much on the amount of anthocyanins as on which anthocyanin is abundant in the petals (Wilbert et al. 1997). The Zufall and Rausher conclusions may apply better to penstemons, where red flower colors appear to have evolved by changing the kinds of anthocyanins (Rausher 2008).

In *Ipomoea*, color may be particularly important in pollinator shifts, but as we have argued previously, differences in nectar rewards are likely to be important too. *Ipomoea* species variously appeal to bees, hummingbirds, and hawk moths, but their conformity to multicharacter syndromes may be less coherent than in monkeyflowers and penstemons. In six Argentine species of *Ipomoea*, differences in nectar seem to be explained more by differences in flower size than by the type of pollinator (Galetto and Bernardello 2004). The bird-pollinated species produce more dilute nectar less gradually, but a large-flowered sphingophile produces the most nectar, and several of the bee-pollinated species have sucrose-rich nectar. Morning glory flowers are radially symmetric, which may provide less opportunity for conformity to multicharacter syndromes. In addition, the center of radiation for morning glories is tropical, where there is much more variety in the hummingbird fauna than in the temperate montane biomes we are more familiar with. Additional information on the phylogeny (Miller et al. 2004) will be especially informative when the pollination biology of the species included in the phylogeny has been studied.

Costus. There are ca. 51 species of Neotropical *Costus* (fig. 1G, 1H). They are pollinated either by euglossine orchid bees or by hummingbirds, and pollination syndrome characters are reliable predictors of which kind of animal actually visits the flowers (Kay and Schemske 2003). The New World clade is nested within Old World *Costus*, and molecular clock data indicate a 1.5–7.1-million-year-old radiation. Hummingbird pollination is inferred to have arisen about seven times, with no compelling case for any reversals (Kay et al. 2005). Again, ornithophily appears to be a one-way attractor. *Costus* arrived in the Neotropics after hummingbirds were established. It already had large flowers with abundant nectar. It adapted to orchid bees first, and in its subsequent diversification, a high proportion of the euglossine-pollinated lineages have spun off hummingbird-pollinated lineages. *Costus* plants are typically very widely spaced and thus are appropriately pollinated by orchid bees and hummingbirds, both of which remember where resources are and forage over long distances. This constraint may prevent *Costus* from switching over to smaller pollinators. The pollination syndromes in *Costus* are based more on signaling than on rewards. Nectar offerings are high in both orchid bee–pollinated species and hummingbird-pollinated species.

Within a pollination syndrome, *Costus* species are maintained by barriers such as where pollen is placed on the body of the pollinator. Kay (2006) reported on the reproductive isolating barriers between two closely related species, *Costus pulverulentus* and *Costus scaber*. Postzygotic barriers are probably fairly weak, and there are only modest barriers because of geography, microhabitat, and phenology. Both these flower species mostly use the long-tailed hermit hummingbird, but the anthers and stigma of *C. pulverulentus* are exserted, so that pollen is carried on the bird's forehead, whereas in *C. scaber* the anthers and stigma are more included in the floral tube, so that pollen is carried on the beak. This produces reproductive isolating barriers in both directions, though more so for *C. scaber* pollen donors by *C. pulverulentus* recipients than vice versa. There are also strong barriers to interbreeding at the stage of pollen germination and growth down the style, but we find Kay's results for pollen movement noteworthy. We would expect that more pollen would be lost from beaks than from feathered foreheads (Castellanos et al. 2003), which would lead us to expect that male-male competition might select for more exserted sex organs in *C. scaber*. Perhaps our intuition about pollen wasting is wrong; alternatively, the greater wastage entailed by beak pollination could be offset by improved precision of pollen placement. Details of pollen transfer by the euglossine bees would be valuable.

Columbines. There are 25 North American species of *Aquilegia* (fig. 1*I*, 1*J*). Whittall and Hodges (2007) studied them all, revealing that much of floral evolution occurred through pollinator shifts. An ordination based on floral traits displays three distinct pollination syndromes. The melittophilous species are blue-purple with spurs of ca. 1 cm. The ornithophiles are scarlet with yellow and have slightly longer, often straighter spurs. The sphingophiles are yellow or pale violet, with much longer spurs and upright flowers. Melittophily is the ancestral condition, from which ornithophily arose twice. Then sphingophily arose from ornithophily five times. This directionality is statistically significant, and the shifts are significantly associated with increased spur length, which occurred mainly in one of the two lineages that emerged from points of cladogenesis where there was a pollinator shift. Sphingophily never arose directly from melittophily, and no Eurasian columbines have gone over to hawk moth pollination.

Ornithophilous and sphingophilous columbines are still visited and pollinated by bees, even though the nectar is inaccessible to them unless they rob from the spurs, which effects no pollination. Pollen can be transferred by bees when they collect pollen, and we predict that they remove large quantities of pollen per visit, of which a relatively low proportion is moved to stigmas compared with amounts transferred by hummingbirds and hawk moths. In some populations in some years, however, the "legitimate" pollinators are so rare that attracting bees is probably adaptive, averaged over evolutionary time and space (Miller 1978, 1981).

Intriguingly, the melittophilous columbines seem to present their pollen gradually as waves of anthers mature, whereas the ornithophiles and sphingophiles seem to dehisce anthers in a less orderly manner (J. Whittall, personal communication). If verified, this would be consistent with our finding that melittophilous penstemons have more metered pollen presentation than related ornithophiles (Castellanos et al. 2006).

Much research on pollinator shifts in columbines has focused on the comparison of ornithophiles to sphingophiles (Fulton and Hodges 1999), which we will not abstract here. We suggest, however, that *Aquilegia* differs from *Penstemon* and the other genera we have mentioned in having many anthers, which might predispose it to sphingophily. We can further extend speculation along these lines to explain the lack of shifts in *Penstemon* and *Mimulus* to other syndromes. If we are right, then perhaps there are no shifts to hawk moth pollination in these genera because hawk moths (it may be presumed) remove very little pollen in a visit to a flower with four anthers. Other explanations of the same general structure but with differing details could also be offered. It seems likely that the nectar spurs of columbines can easily respond to selection favoring greater length without compromising their simple function. This may not be true for penstemons where nectar is secreted by the stamens near their base and whose normal floral function seems to depend on stiff stamens and styles being positioned for nototribic pollen transfer. If selection pulled penstemon corolla tubes out to the length and thinness required to enforce sphingophily, the filaments and style might become so attenuated and floppy as to lower the precision with which they transfer pollen via birds or moths.

Multiple Traits

The concept of a few pollination syndrome vortices of different strengths and sizes encompasses the three elements of transitions that we set out to explain: (1) clusters of phenotypically similar plants that have converged on one of two alternative states, (2) occasional vortex-escaping transitions from one state to the other, and (3) biased directionality of those transitions. But the abstract notion of vortices needs to be fleshed out. What factors give a vortex its distinctive character?

One important factor, recognized by Faegri and van der Pijl (1979), must lie in the cognitive characteristics of the pollinator. They wrote, "In reality the 'why' of pollination ecology is largely animal psychology" (p. 5). It seems likely that the visual system of the pollinator and some amount of sensory drive (sensu Endler 1992) underlie the strong correlation of flower color with pollination mode, although it is very risky to ascribe observed color preferences to innate predispositions of pollinators (Chittka and Waser 1997). Differences in the odors that attract various animals add another dimension to differences in pollination syndrome. Other important cognitive factors would include how pollinators sample, perceive, and react to nectar or pollen rewards. For example, bumblebees strongly prefer smaller volumes of more concentrated nectar to energetically equivalent, larger volumes of more dilute nectar (Cnaani et al. 2006). Although we lack parallel data for hummingbirds, this behavioral predisposition of bees may open an opportunity for plants to discourage those bees (by providing copious but dilute nectar) without discouraging birds to the same degree. This example reinforces the necessity to consider the question raised by Faegri and van der Pijl

(1979) and stressed by Castellanos et al. (2004): When a floral character changes during a pollinator transition, does the change serve to improve pollination service by the new pollinator or to deter wasteful visits by the old one? It is not always easy to tell, but before we can make a meaningful interpretation of the character as a response to an animal's cognitive properties, we must answer the question of which animal we are talking about.

In addition to the cognitive properties of different pollinators, floral phenotypes should also respond adaptively to the principal pollinator's shape, size, and characteristic posture. Obviously, selection will tend to place anthers and stigmas so that pollen transfer is achieved, but again there are subtleties. For example, the exserted sex organs and the narrow corolla tube of ornithophilous penstemons may work well only in concert (Castellanos et al. 2004). If the corolla tube is narrow, birds must approach straight along its axis, and exserted sex organs on the flower's midline will contact the forehead. If the organs are exserted but the corolla remains widely flared, birds can approach laterally and may not pick up or deposit pollen. These considerations suggest that there may be a characteristic order in which character changes are introduced during a shift between syndromes (Wilson et al. 2006). Some may be prerequisites for others. We suspect that reward and signaling characters are likely to change first because they relate directly to enforcing the change from old pollinator to new. Changes in morphology and pollen presentation would tend to follow as a part of improving the effectiveness of the new pollinator.

Explanations of specialization based on distinct syndromes (as opposed to wandering) have typically assumed the existence of some kind of trade-off in which being adapted to one kind of pollinator (bees) makes flowers maladapted to another kind of pollinator (birds; see the many citations in Aigner 2006; Muchhala 2006). A flower cannot be in two vortices, or else the vortices would not be distinct. Previously, we drew attention to heterogeneities in the slopes of selection gradients (Wilson and Thomson 1996). For example, pollinator-driven divergence could result if shorter stamens and styles work well with bees whereas longer stamens and styles work well with birds. Such differences in the direction of the slopes have eluded empirical study, although they seem plausible enough when we observe the end products of divergence. Aigner (2001, 2004, 2006) has carefully discussed the trade-off assumption, studying natural variation, doing experiments with manipulated phenotypes, and using graphical models. He points to a situation in which trade-offs are not manifest; instead, a flower works well with one kind of pollinator (bees) regardless of phenotype, but with another kind of pollinator (birds), the flower has a fitness-on-phenotype function that is narrow and peaked. When this is the case, the flower ought to master the use of birds, acquiring long stamens and styles, while still using bees. The flowers will become more generalized, at least in the sense of lengthening their list of effective pollinators. In Schemske and Bradshaw's (1999) study (fig. 3), when one character at a time was looked at, only anthocyanin concentration showed a significant fitness trade-off: bees preferred more and birds preferred less anthocyanin. Other single characters were preferred by bees or by birds, with the other kind of animal being indifferent. Aigner was cautious about even

recognizing syndromes as being vortices because his attention was focused on single characters. Our view, focusing on pollinator quality, is that bees can become conditional parasites. This allows for another kind of trade-off. Once a population is past the threshold, any character that discourages bees is selected for, as is any character that improves the pollen transfer effectiveness of birds. The stability of the bird vortex becomes multivariate as more characters are progressively recruited.

The stability of the bee vortex derives from its evolutionary history: multiple characters, which were presumably recruited sequentially over evolutionary time, function together in an integrated fashion. Indeed, it is the multivariate nature of the syndrome that makes the syndrome a vortex—that coupled with the social way in which flowers work as a team (see Wilson et al. 2006). In a large patch of a melittophilous penstemon, an individual that makes unusually dilute copious nectar is unlikely to attract hummingbirds, and even if a hummingbird did visit, its potential effectiveness as a pollinator would probably not be realized. The mutant individual's flower color, landing platform, vestibule size, and anther and stigma exsertion would all be suboptimal for hummingbird pollination, and all the other individuals in the patch would be more attractive to bees than to hummingbirds. Thus, the status quo perpetuates itself, effecting what might be called multiple-trait centripetal stasis. The adaptive stable state of melittophily ensures that most of the visits are by bees, and that, in turn, ensures that most of the pollination is by bees. In addition, the geometry of the flower is already adapted to bees, so even if a mutation does attract more birds quantitatively, their qualitative superiority is unlikely to be fully manifest. Therefore, selection tends to be purifying and to disfavor deviations from the quantitative norm, even when another pollinator would have the potential to be qualitatively better on a per-visit basis (see similar argument by Kirkpatrick and Barton [1997] for how adaptation at the center of a range prevents adaptation at the edge of a species' range).

Functional interactions among syndrome characters may channel and constrain evolution (Armbruster 1990; Wilson et al. 2007). Syndromes and other such strategic concepts are fundamentally about how characters work well together (Fenster et al. 2004), but there has been almost no empirical study of multiple-trait synergism of the characters involved in bee-to-bird transitions. In lieu of such work, we review two empirical studies on individual characters that are important in the bee and bird pollination syndromes and whose explanation involves multiple traits, some affecting quality and some affecting quantity.

Campions. In eastern North America (where there is only one species of hummingbird, the ruby-throated hummingbird), *Silene virginica* conforms to the hummingbird syndrome (fig. 1K, 1L). Close relatives are pollinated by nocturnal moths and by diurnal bees and Lepidoptera (Fenster et al. 2006). Fenster and Dudash (2001) caged plants with mesh that kept out birds but allowed bees to visit. They found that the caged plants produced fewer seeds each year for several years at both a woodland site and a meadow site. Therefore, the plants benefited from the pollinators that syndrome characters suggest they are adapted to. The degree of benefit varied with year and site. However, the plants may well have also benefited from "illegitimate" pollinators, particularly when and where hummingbirds were scarce. Subsequently, Fenster et al. (2006) artificially decoupled petal size and corolla tube width from nectar offerings and found that hummingbirds prefer flowers with larger petals, but they do not generally prefer narrower tubes. This suggests that the narrowness of the floral tubes in *S. virginica* is an adaptation not to improve the quantity of hummingbird visits but to improve the quality of bird visits or to exclude pollinators of inferior quality.

Sages. Many of the features that make the penstemons interesting are mirrored among the sages, and *Salvia* could provide even further scope for comparative studies because it includes both passerine and hummingbird ornithophiles. In addition to classic work reviewed by Faegri and van der Pijl (1979, pp. 147–148), much phylogenetic and functional research is under way, and we expect new insights about pollinator transitions to emerge. Wester and Claßen-Bockhoff (2006) have studied pollen presentation. Many melittophilous *Salvia* have a well-known "staminal lever" mechanism that has generally been interpreted as a device for the precise nototribic placement of pollen on bees. Typically, the anthers are held against the roof of the flower until a bee enters deeply to probe for nectar, at which point the anthers are indirectly pushed down to touch the bee's notum. This mechanism is frequently inactivated in ornithophilous species (fig. 5); in a study of anther function in the hummingbird-pollinated *Salvia haenkei*, Wester and Claßen-Bockhoff (2006, p. 142) mention that at least 50 similar ornithophilous species show "reduced or stiffened staminal levers." Although phylogenetic information is not presented, it seems likely that some of these represent independent convergence toward bird pollination, causing deactivation of the lever mechanism. Although these authors are inclined to view the loss of levering as a side effect of elongation and narrowing of the corolla tube to better fit birds or exclude bees, an alternative explanation is possible (such as the pollen-dosing explanation by Castellanos et al. [2006]). Perhaps staminal levering is advantageous in bee flowers because it achieves restricted pollen dosing in addition to precise placement. Then, as in penstemons and perhaps in columbines, adaptation to birds relaxes selection for dosing. Although we

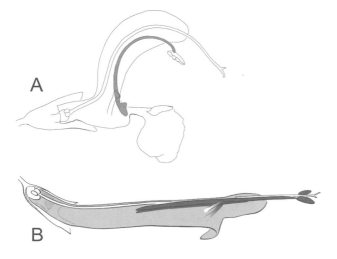

Fig. 5 Loss of the staminal lever in a hummingbird-pollinated sage. *A,* Melittophilous *Salvia pratensis. B,* Ornithophilous *Salvia haenkei* (artist: D. Franke).

need more phylogenetic information to judge the strength of the correlation between lever loss and ornithophily, and more information about pollen transfer to judge the applicability of the pollen-dosing hypothesis, the putative correlation is highly suggestive of some link between pollen transfer and the transition to ornithophily.

Broadening the Applicability of Our Ideas

The salient challenge for explaining bee-to-bird pollination shifts that are replicated more or less faithfully across different genera is to explain what factors destabilize established plant-pollinator relationships. One would expect established relationships to continue under purifying selection unless some special circumstance turns wandering evolution to revolution. A sufficient but easily rejected possibility is that bees simply disappeared from large geographic regions for large spans of evolutionary time. Some melittophilous species would then become extinct, others would become autogamous, and some might adapt to other pollinators that would visit them without immediate evolution. We consider such a regional bee extirpation to be unbelievable; a good explanation must explain how bee pollination can become destabilized while bees remain a component of the regional community. We can envision sustained depressions of numbers or even transient disappearances but not a sustained wholesale disappearance of bee pollinators. Perhaps more palatable is to invoke a depression in bee visitation during certain times of day in certain habitats (Stebbins 1989). By this view, a plant lineage would first evolve to grow in a different habitat than its progenitors, and this would impose on it a changed pollinator regime that would then disrupt the status quo (Sargent and Vamosi 2008).

In what other ways could the usual neglect by hummingbirds of a melittophilous flower be disrupted? Changes in plant community composition can upset prevailing patterns of floral use. Faegri and van der Pijl (1979, p. 51) recount that "a breakdown of blossom constancy ... is easily seen when, for example, *Trifolium pratense* comes into flower: owing to its greater nectar production it suddenly attracts all long-tongued bumblebees that are able to reach the nectar, causing an immediate neglect of other bee blossoms." A novel invasion of such a super-rich species could disrupt traditional allegiances; if bees were to abandon a penstemon for a clover, for example, the penstemon might in turn begin receiving more attention from birds. Alternatively, an important ornithophile might disappear from a local community, forcing birds to forage on a melittophilous species that they would normally disdain. Substantial spatiotemporal variation in bee : bird ratios has been observed in the ornithophilous *Ipomopsis aggregata* (Price et al. 2005). Abandonment of a host plant by pollinators does not require the disappearance of the plant itself, only the reward. For example, in Irwin, Colorado, in 2002, a drought caused *I. aggregata* to cease producing nectar, which in turn increased hummingbird attendance at artificial feeders (J. D. Thomson, unpublished data).

These possible changes in situation seem plausible in the short term, but for the situation to be sustained over evolutionary time would be much rarer (as well as practically impossible to study). An interesting moment in evolutionary time may occur at the end of a glacial period when range-extending hummingbirds first reach communities that lack ornithophilous plants. In most contemporary communities in which birds are common, they can afford to bypass low-nectar melittophilous species in favor of high-nectar species adapted to birds. If ornithophiles have not yet evolved or migrated into communities, however, newly arrived birds will need to choose the best foraging options from the existing flora, which will probably be dominated by melittophiles. The best species available may be those that, like some penstemons, replenish nectar rapidly after draining (Cruden et al. 1983; Castellanos et al. 2002); such characteristics may predispose certain genera to shifts toward ornithophily. The first wave of birds may provide a unique opportunity for melittophilous species to recruit bird visitors and thereby cross the efficiency threshold. After ornithophiles that offer high volumes of nectar have become abundant in a community, such transitions may become much harder.

We offer these speculations about ecological changes not because we think we can know which of these or others allowed for bee-to-bird shifts but to illustrate the role of ecological changes. That such moderately prolonged local rare extrinsic changes set the stage for pollinator shifts is a general concept. It is analogous to the familiar scenario for allopatric speciation in which a peripheral, isolated population is first extrinsically isolated by geography and then selection directly or indirectly changes its reproductive system so that it is intrinsically isolated (Futuyma 1998, p. 482). Here, we are not necessarily dealing with the origin of reproductive isolating mechanisms (which may come before a shift in syndromes or during a shift in syndromes), but the same general idea applies: shifts may occur in isolated lineages that experience a novel selective regime.

Although we view ecological changes as the most likely initiators of pollinator shifts, a very special mutation or coincidence of special mutations could in principle cause a shift. It would have to be a mutation that caused a drastic change in one character (e.g., nectar production) or just the right pleiotropic changes in multiple characters (e.g., a change from pink to red coupled with a change from low to high nectar production). Aigner (2006) seems to suggest that such pleiotropic variation could be responsible for differences in specialization on bees versus on birds. A pleiotropic mutation of just the right sort provides a multicharacter way to obviate the need for single-character trade-offs. Evidence of such a pleiotropic mutation would be hard to find. If such a genetic correlation were found between two established species (a melittophile and an ornithophile), it would not necessarily prove that the two character changes arose simultaneously through the same pleiotropic mutation; the mutation for copious nectar production could have arisen first and swept through the population, and then the mutation for red could have arisen in a closely linked gene and swept through the population, leaving no gene copies that had the first but not the second mutation. We remain skeptical of shifts without changes in ecology, simply because the selective sweep would spread throughout the species and not leave behind a melittophilous ancestral form for us to observe. The reason for erecting all our elaborate explanations is to explain the multifaceted nature of syndromes.

Nevertheless, Bradshaw and Schemske's work still suggests that large mutations may promote shifts, moving lineages past the threshold beyond which an evolutionary shift no longer requires special ecology. Because key syndrome characters in *Mimulus* are indeed controlled by major genes, it is likely that the pollinator shift in *Mimulus* involved "special genetics," i.e., large novel mutations rather than standing variation. It is certainly conceivable that similar effects influence transitions in other clades, too. In our minds, however, mutations of large effect are probably not essential for pollinator transitions. Normal variation drawn from many genes of small effect throughout the genome can be recombined through sexual reproduction and selection to cause very large changes in quantitative traits (Futuyma 1998, pp. 27, 283–285; Barton and Partridge 2000; see also Mitchell and Shaw 1993 for traits in *Penstemon*). For the simple biochemical pathways involved in pigment production, perhaps a few loci will play prominent roles, but for many other traits, there is no reason to think that new mutations are needed for a shift between pollination syndromes, that is, to get the lineage past the threshold. Notice that, regardless of the genetic basis for character variation, a pollinator shift still takes many generations and in this sense is always genetics limited (Wilson et al. 2006). Whether it proceeds by the fixation of new macromutations or by bringing together numerous alleles of small effect, the substitution of alleles by selection takes time.

The phenomenon we have focused on—bee-to-bird shifts of distinct syndromes—does invoke some biology specific to bees and birds. Most notably, bees feed pollen to their brood and prefer concentrated nectar, whereas birds have no interest in pollen and can better transport and process copious amounts of dilute nectar. Schemske and Horvitz (1984) made pollination biologists appreciate differences in quality of pollination and how those differences in quality are important in understanding floral diversification (Herrera 1987; Wilson and Thomson 1996). We suggest that in the absence of quality differences—i.e., if all kinds of pollinators were to transfer the same fraction of the pollen they removed—the phenomenon of pollination syndromes would be greatly diminished. Syndromes would have no basis in effect, only in the trade-offs that cause differences in quantity of pollinator visits. Such "syndromes" would be simpler than the multicharacter vortices that we have been postulating.

Basically, the broader applicability of our ideas to pollination systems depends on the extent to which syndromes exist and are distinct. When diversity is syndromelike, the factors we have considered would seem to apply. One would just have to know how the two types of pollinators differ in their responses to various attractants and in their potential pollen-transfer efficiencies. How do long-tongued flies differ in psychology and in pollen-transfer effectiveness from bees?

Although we have tried to discuss several aspects of evolutionary transitions between well-defined pollination syndromes, our principal thesis is that such transitions involve thresholds. Something must propel a plant lineage out of one pollinator's sphere of influence and into another's, and we think that pollen-transfer efficiencies may well determine where the critical thresholds lie. Thinking about pollinator efficiencies further suggests reasons why shifts ought to be easier in one direction than in the reverse, why shifts from syndrome X to Y should be easier than from X to Z, and how some lineages can be predisposed to subsequently shifting from Y to Z. Such shifts should generally be hard but will be easiest when the efficiency of an "illegitimate" pollinator is fortuitously quite good and can be improved on by some simple changes in floral dimensions. Of course, these ideas have much less force in explaining diversifications based on adaptive wandering within one broad flat vortex (generalized entomophily) or even being drawn from a generalized relationship with many pollinators into a specialized relationship with a few. Threshold mechanisms become essential only when escape from one vortex must precede entry into another one.

Does our explanatory framework have any applicability outside pollination biology? There may well be parallel, vortexlike phenomena in other transitions discussed in this special issue, such as shifts between outcrossing and selfing, cosexuality and dioecy, or animal pollination and wind pollination. If so, the study of those transitions would benefit from explicit consideration of how the alternative states are generally stabilized, what the defining characters are, and how those characters interact with ecological circumstances to drive lineages over thresholds. But these other types of evolutionary transitions also require other considerations. Many of them would change the genetic system much more radically than would a pollinator shift from bees to birds. The change in the genetic system, for example, a large increase in homozygosity, could itself enter into the evolutionary dynamics of the transition (see Wright et al. 2008). We have not considered genetic ramifications of this sort because we expect them to be minor in pollinator shifts. For similar reasons, we have also ignored reallocation of resources from one function to another, although such reallocation may be central to other transitions such as sexual systems (see Sakai et al. 1997). Although it is possible that increasing nectar output associated with ornithophily involves a reallocation of energy or of water from some other function, we judged such effects to be small enough to be excluded from our main arguments. Thus, our narrow focus led us to ignore some phenomena that might play key roles in other transitions. However, the elements that we chose to develop at length may indeed have analogs in other transitions.

Acknowledgments

We thank Toby Bradshaw, Doug Schemske, Chris Caruso, Kathleen Kay, Scott Hodges, Mark Rausher, Petra Wester, Justen Whittall, and two anonymous reviewers for data, ideas, corrections, and unpublished work. Maria Clara Castellanos made invaluable contributions to our understanding of penstemons. The penstemon work was supported by a National Science Foundation grant to J. D. Thomson, P. Wilson, W. S. Armbruster, and A. Wolfe, plus Natural Sciences and Engineering Research Council funding to J. D. Thomson.

Literature Cited

Aigner PA 2001 Optimality modeling and fitness trade-offs: when should plants become pollinator specialists? Oikos 95:177–184.

——— 2004 Floral specialization without trade-offs: optimal corolla flare in contrasting pollination environments. Ecology 85:2560–2569.

——— 2006 The evolution of specialized floral phenotypes in a fine-grained pollination environment. Pages 23–46 in NM Waser, J Ollerton, eds. Plant-pollinator interactions from specialization to generalization. University of Chicago Press, Chicago.

Armbruster WS 1990 Estimating and testing the shapes of adaptive surfaces: the morphology and pollination of Dalechampia blossoms. Am Nat 135:14–31.

Armbruster WS, CB Fenster, MR Dudash 2000 Pollination "principles" revisited: specialization, pollination syndromes, and the evolution of flowers. Det Nor Vidensk Acad I Mat Natur Lk Skr Ny Ser 39:179–200.

Baker HG 1963 Evolutionary mechanisms in pollination biology. Science 139:877–883.

Barton N, L Partridge 2000 Limits to natural selection. BioEssays 22:1075–1084.

Beardsley PM, A Yen, RG Olmstead 2003 AFLP phylogeny of Mimulus section Erythranthe and the evolution of hummingbird pollination. Evolution 57:1397–1410.

Beisner BE, DT Haydon, K Cuddington 2003 Alternative stable states in ecology. Front Ecol Environ 1:376–382.

Bradshaw HD, KG Otto, BE Frewen, JK McKay, DW Schemske 1999 Quantitative trait loci affecting differences in floral morphology between two species of monkeyflower (Mimulus). Genetics 149:367–382.

Bradshaw HD, DW Schemske 2003 Allele substitution at a flower colour locus produces a pollinator shift in monkeyflowers. Nature 426:176–178.

Bradshaw HD, SM Wilbert, KG Otto, DW Schemske 1995 Genetic mapping of floral traits associated with reproductive isolation in monkeyflowers (Mimulus). Nature 376:762–765.

Castellanos MC, P Wilson, S Keller, AD Wolfe, JD Thomson 2006 Anther dehiscence and pollen production in Penstemon and relatives. Am Nat 167:288–296.

Castellanos MC, P Wilson, JD Thomson 2002 Dynamic nectar replenishment in flowers of Penstemon (Scrophulariaceae). Am J Bot 89:111–118.

——— 2003 Pollen transfer by hummingbirds and bumblebees, and the divergence of pollination modes in Penstemon. Evolution 57:2742–2752.

——— 2004 "Anti-bee" and "pro-bird" changes during the evolution of hummingbird pollination in Penstemon flowers. J Evol Biol 17:876–885.

Chittka L, NM Waser 1997 Why red flowers are not invisible to bees. Isr J Plant Sci 45:169–183.

Cnaani JC, JD Thomson, DR Papaj 2006 The effect of reward properties on learning and choice in foraging bumblebees. Ethology 112:278–285.

Corbet SA 2006 A typology of pollination systems: implications for crop management and the conservation of wild plants. Pages 315–340 in NM Waser, J Ollerton, eds. Plant-pollinator interactions: from specialization to generalization. University of Chicago Press, Chicago.

Coyne JA 1995 Speciation in monkeyflowers. Nature 376:726–727.

Cruden RW 1972 Pollinators in high-elevation ecosystems: relative effectiveness of bees and birds. Science 176:1439–1440.

Cruden RW, SM Hermann, S Peterson 1983 Patterns of nectar production and plant-pollinator coevolution. Pages 80–125 in BL Bentley, TS Elias, eds. The biology of nectaries. Columbia University Press, New York.

Darwin C 1862 On the various contrivances by which British and foreign orchids are fertilized. J Murray, London.

Dilley J, P Wilson, MR Mesler 2000 The radiation of Calochortus: generalist flowers moving through a mosaic of potential pollinators. Oikos 89:209–222.

Endler JA 1992 Signals, signal conditions, and the direction of evolution. Am Nat 139(suppl):S125–S153.

Faegri K, L van der Pijl 1979 Principles of pollination ecology. 3rd ed. Pergamon, Oxford.

Fenster CB, WS Armbruster, MR Dudash, P Wilson, JD Thomson 2004 Floral specialization and pollination syndromes. Annu Rev Ecol Evol Syst 35:375–403.

Fenster CB, G Cheely, MR Dudash, RJ Reynolds 2006 Nectar reward and advertisement in hummingbird-pollinated *Silene virginica* (Caryophyllaceae). Am J Bot 93:1800–1807.

Fenster CB, MR Dudash 2001 Spatiotemporal variation in the role of hummingbirds as pollinators of *Silene virginica*. Ecology 82:844–851.

Fisher RA 1930 The genetical theory of natural selection. Clarendon, Oxford.

Fulton M, SA Hodges 1999 Floral isolation between *Aquilegia formosa* and *Aquilegia pubescens*. Proc R Soc B 266:2247–2252.

Futuyma DJ 1998 Evolutionary biology. 3rd ed. Sinauer, Sunderland, MA.

Galetto L, G Bernardello 2004 Floral nectaries, nectar production dynamics and chemical composition in six *Ipomoea* species (Convolvulaceae) in relation to pollinators. Ann Bot 94:269–280.

Gegear RJ, JG Burns 2007 The birds, the bees, and the virtual flowers: can pollinator behavior drive ecological speciation in flowering plants? Am Nat 170:551–556.

Gilpin ME, ME Soulé 1986 Minimum viable populations: processes of species extinction. Pages 19–34 *in* ME Soulé, ed. Conservation biology. Sinauer, Sunderland, MA.

Gottlieb LD 1984 Genetics and morphological evolution in plants. Am Nat 123:681–709.

Grant V 1994 Historical development of ornithophily in the western North American flora. Proc Natl Acad Sci USA 91:10407–10411.

Herrera CM 1987 Components of pollination "quality": comparative analysis of a diverse insect assemblage. Oikos 50:79–90.

——— 1996 Floral traits and plant adaptation to insect pollinators: a devil's advocate approach. Pages 65–87 *in* DG Lloyd, SCH Barrett, eds. Floral biology: studies on floral evolution in animal-pollinated plants. Chapman & Hall, New York.

Hiesey WM, MA Nobs, O Björkman 1971 Experimental studies on the nature of species. V. Biosystematics, genetics, and physiological ecology of the *Erythranthe* section of *Mimulus*. Carnegie Inst Wash Publ 628.

Jordan E 2004 Inheritance patterns in floral characters of *Penstemon* and pollinator preference. MS thesis. California State University, Northridge.

Kay KM 2006 Reproductive isolation between two closely-related hummingbird-pollinated Neotropical gingers. Evolution 60:538–552.

Kay KM, PA Reeves, RG Olmstead, DW Schemske 2005 Rapid speciation and the evolution of hummingbird pollination in Neotropical *Costus* subgenus *Costus* (Costaceae): evidence from nrDNA ITS and ETS sequences. Am J Bot 92:1899–1910.

Kay KM, DW Schemske 2003 Pollinator assemblages and visitation rates for 11 species of Neotropical *Costus* (Costaceae). Biotropica 35:198–207.

Kirkpatrick M, NH Barton 1997 Evolution of a species' range. Am Nat 150:1–23.

Kölreuter JG 1761–1766 Vorläufige Nachricht von einigen das Geschlecht der Pflanzen betreffenden Versuchen und Beobachten. Gleditschischen Handlung, Leipzig.

Losos JB 1992 The evolution of convergent structure in Caribbean *Anolis* communities. Syst Biol 41:403–420.

Mayfield M, NM Waser, MV Price 2001 Exploring the "most effective pollinator principle" with complex flowers: bumblebees and *Ipomopsis aggregata*. Ann Bot 88:591–596.

Miller RB 1978 The pollination ecology of *Aquilegia elegantula* and *A. caerulea* (Ranunculaceae) in Colorado. Am J Bot 65:406–414.

——— 1981 Hawkmoths and the geographic patterns of floral variation in *Aquilegia caerulea*. Evolution 35:763–774.

Miller RE, JA McDonald, PS Manos 2004 Systematics of *Ipomoea* subgenus *Quamoclit* (Convolvulaceae) based on its sequence data and a Bayesian phylogenetic analysis. Am J Bot 91:1208–1218.

Mitchell RJ, RG Shaw 1993 Heritability of floral traits for the perennial wild flower *Penstemon centranthifolius* (Scrophulariaceae): clones and crosses. Heredity 71:185–192.

Muchhala N 2006 The pollination biology of *Burmeistera* (Campanulaceae): specialization and syndromes. Am J Bot 93:1081–1089.

Müller H, F Delpino 1871 Application of the Darwinian theory to flowers and the insects which visit them. Am Nat 5:271–297.

Ollerton J 1996 Reconciling ecological processes with phylogenetic patterns: the apparent paradox of plant-pollinator systems. J Ecol 84:767–769.

Orr HA, JA Coyne 1992 The genetics of adaptation: a reassessment. Am Nat 140:725–742.

Price MV, NM Waser, RE Irwin, DR Campbell, AK Brody 2005 Temporal and spatial variation in pollination of a montane herb: a seven-year study. Ecology 86:2106–2116.

Rausher MD 2008 Evolutionary transitions in floral color. Int J Plant Sci 169:7–21.

Raven PH 1972 Why are bird-visited flowers predominantly red? Evolution 26:674.

Rodríguez-Gironés MA, L Santamaría 2004 Why are so many bird flowers red? PLoS Biol 2:1515–1519.

Sakai AK, SG Weller, M-L Chen, S-Y Chou, C Tasanont 1997 Evolution of gynodioecy and maintenance of females: the role of inbreeding depression, outcrossing rates, and resource allocation in *Schiedea adamantis* (Caryophyllaceae). Evolution 51:724–736.

Sargent RD, JC Vamosi 2008 The influence of canopy position, pollinator syndrome, and region on evolutionary transitions in pollinator guild size. Int J Plant Sci 169:39–47.

Schemske DW, HD Bradshaw 1999 Pollinator preference and the evolution of floral traits in monkeyflowers (*Mimulus*). Proc Natl Acad Sci USA 96:11910–11915.

Schemske DW, CC Horvitz 1984 Variation among floral visitors in pollination ability: a precondition for mutualism specialization. Science 225:519–521.

Simpson MG 2006 Plant systematics. Elsevier, Oxford.

Sprengel CK 1793 Das entdeckte Geheimnis der Natur im Bau und in der Befruchtung der Blumen. Friedrich Vieweg dem aeltern, Berlin.

Stebbins GL 1970 Adaptive radiation of reproductive characteristics in angiosperms. I. Pollination mechanisms. Annu Rev Ecol Syst 1:307–326.

——— 1989 Adaptive shifts toward hummingbird pollination. Pages 39–60 *in* JH Bock, YB Linhart, eds. The evolutionary ecology of plants. Westview, Boulder, CO.

Thompson JN 1994 The coevolutionary process. University of Chicago Press, Chicago.

Thomson JD, P Wilson, M Valenzuela, M Malzone 2000 Pollen presentation and pollinator syndromes, with special reference to *Penstemon*. Plant Species Biol 15:11–29.

Thomson JD 1978 Effects of stand composition on insect visitation in two-species mixtures of *Hieracium*. Am Midl Nat 100:431–440.

——— 2003 When is it mutualism? Am Nat 162(suppl):S1–S9.

Thomson JD, BA Thomson 1992 Pollen presentation and viability schedules in animal-pollinated plants: consequences for reproductive success. Pages 1–24 *in* R Wyatt, ed. Ecology and evolution of plant reproduction: new approaches. Chapman & Hall, New York.

Waser NM, L Chittka, MV Price, NM Williams, J Ollerton 1996 Generalization in pollination systems, and why it matters. Ecology 77:1043–1060.

Wester P, R Claßen-Bockhoff 2006 Hummingbird pollination in

Salvia haenkei (Lamiaceae) lacking the typical lever mechanism. Plant Syst Evol 257:133–146.

Whittall JB, SA Hodges 2007 Pollinator shifts drive increasingly long nectar spurs in columbine flowers. Nature 447:706–709.

Wilbert SM, DW Schemske, HD Bradshaw 1997 Floral anthocyanins from two monkeyflower species with different pollinators. Biochem Syst Ecol 25:437–443.

Williams NM, JD Thomson 1998 Trapline foraging by bumble bees. III. Temporal patterning of visits. Behav Ecol 9:612–621.

Wilson P, MC Castellanos, JN Hogue, JD Thomson, WS Armbruster 2004 A multivariate search for pollination syndromes among penstemons. Oikos 104:345–361.

Wilson P, MC Castellanos, AD Wolfe, JD Thomson 2006 Shifts between bee- and bird-pollination among penstemons. Pages 47–68 *in* NM Waser, J Ollerton, eds. Plant-pollinator interactions: from specialization to generalization. University of Chicago Press, Chicago.

Wilson P, JD Thomson 1996 How do flowers diverge? Pages 88–111

in DG Lloyd, SCH Barrett, eds. Floral biology: studies on floral evolution in animal-pollinated plants. Chapman & Hall, New York.

Wilson P, AD Wolfe, WS Armbruster, JD Thomson 2007 Constrained lability in floral evolution: counting convergent origins of hummingbird pollination in *Penstemon* and *Keckiella*. New Phytol 176: 883–890.

Wolfe AD, SL Datwyler, CP Randle 2002 A phylogenetic and biogeographic analysis of the Cheloneae (Scrophulariaceae) based on ITS and matK sequence data. Syst Bot 27:138–148.

Wolfe AD, CP Randle, SL Datwyler, JJ Morawetz, N Arguedas, J Diaz 2006 Phylogeny, taxonomic affinities, and biogeography of *Penstemon* (Plantaginaceae) based on ITS and cpDNA sequence data. Am J Bot 94:1699–1713.

Wright SI, RW Ness, JP Foxe, SCH Barrett 2008 Genomic consequences of outcrossing and selfing in plants. Int J Plant Sci 169: 105–118.

Zufall RA, MD Rausher 2004 Genetic changes associated with floral adaptation restrict future evolutionary potential. Nature 428:847–850.

Int. J. Plant Sci. 169(1):39–47. 2008.
1058-5893/2008/16901-0004$15.00   DOI: 10.1086/523359

THE INFLUENCE OF CANOPY POSITION, POLLINATOR SYNDROME, AND REGION ON EVOLUTIONARY TRANSITIONS IN POLLINATOR GUILD SIZE

R. D. Sargent* and J. C. Vamosi[†]

*Department of Integrative Biology, University of California, Berkeley, California 94720-3140, U.S.A.; and
[†]Department of Biological Sciences, University of Calgary, Calgary, Alberta T2N 1N4, Canada

Little is known about how ecological context influences the probability of transitions in the extent of pollinator specialization. One unexplored hypothesis suggests that transitions to environments with different light conditions should be accompanied by transitions in pollinator guild because of the combined effects of exposure to a new pollinator community and the different relative costs associated with pollinator attraction in the understory and the canopy. Using data from literature surveys, we compiled a data set of habitat light availability (e.g., canopy vs. understory), pollinator identity, and guild size for 481 angiosperm species representing four broad regions (India, Neotropics, paleotropics, and Canada). Phylogenetic independent contrasts were used to assess the degree to which transitions in canopy position are associated with transitions in the usage of particular pollinators and pollinator guild size. We further examined the degree to which each of these traits tends to be evolutionarily labile versus evolutionarily conserved. Our analysis demonstrates that species that tend to occupy the same position in the canopy are more closely related than expected by chance, as are species that follow traditional pollinator syndromes (e.g., bee or bird), but species that have exceptionally wide pollinator guilds (e.g., are visited by bees, flies, and moths) are widely scattered across the angiosperm phylogeny. Transitions to generalist pollination appeared to be strongly associated with beetle and fly pollination and with position in the canopy above the forest floor.

Keywords: pollination, phylogenetic niche conservatism, phylogenetic independent contrasts, light conditions.

Introduction

Plant evolutionary biologists have a particular interest in the frequency (Wilson et al. 2006), directionality (Armbruster and Baldwin 1998; Castellanos et al. 2004; Wilson et al. 2006), and ecological context (Sargent and Otto 2006; Vamosi et al. 2006; van der Niet et al. 2006) of evolutionary transitions in plant-pollinator interactions. Such interactions are fundamental to understanding how selection operates on floral traits, mating systems, and inflorescence architecture. Furthermore, plant-pollinator interactions are hypothesized to play an integral role in angiosperm speciation (Grant 1994; Dodd et al. 1999; Fulton and Hodges 1999; Bradshaw and Schemske 2003; Johnson 2006). One unknown and virtually unexplored area is the degree to which the size of a plant's pollinator community (or guild) is determined by heritable traits that can evolve versus a highly labile state that depends mainly on the environment in which a plant lives (e.g., Waser et al. 1996; Fenster and Dudash 2001).

The argument that pollination is labile with respect to environmental influences suggests that the environment itself is labile with respect to phylogeny (i.e., sister species or populations are likely to experience differing environments and thus different pollinator communities). Whether this is true has not been adequately explored.

Certainly, habitat specialization is widely held to have influenced the diversification of angiosperm species and traits (e.g., Fine et al. 2004; Rajakaruna 2004). Shifts in habitat specialization and pollinator guild size may be evolutionarily correlated because of selection pressure imposed by a new pollinator community upon transition to a novel habitat (Grant and Grant 1965; Johnson 1996; von Hagen and Kadereit 2003; van der Niet et al. 2006).

As part of their evolutionary history, or indeed, even within a few generations, plant populations are likely to face transitions to conditions with differences in light availability. Because of its effect on the rate of photosynthesis, light availability affects the amount of resources available to the plant, and, of particular importance to pollination biologists, can affect the amount of resources available for floral rewards. Not surprisingly, plants growing in low-light conditions exhibit a reduction in floral nectar production (Boose 1997), fruit set (Niesenbaum 1993), and pollination (Bertin and Sholes 1993; Larson and Barrett 2000). However, it remains unclear whether transitions to low-light environments should be accompanied by transitions to plant specialization or generalization. One hypothesis suggests that plants that are adapted to understory habitats will experience strong selection to restrict their costly floral rewards to only a small number of faithful (i.e., specialized) pollinators (Feinsinger 1983). However, whether such faithful pollinators exist in low-light environments remains questionable. Low light availability affects the pollinator community and visitation rates (Herrera 1995) and is associated with high pollen limitation (Larson and Barrett 2000). Furthermore, forest understory habitats may contain disproportionately more generalist (indiscriminate) than specialist pollinators (Bawa 1990). Thus, a second hypothesis purports that low-light environments may constrain the pollination of species to be rather generalized. Interestingly, both hypotheses suggest that transitions to a habitat with a different light availability will be accompanied by shifts in a plant's pollinator guild. Accordingly, community surveys of the pollination ecology of understory plants provide support for a correlation between habitat and plant-pollinator interactions (e.g., Schemske 1981; Sakai et al. 1999; Kay and Schemske 2003; Mayfield et al. 2006). However, no study has tested whether there is a correlation between transitions in habitat light availability and transitions in pollinator guild size.

The evolutionary history of a trait has important implications for the assessment of across-species ecological patterns. When traits are phylogenetically clustered or conserved, traditional statistical tests for associations between traits among taxa will fail the assumption of independence. Thus, when conducting tests of ecological patterns, it is vital to take phylogenetic relatedness into account (Felsenstein 1985). However, the importance of correcting for phylogeny depends on how likely it is that the traits influencing pollinator composition are species or clade specific. For example, if pollinator composition changes with light environment and light environment is so labile as to change from population to population or from sister species to sister species, then correcting for phylogeny loses importance. Finally, difficulties arise in assigning causal relationships to global patterns. For instance, the composition of pollinators within the guild may be correlated with guild size (e.g., a species that includes birds as pollinators may have smaller guild size than one that includes flies because of the nectar quantities required to attract each), yet ultimately, guild size and composition may depend on regional affinities of the pollinators themselves (e.g., bird pollination may be more common in Neotropical regions than in paleotropical regions; Momose et al. 1998). Identifying and dealing with such confounding influences remains a central goal of comparative biology.

In addition to the statistical considerations, information about phylogenetic distribution can give us clues about the evolutionary constraints inherent to a particular trait or set of traits. The tendency for closely related species to have similar sets of ecological characteristics is known as phylogenetic niche conservatism (Wiens and Graham 2005). When ecological traits such as pollinator guild size and habitat associations are phylogenetically conserved, species may be more limited in their ability to adapt to novel habitats or, in the case of pollination mode, novel biotic interactions. Reproductive traits such as sexual system and pollination syndrome have previously been demonstrated to exhibit some phylogenetic niche conservatism (Fox 1985; Chazdon et al. 2003; Wilson et al. 2006). Conversely, the extent to which pollinator-mediated speciation is responsible for the diversification of angiosperms relies on plant-pollinator interactions being relatively labile with respect to phylogeny (Goldblatt and Manning 2006; Thomson and Wilson 2008), yet we are unaware of any studies that have set out to specifically examine the degree of phylogenetic conservatism in the size of a species' pollination guild.

Using literature surveys, we assembled information about primary habitat and pollination guild size for 481 angiosperm species. Most available studies of pollinator guild size were on tropical plant species. We used phylogenetic independent contrasts to address (1) the degree to which plant habit, pollinator syndrome, and plant-pollinator interactions tend to be phylogenetically labile versus conserved and (2) the evidence for evolutionary associations among pollinator specialization, canopy position, pollinator syndrome, and region.

Methods

Data Collection

We used the Web of Science (http://apps.newisiknowledge .com)to search for articles containing the search terms "pollinat* and understory" or "pollinat* and canopy or gap." We found a total of 39 articles containing information pertinent to our study. Most of these studies were performed in the tropics and can be pooled into four main regions: paleotropical ($N = 285$ plant species), Neotropical ($N = 73$), tropical India ($N = 105$), and temperate (largely Canadian; $N = 15$). We were interested in species that clearly inhabited a particular light environment (e.g., understory, forest floor, gap, canopy, etc.). We accomplished this by assigning the following light regime values to the habitats listed in published studies. Forest floor (very low light) = 1; understory (low light) = 2; midrange canopy, such as species that are epiphytes, emergents in the subcanopy (moderate light) = 3; and canopy/gap (high light) = 4.

We then determined the extent of pollination specialization by surveying studies that monitored the identity of animal species that visited a particular plant species. Regrettably, much variation exists in the literature in the precision of pollinator identification. Some studies identified pollinators to the species level, whereas many others listed pollinators as, e.g., "Coleoptera" or, more vaguely, "diverse insects." In an attempt to merge this variation into a coherent meta-analysis, we considered each insect order as a separate group of pollinators. Our designation of pollinator guild size was designed to adhere as closely as possible, given the constraints of the data, to the "functional group" definition of pollinator specialization outlined by Fenster et al. (2004). For example, if a species is pollinated by only four species of flies, it was categorized as having a pollinator guild size of 1, whereas a species pollinated by bees and flies was categorized as having a guild size of 2. Plants that had been recorded as visited by diverse insects were given a guild size of 3 because that tended to be the highest number of insect orders pollinating a single plant species. In cases where pollinators were identified only as "insect species," we assumed that each species was a different order. Wind-pollinated plant species were excluded. Because of the difficulties associated with determining the degree to which an animal visiting a flower is actually involved in pollination (i.e., transfer of self or outcross pollen to the stigma of a conspecific flower), we considered a species to be a pollinator if it was reported to visit the flowers of a particular plant species. With this data set we performed ANOVA tests to determine (1) whether a cross-species (i.e., with no phylogenetic correction) relationship existed between light index and guild size and (2) whether particular pollinator types were associated with species that had many or few other pollinating types visit. Finally, we also performed Pearson contingency tests to determine (3) whether particular orders of pollinators (which may have a higher likelihood of specializing) predominated in certain habitats and regions.

Phylogenetic Independent Contrasts

A backbone phylogeny of the species in the database was created using the program Phylomatic (Webb and Donoghue 2005). Species were placed on the backbone using the angiosperm maximally resolved seed plant tree (Stevens 2001). Generic relationships within families were further resolved where possible (a list of phylogeny sources is available from the authors). We used the function BLADJ, implemented in the program PHYLOCOM (Webb et al. 2006), to estimate branch lengths from fossil-calibrated node ages on the tree (Wikstrom et al. 2001). Data and phylogeny are available from the authors upon request.

To determine whether there is phylogenetic conservatism in our two traits, we examined whether related species were more similar in terms of light regime and pollinator breadth than expected by random chance. We used a commonly employed method for detecting phylogenetic signal that quantifies the similarity of close relatives but does not make any statement or assumption about the processes that lead to phylogenetic signal (Blomberg and Garland 2002). We used standardized phylogenetic independent contrasts (PICs), implemented using the Analysis of Traits (AOT) tool in the program PHYLOCOM (Webb et al. 2006), to test the hypothesis that transitions in the size of a plant species' pollination guild vary predictably with transitions in light regime. Although maximum likelihood and Bayesian methods are available to address correlated evolution and may be more powerful for small data sets, recent studies have found these methods to be very sensitive to taxon sampling (Nosil and Mooers 2005) and differential diversification (i.e., if the traits in question influence speciation or extinction; Vamosi et al. 2003; Maddison 2006). PICs, on the other hand, have been found to be relatively robust and conservative (Oakley and Cunningham 2000; Martins et al. 2002). Our estimate of phylogenetic signal calculates the average magnitude of contrasts across the tree and compares it to a null expectation generated using 10,000 randomizations of trait values across the tips of the tree. Tests were conducted coding the data as continuous variables (see light and guild index in "Data Collection"). Because the cross-species comparison revealed a trend for forest floor species to have specialized pollinators (only one visiting order), we include the results of an a posteriori Wilcoxon signed-rank test that examines whether transitions to the forest floor habitat (i.e., with binary coding of the habitat as forest floor or not) results in consistent decreases in the pollination guild index.

Results

Our literature survey included a total of 481 species in 275 genera and 91 families. To give an overview of the families and trait distribution in our study, we illustrate average trait values by family in figure 1. Ten families account for 238 (49.2%) of species in the data set: Zingiberaceae (40 species), Rubiaceae (37), Euphorbiaceae (33), Dipterocarpaceae (28), Annonaceae (24), Fabaceae (17), Lauraceae (17), Malvaceae

(17), Costaceae (13), and Verbenaceae (12). We obtained primary habitat and pollinating species (based on visitation) data for all species in the data set. The majority of the species are tropical in distribution.

Analyses of Uncorrected Data

Most species (81.2%) were pollinated by only one functional group, confirming the results of previous studies demonstrating that tropical plant species tend to exhibit a high degree of specialization on a particular species or related group of pollinators (reviewed in Ollerton et al. 2006). It is important to note that finding a high proportion of specialists reflects the method of categorizing specialization by functional group (here, approximated as taxonomic order) rather than by individual pollinating species (Fenster et al. 2004). Nearly half the species in the data set were recorded as being found in forest floor or understory habitats. The remaining species were found in some sort of open habitat (table 1). The number of pollinating orders tended to increase with the amount of light in the environment, with forest floor species displaying the highest incidence of specialization for pollination ($F_{3,477} = 3.53$, $P = 0.015$). A Tukey test identified the key significant difference to be between canopy/gap species and forest floor species. The probability of encountering certain pollinators in the different light environments could have profound effects on our perception of pollinator specialization. Table 2 displays the major pollinating taxa represented in our data set. Hymenoptera, Lepidoptera, Coleoptera, and Diptera represented 87% of the 613 recorded visitors to the 481 species. Note that more than one species within a given order may have visited but that this is recorded as only one visiting order in our data set. We also display the expected and observed frequencies of these pollinators in each light environment. Lepidoptera, for example, appear as more common visitors to the forest floor species (light environment 1 in table 2), while birds were observed more than expected in the understory (light environment 2). Canopy position preferences of the pollinators themselves may create bias for where pollinator specialization occurs (i.e., no pollination by Thysanoptera, Hemiptera, birds, or mammals occurred in forest floor species; forest floor species can have a maximum guild index of 4, while species in other habitats can have a guild index of up to 8). Finally, plant species that were observed to employ certain pollinators were also seen to have varying degrees of specialization. Plant species employing birds had fewer other pollinators ($F_{1,479} = 6.35$, $P = 0.01$), while the opposite was true for plant species visited by Hymenoptera ($F_{1,479} = 22.15$, $P < 0.001$), Lepidoptera ($F_{1,479} = 18.91$, $P < 0.001$), Diptera ($F_{1,479} = 296.16$, $P < 0.001$), Coleoptera ($F_{1,479} = 36.19$, $P < 0.001$), and Hemiptera ($F_{1,479} = 17.19$, $P < 0.001$). Plant species that included mammals and thrips in their pollinator guild were not significantly different in their specialization from those that did not. Contingency tests of all pollinator types revealed highly significant differences in frequency between the regions ($P < 0.001$ for all; fig. 2). Pollinator breadth was significantly higher (i.e., more generalized) in India than in Neotropical and paleotropical regions ($F_{3,470} = 7.75$, $P < 0.001$) and was surprisingly higher than that of temperate regions, although not significantly so.

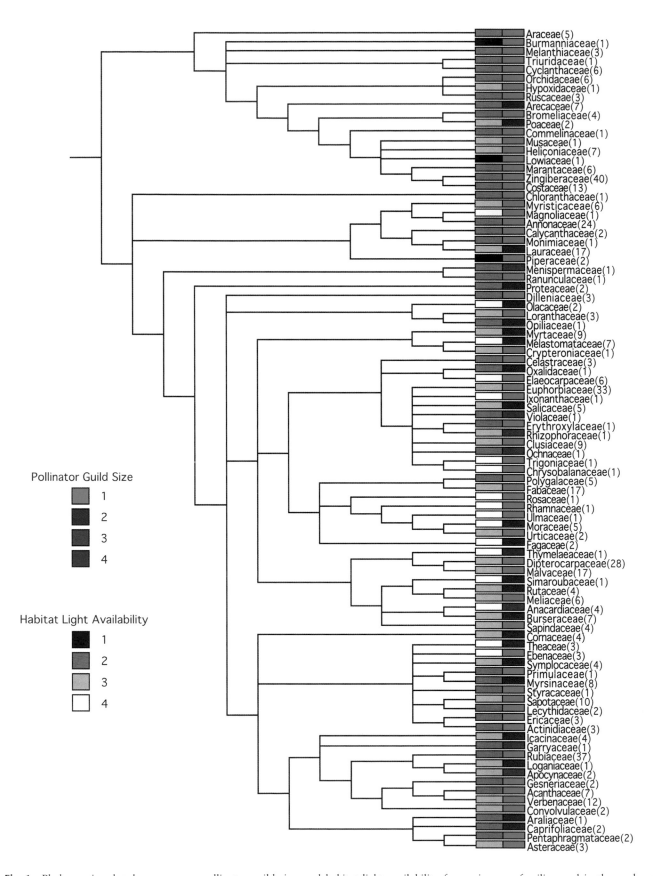

Fig. 1 Phylogenetic relatedness, average pollinator guild size, and habitat light availability for angiosperm families used in the analyses. Numbers in parentheses indicate the number of species included in the analysis.

Table 1

Summary of the 481 Species Used in Phylogenetic Analysis

Canopy layer	Light index	N	No. pollinator orders ($\pm$SE)
Forest floor	1	41	1.02 ± .11
Understory	2	208	1.32 ± .05
Liana	3	21	1.23 ± .16
Epiphyte	3	5	1.00 ± .33
Subcanopy	3	47	1.38 ± .11
Emergent	3	33	1.14 ± .16
Gap	4	52	1.33 ± .10
Canopy	4	74	1.53 ± .09

Note. Canopy layers were divided into four ordinal light environments (light index). Because the majority of information regarding pollinator identification "on the wing" is in the form of, for example, "stingless bees/beetles," we necessarily had to reduce our specialization index to the ordinal level (and in the cases of birds and mammals, to the class level). Cross-species analysis reveals significant differences in the pollinator breadth of the different light levels ($F_{3,477} = 3.53$, $P = 0.015$), with a Tukey test identifying the key significant difference to be between canopy/gap species and forest floor species. The mean pollination guild index for all species in the analysis is 1.32 ± 0.74.

Phylogenetic Comparative Tests

The significant result of the cross-species analysis above may be to some degree caused by the nonindependence of related species. We tested the association between habitat and size of pollinator guild using phylogenetic independent contrasts and found no correlation between these two traits (fig. 2; $F_{1,311} = 0.589$, $P = 0.443$). The results of phylogenetic tests for correlations between guild size and pollinator types were similar to those from the nonphylogenetic analysis (table 3). Inclusion of Hymenoptera, Lepidoptera, and Diptera in a plant's pollinator pool was associated with increases in pollinator guild size, while the inclusion of birds displayed a nonsignificant trend toward decreases in pollinator guild size. By contrast, none of the pollinator types showed a phylogenetic association with canopy position. The trend for increased pollinator breadth in India was marginally significant in a phylogenetic analysis ($P = 0.05$), possibly because of a dearth of bird pollination ($P =$

0.01). Canopy position exhibited strong phylogenetic conservatism ($P = 0.001$), while size of pollination guild did not ($P = 0.112$). However, all pollination syndromes tested (table 3), with the exception of Diptera, displayed strong phylogenetic signal ($P < 0.001$). Habitation of the forest floor also appeared to be phylogenetically conserved ($P = 0.020$), and the forest floor habitat was marginally associated with fewer pollinators (one-tailed Wilcoxon-signed rank test, $P = 0.059$, df = 31).

Discussion

Examining the geographical and ecological context under which transitions to pollinator specialization are favored can provide insight into the selective forces contributing to pollinator specialization. We found little support for the hypothesis that a plant's primary light environment is correlated with the size of a plant's pollination guild, especially when phylogeny is accounted for. Instead, we found strong patterns of association between generalization, region, and pollinator identity. The finding that the regional pollinator pool is an important determinant of pollinator guild size has consequences for plant evolution. For instance, the selection pressures imposed by generalized pollination syndromes are considered to be too diffuse to precipitate speciation (Waser et al. 1996), and therefore we might predict that clades with traits (e.g., in terms of pollinator identity or region) associated with generalist pollination will have fewer species per clade.

Our results suggest that forest floor species tend to have a smaller number of pollinators than canopy or gap species, providing some support for the idea that extremely shaded conditions select for specialization. Plant species in shaded environments have been hypothesized be more specialized due to selective pressures to reduce wastage of floral resources on inefficient pollinators (Feinsinger 1983). Conversely, plants in shady habitats may tend to be more generalized because such habitats tend to be dominated by generalist insect pollinators (Bawa 1990). Thus, we provide some evidence consistent with Feinsinger's (1983) hypothesis. However, because we did not observe strong correlations between canopy position and pollination guild size, a variety of light environments appear to provide the necessary conditions for transitions toward greater or smaller pollinator breadth.

Table 2

Presence of the Major Pollinating Taxa in Canopy Layers 1–4 and the Mean Guild Index in Plant Species That Employ Each Taxon

Pollinator	N	Layer 1 (8.5%)	Layer 2 (43.2%)	Layer 3 (22.0%)	Layer 4 (26.2%)	Guild index
Hymenoptera	257	22 (8.6%)	114 (44.4%)	46 (17.9%)	75 (29.2%)	1.46 ± .05[**]
Lepidoptera[*]	84	9 (10.7%)	26 (30.9%)	27 (33.3%)	21 (25.0%)	1.63 ± .08[**]
Coleoptera[*]	127	9 (7.1%)	45 (36.0%)	41 (32.3%)	32 (25.2%)	1.65 ± .06[**]
Diptera	73	2 (2.7%)	40 (54.8%)	13 (17.8%)	18 (24.7%)	2.39 ± .07[**]
Thysanoptera	12	0	6 (50.0%)	1 (8.3%)	5 (41.7%)	1.42 ± .21
Hemiptera	8	0	3 (37.5%)	4 (50.0%)	1 (12.5%)	2.38 ± .26[**]
Birds[*]	57	0	30 (52.6%)	16 (28.1%)	11 (19.3%)	1.09 ± .10[*]
Mammals	10	0	4 (40.0%)	4 (40.0%)	2 (20.0%)	1.00 ± .23

Note. Mean guild index refers to the number of pollinating orders utilized. Guild indexes differing significantly when the pollinator was present versus absent are marked with asterisks. See table 1 for explanation of canopy layers.

[*] $P < 0.05$ (Pearson contingency test).

[**] $P < 0.001$.

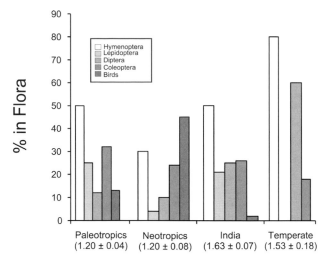

Fig. 2 Distribution of pollinator visitation in geographical regions (percentage of species visited by particular pollinating orders). Contingency tests of all pollinator types revealed highly significant differences in frequency between the regions ($P < 0.001$). Numbers below the X-axis indicate the mean pollinator guild size ($\pm$SE) of species for the region.

Pollinator Guild Size, Identity, and Region

It is important to note that we do find that certain light environments are depauperate in pollinator diversity, possibly because of pollinator preferences for certain light intensities (Bawa 1990; Bertin and Sholes 1993; Bishop and Armbruster 1999). Thus, forest floor species could appear to attract only a small number of pollinators, not because of floral adaptations but because a smaller suite of pollinators is available at this low light level. This finding is supported by the observation that shaded habitats have lower pollinator fauna species richness than unshaded habitats in Alaska (Bishop and Armbruster 1999) but is in contrast to the findings of other studies (e.g., Bawa et al. 1985) that have found a greater diversity of pollinators at low levels of the canopy strata.

Interestingly, our results suggest that pollinator identity strongly influenced pollinator guild size, with bee, moth, beetle,

and fly pollination being associated with increased pollinator guild size and bird pollination with decreased pollinator guild size. Region also exerted a strong influence, with the tropical species of India displaying significantly higher guild size than Neotropical and other paleotropical regions, perhaps because of a reduction in the amount of pollination by birds. Incidentally, our phylogenetic global analysis finds little evidence to substantiate the claims of Devy and Davidar (2003) that Neotropical countries such as Venezuela and Jamaica (two countries that figure heavily in our data set) show a paucity in the number of pollinator guilds and a high proportion of species skewed toward generalization.

Somewhat paradoxically, the pollinators of forest floor species often included Lepidoptera—a group that was associated with relatively generalist species in our study—but not birds, which are associated with specialization. This finding is consistent with the hypothesis that forest floor plants are less likely to be bird pollinated because high amounts of light are needed to produce the copious amounts of nectar required to attract birds (Stiles 1978). However, our phylogenetic analysis revealed reduced significance of the association between Lepidoptera pollination and low canopy position, indicating that the pattern can most likely be attributed to disproportionate representation in the data set of certain families with Lepidoptera-pollinated forest floor species. Nevertheless, further studies that combine the knowledge that certain pollinators (1) may prefer particular light intensities and (2) may be more likely to influence plant traits in such a way as to exclude other pollinators (Kato 1996, 2005) are warranted. In particular, the similarity of the association of beetles and flies with generalization is intriguing because there are few common foraging characteristics between these pollinators other than their propensity to visit flowers with an open floral form (Devy and Davidar 2003). However, having an open floral form may reduce the potential for adaptive trade-offs in floral morphology that have been shown to contribute to the evolution of specialization (Muchhala 2007), which might help to explain why the possession of radially symmetrical flowers is associated with reduced species richness (Sargent 2004).

Phylogenetic Conservatism

Plant canopy position exhibited significant phylogenetic conservatism. Thus, closely related species in our sample were more

Table 3

Phylogenetic Signal and Correlations with Habitat and Guild Size for Pollinator Types with $N > 20$

Pollinator	Signal (P)	Correlation with canopy position			Correlation with guild size		
		C	df	P	C	df	P
Hymenoptera	.001	.08	90	.623	.22	90	.025
Lepidoptera	.001	−.23	48	.186	.52	48	.001
Coleoptera	.001	.10	64	.327	.78	64	<.0001
Diptera	.158	−.10	60	.438	1.16	60	<.0001
Bird	.001	.29	25	.109	−.13	25	.117

Note.   Positive correlations indicate that the inclusion of a particular pollinator is associated with higher levels in the canopy or increases in guild size (e.g., plant species that are visited by bees are more likely to be visited by other pollinators as well compared to species that are not visited by bees). C = mean contrast value (e.g., the mean difference in trait X [e.g., canopy position] between sister groups that differ in whether bees act as pollinators). In two-tailed Wilcoxon signed rank tests, C was significantly different from zero (Vamosi and Vamosi 2005).

likely to have similarities in their light environment than distantly related species. This result is consistent with previous research demonstrating that a plant species' light environment and shade tolerance tend to be evolutionarily conserved rather than labile (Prinzing et al. 2001; Niinemets and Valladares 2006). The degree to which plant ecological traits tend to be conservative in their evolution is an important and somewhat controversial topic in plant ecology (reviewed in Webb et al. 2002; see also Silvertown et al. 2006) and may have implications for the mode and pace of plant speciation (reviewed in Wiens and Graham 2005). It is somewhat surprising that light environment would tend to be phylogenetically conserved because the amount of light that a plant receives tends to vary at a much smaller spatial scale than other environmental gradients (Prinzing et al. 2001). Our result may be a reflection of the fact that a species' position in the canopy most likely reflects ecological gradients beyond simple differences in the amount of light received. For example, particularly in tropical species, an epiphytic habit is correlated with occurrence in the canopy layer, and if epiphyte habitat itself is phylogenetically conserved, it would tend to inflate the degree to which light environment exhibits phylogenetic signal.

In contrast to canopy position, we found no evidence that the size of a plant's pollinator guild is phylogenetically conserved; closely related species were as likely to have similar pollinator guild sizes as distantly related species. To our knowledge, this is the first statistical test of the phylogenetic conservatism of pollinator guild size. We did, however, find that the inclusion of a particular pollinator order was strongly conserved, which is consistent with the idea that pollination systems exhibit niche conservatism (Wilson et al. 2006; Thomson and Wilson 2008). Our results suggest that lineages with specialized pollination systems may experience switches in the identity of their main pollinators rarely when compared to the frequency with which plant species increase or decrease their pollinator breadth.

This finding is significant because theories of pollinator-driven speciation rest on the assumption that pollinator syndromes are evolutionarily labile and are prone to change relatively rapidly upon exposure to different pollinator environments (reviewed in Johnson 2006). Paradoxically, our results, in conjunction with results reported by Wilson et al. (2006), suggest that, at least for the species represented in our study, pollinator syndromes may be relatively conserved. Because the evolutionary lability of pollinator syndromes has implications for our understanding of the mechanisms of plant speciation, we suggest that further investigations of the degree to which pollinator syndromes tend to exhibit phylogenetic conservatism are warranted.

Finally, it is important to note that the method that we used to define the size of a plant species' pollinator guild is only one of several possible methods (reviewed in Fenster et al. 2004). Our pollinator guild data do not distinguish between pollination and visitation, which may have different implications for natural selection and speciation (Fenster et al. 2004; Maad and Nilsson 2004). Other workers in this field (e.g., Waser et al.

1996) favor an approach that would have many of our "specialized" species categorized as generalists, which would have influenced the phylogenetic pattern of specialization. Most comparative studies of pollinator specialization to date have been forced to interpret visitation data as pollination data because of the scarcity of data sets that have recorded pollination (e.g., Ollerton et al. 2006; Wilson et al. 2006). Our approach represents a middle ground between methods advocated by Waser et al. (1996) and Fenster et al. (2004). Unfortunately, a methodology for classifying a plant's pollinator guild that avoids the pitfalls of lumping together some pollinators that may in fact influence floral evolution differently (sensu Fenster et al. 2004) while separating others that may not (sensu Waser et al. 1996) has yet to be described. We feel that our methodology reduces the subjectivity associated with classifying pollinators into functional groups while maintaining the philosophy behind it.

Controversy surrounds the role of pollinator interactions in the diversification of the angiosperms. Our results suggest that evolutionary transitions between specialized and generalized interactions are not uncommon, are not subject to phylogenetic constraint, and are (largely) not limited to certain light environments. However, certain pollinators were associated with more specialized pollination than others, and whether a particular pollinator type was included within a plant's suite of visitors did have a strong phylogenetic signal. These findings may have implications for plant speciation. Undoubtedly, speciation accompanied by floral shifts for specialized pollinator types (e.g., bird vs. bee) has occurred (reviewed in Thomson and Wilson 2008). Our data suggests that (1) the majority of plant speciation has occurred without shifts in pollinator types (as evidenced by the strong phylogenetic signal for the majority of pollinators) and (2) further instances where pollinator-mediated speciation has occurred will most likely be found within clades that include bird pollination and do not include pollination by beetles and flies. To our knowledge, although isolated cases exist to substantiate the claim that transitions to certain pollinators or positions in the canopy may subsequently influence the diversification rates within lineages or regions (e.g., Kay et al. [2005] found high rates of speciation in hummingbird-pollinated *Costus* lineages), large-scale phylogenetic examinations of speciation rates among clades with contrasting pollinators and/or locales have yet to be conducted. We assert that formal phylogenetic tests of the association between specialized pollination and angiosperm speciation are an important goal of this field.

Acknowledgments

We thank S. Armbruster, S. Barrett, R. Ree, and one anonymous reviewer for their insightful comments on an earlier draft of the manuscript. We thank S. Kembel for his advice on configuring PHYLOCOM. R. D. Sargent was supported by a Natural Sciences and Engineering Research Council of Canada postdoctoral fellowship.

Literature Cited

Armbruster WS, BG Baldwin 1998 Switch from specialized to generalized pollination. Nature 394:632.

Bawa KS 1990 Plant-pollinator interactions in tropical rain forests. Annu Rev Ecol Syst 21:399–422.

Bawa KS, SH Bullock, DR Perry 1985 Reproductive biology of tropical lowland rain forest trees. 2. Pollination systems. Am J Bot 72: 346–356.

Bertin RI, ODV Sholes 1993 Weather, pollination and the phenology of *Geranium maculatum*. Am Midl Nat 129:52–66.

Bishop JA, WS Armbruster 1999 Thermoregulatory abilities of Alaskan bees: effects of size, phylogeny and ecology. Funct Ecol 13:711–724.

Blomberg SP, T Garland 2002 Tempo and mode in evolution: phylogenetic inertia, adaptation and comparative methods. J Evol Biol 15:899–910.

Boose DL 1997 Sources of variation in floral nectar production rate in *Epilobium canum* (Onagraceae): implications for natural selection. Oecologia 110:493–500.

Bradshaw HD, DW Schemske 2003 Allele substitution at a flower colour locus produces a pollinator shift in monkeyflowers. Nature 426:176–178.

Castellanos MC, P Wilson, JD Thomson 2004 "Anti-bee" and "pro-bird" changes during the evolution of hummingbird pollination in *Penstemon* flowers. J Evol Biol 17:876–886.

Chazdon RL, S Careaga, C Webb, O Vargas 2003 Community and phylogenetic structure of reproductive traits of woody species in wet tropical forests. Ecology 73:331–348.

Devy MS, P Davidar 2003 Pollination systems of trees in Kakachi, a mid-elevation wet evergreen forest in Western Ghats, India. Am J Bot 90:650–657.

Dodd ME, J Silvertown, MW Chase 1999 Phylogenetic analysis of trait evolution and species diversity variation among angiosperm families. Evolution 53:732–744.

Feinsinger P 1983 Coevolution and pollination. Pages 283–310 *in* DJ Futuyma, M Slatkin, eds. Coevolution. Sinauer, Sunderland, MA.

Felsenstein J 1985 Phylogenies and the comparative method. Am Nat 125:1–15.

Fenster CB, WS Armbruster, P Wilson, MR Dudash, JD Thomson 2004 Pollination syndromes and floral specialization. Annu Rev Ecol Evol Syst 35:375–403.

Fenster CB, MR Dudash 2001 Spatiotemporal variation in the role of hummingbirds as pollinators of *Silene virginica*. Ecology 82: 844–851.

Fine PVA, I Mesones, PD Coley 2004 Herbivores promote habitat specialization by trees in Amazonian forests. Science 305:663–665.

Fox JF 1985 Incidence of dioecy in relation to growth form, pollination and dispersal. Oecologia 67:244–249.

Fulton M, SA Hodges 1999 Floral isolation between *Aquilegia formosa* and *Aquilegia pubescens*. Proc R Soc B 266:2247–2252.

Goldblatt P, JC Manning 2006 Radiation of pollination systems in the Iridaceae of sub-Saharan Africa. Ann Bot 97:317–344.

Grant V 1994 Modes and origins of mechanical and ethological isolation in angiosperms. Proc Natl Acad Sci USA 91:3–10.

Grant V, KA Grant 1965 Flower pollination in the phlox family. Columbia University Press, New York.

Herrera CM 1995 Microclimate and individual variation in pollinators: flower plants are more than their flowers. Ecology 76:1518–1524.

Johnson SD 1996 Pollination, adaptation and speciation models in the Cape flora of South Africa. Taxon 45:59–66.

——— 2006 Pollinator-driven speciation in plants. Pages 295–309 *in* LD Harder, SCH Barrett, eds. Ecology and evolution of flowers. Oxford University Press, New York.

Kato M 1996 Plant-pollinator interactions in the understory of a lowland mixed dipterocarp forest in Sarawak. Am J Bot 83:732–743.

——— 2005 Ecology of traplining bees and understory pollinators in ecological studies. Pages 128–133 *in* DW Roubik, S Sakai, AAH Hamid Karim, eds. Pollination ecology and the rain forest. Vol 174. Springer, New York.

Kay KM, PA Reeves, RG Olmstead, DW Schemske 2005 Rapid speciation and the evolution of hummingbird pollination in Neotropical *Costus* subgenus *Costus* (Costaceae): evidence from nrDNA ITS and ETS sequences. Am J Bot 92:1899–1910.

Kay KM, DW Schemske 2003 Pollinator assemblages and visitation rates for 11 species of Neotropical *Costus* (Costaceae). Biotropica 35:198–207.

Larson BMH, SCH Barrett 2000 A comparative analysis of pollen limitation in flowering plants. Biol J Linn Soc 69:503–520.

Maad J, LA Nilsson 2004 On the mechanism of floral shifts in speciation: gained pollination efficiency from tongue- to eye-attachment of pollinia in *Platanthera* (Orchidaceae). Biol J Linn Soc 83:481–495.

Maddison WP 2006 Confounding asymmetries in evolutionary diversification and character change. Evolution 60:1743–1746.

Martins EP, JAF Diniz, EA Houseworth 2002 Adaptive constraints and the phylogenetic comparative method: a computer simulation test. Evolution 56:1–13.

Mayfield MM, DD Ackerly, GC Daily 2006 The diversity and conservation of plant reproductive and dispersal functional traits in human-dominated tropical landscapes. J Ecol 94:522–536.

Momose K, T Yumoto, T Nagamitsu, M Kato, H Nagamasu, S Sakai, RD Harrison, T Itioka, AA Hamid, T Inoue 1998 Pollination biology in a lowland dipterocarp forest in Sarawak, Malaysia. I. Characteristics of the plant-pollinator community in a lowland dipterocarp forest. Am J Bot 85:1477–1501.

Muchhala N 2007 Adaptive trade-off in floral morphology mediates specialization for flowers pollinated by bats and hummingbirds. Am Nat 169:494–504.

Niesenbaum RA 1993 Light or pollen: seasonal limitations on female reproductive success in the understory shrub *Lindera benzoin*. J Ecol 81:315–323.

Niinemets U, F Valladares 2006 Tolerance to shade, drought, and waterlogging of temperate northern hemisphere trees and shrubs. Ecol Monogr 76:521–547.

Nosil P, AO Mooers 2005 Testing hypotheses about ecological specialization using phylogenetic trees. Evolution 59:2256–2263.

Oakley TH, CW Cunningham 2000 Independent contrasts succeed where ancestor reconstruction fails in a known bacteriophage phylogeny. Evolution 54:397–405.

Ollerton J, SD Johnson, AB Hingston 2006 Geographical variation in diversity and specificity of pollination systems. Pages 283–308 *in* NM Waser, J Ollerton, eds. Plant-pollinator interactions: from specialization to generalization. University of Chicago Press, Chicago.

Prinzing A, W Durka, S Klotz, R Brandl 2001 The niche of higher plants: evidence for phylogenetic conservatism. Proc R Soc B 268: 2383–2389.

Rajakaruna N 2004 The edaphic factor in the origin of species. Int Geol Rev 46:471–478.

Sakai S, M Kato, T Inoue 1999 Three pollination guilds and variation in floral characteristics of Bornean gingers (Zingiberaceae and Costaceae). Am J Bot 86:646–658.

Sargent RD 2004 Floral symmetry affects speciation rates in angiosperms. Proc R Soc B 271:603–608.

Sargent RD, SP Otto 2006 The role of local species abundance in the evolution of pollinator attraction in flowering plants. Am Nat 167: 67–80.

Schemske DW 1981 Floral convergence and pollinator sharing in two bee-pollinated tropical herbs. Ecology 62:946–954.

Silvertown J, K McConway, D Gowing, M Dodd, MF Fay, JA Joseph, K Dolphin 2006 Absence of phylogenetic signal in the niche structure of meadow plant communities. Proc R Soc B 273:39–44.

Stevens PF 2001 Angiosperm phylogeny website, version 7. http://www.mobot.org/MOBOT/research/APweb/.

Stiles FG 1978 Ecological and evolutionary implications of bird pollination. Am Zool 18:715–727.

Thomson JD, P Wilson 2008 Explaining evolutionary shifts between bee and hummingbird pollination: convergence, divergence, and directionality. Int J Plant Sci 169:23–38.

Vamosi JC, TM Knight, JA Steets, SJ Mazer, M Burd, T-L Ashman 2006 Pollination decays in biodiversity hotspots. Proc Natl Acad Sci USA 103:956–961.

Vamosi JC, SP Otto, SCH Barrett 2003 Phylogenetic analysis of the ecological correlates of dioecy in angiosperms. J Evol Biol 16:1006–1018.

Vamosi SM, JC Vamosi 2005 Endless tests: guidelines to analyzing non-nested sister-group comparisons. Evol Ecol Res 7:567–579.

van der Niet T, SD Johnson, HP Linder 2006 Macroevolutionary data suggest a role for reinforcement in pollination system shifts. Evolution 60:1596–1601.

von Hagen KB, JW Kadereit 2003 The diversification of *Halenia* (Gentianaceae): ecological opportunity versus key innovation. Evolution 57:2507–2518.

Waser NM, L Chittka, MV Price, NM Williams, J Ollerton 1996 Generalization in pollination systems, and why it matters. Ecology 77:1043–1060.

Webb CO, DD Ackerly, SW Kembel 2006 Phylocom: software for the analysis of community phylogenetic structure and trait evolution, version 3.40. http://www.phylodiversity.net/phylocom.

Webb CO, DD Ackerly, MA McPeek, MJ Donoghue 2002 Phylogenies and community ecology. Annu Rev Ecol Syst 33:475–505.

Webb CO, MJ Donoghue 2005 Phylomatic. http://www.phylodiversity.net/phylomatic.

Wiens JJ, CH Graham 2005 Niche conservatism: integrating evolution, ecology and conservation biology. Annu Rev Ecol Evol Syst 36:519–539.

Wikström N, V Savolainen, MW Chase 2001 Evolution of the angiosperms: calibrating the family tree. Proc R Soc B 268:2211–2220.

Wilson P, MC Castellanos, AD Wolfe, JD Thomson 2006 Shifts between bee and bird pollination among penstemons. Pages 47–68 *in* NM Waser, J Ollerton, eds. Plant-pollinator interactions: from specialization to generalization. University of Chicago Press, Chicago.

Int. J. Plant Sci. 169(1):49–58. 2008.
1058-5893/2008/16901-0005$15.00   DOI: 10.1086/523365

A PHYLOGENETIC ANALYSIS OF THE EVOLUTION OF WIND POLLINATION IN THE ANGIOSPERMS

Jannice Friedman and Spencer C. H. Barrett

Department of Ecology and Evolutionary Biology, University of Toronto, 25 Willcocks Street, Toronto, Ontario M5S 3B2, Canada

Wind pollination is predominantly a derived condition in angiosperms and is thought to evolve in response to ecological conditions that render animal pollination less advantageous. However, the specific ecological and evolutionary mechanisms responsible for transitions from animal to wind pollination are poorly understood in comparison with other major reproductive transitions in angiosperms, including the evolution of selfing from outcrossing and dioecy from hermaphroditism. To investigate correlations between wind pollination and a range of characters including habitat type, sexual system, floral display size, floral showiness, and ovule number, we used a large-scale molecular phylogeny of the angiosperms and maximum likelihood methods to infer historical patterns of evolution. This approach enabled us to detect correlated evolution and the order of trait acquisition between pollination mode and each of nine characters. Log likelihood ratio tests supported a model of correlated evolution for wind pollination and habitat type, floral sexuality, sexual system, flower size, flower showiness, presence versus absence of nectar, and ovule number. In contrast, wind pollination and geographical distribution and number of flowers per inflorescence evolve independently. We found that in wind-pollinated taxa, nectar is lost more often and ovule number is reduced to one. We also found that wind pollination evolves more frequently in lineages already possessing unisexual flowers and/or unisexual plants. An understanding of the ecological and life-history context in which wind pollination originates is fundamental to further investigation of the microevolutionary forces causing transitions from animal to wind pollination.

Keywords: anemophily, comparative analysis, correlated evolution, ecological correlates, morphological correlates, pollination.

Introduction

Wind pollination (anemophily) has evolved at least 65 times in the angiosperms from biotically pollinated ancestors (Linder 1998). A recent survey estimates that abiotic pollination occurs in at least 18% of angiosperm families (Ackerman 2000), with wind pollination more commonly represented than water pollination. The evolution of wind pollination is thought to occur when environmental conditions render biotic pollination less advantageous (Regal 1982; Cox 1991). For example, a decline in pollinator abundance or changes in the abiotic environment limiting pollinator activity have been invoked to explain why wind pollination has evolved in particular taxa (Berry and Calvo 1989; Weller et al. 1998; Goodwillie 1999). However, the specific ecological mechanisms causing transitions from animal to wind pollination have not been investigated in detail, and, in comparison with other reproductive transitions (e.g., the evolution of selfing from outcrossing and dioecy from hermaphroditism), little is known about the microevolutionary forces responsible for the evolution of wind pollination.

Comparative evidence clearly indicates that wind pollination evolves more frequently in certain clades (Ackerman 2000). As families and genera often share traits as a result of

common ancestry, mapping traits onto phylogenetic trees is our best option for testing correlated evolution without bias from phylogenetic relationships (Felsenstein 1985; Donoghue 1989; Harvey and Pagel 1991). Taking phylogenetic relationships into account, we were interested in finding evidence for the existence of correlations between anemophily and specific morphological and ecological traits and determining the putative evolutionary pathways leading to these associations. Linder (1998) first examined morphological traits and their associations with wind pollination in a comparative context. However, his study was limited to families of the lower rosids and commelinoid monocots and used the concentrated-changes test (Maddison 1990), which is less powerful than current phylogenetic comparative methods (see Schluter et al. 1997; Pagel 1999), to investigate correlations. The recent advent of maximum likelihood phylogenetic comparative methods (Harvey and Pagel 1991; Pagel 1994; Freckleton et al. 2002) provides an opportunity to investigate correlations among life-history traits and ecology and biogeography and to examine the evolutionary history of reproductive associations (e.g., dioecy: Vamosi et al. 2003; dichogamy and self-incompatibility: Routley et al. 2004; protogyny and pollination mode: Sargent and Otto 2004). We were interested in examining the relations between wind pollination and a range of characters that have been proposed to be associated with this condition. We now briefly review the traits chosen and the functional arguments that have been proposed to explain their correlations with anemophily.

Regions of higher latitude, arid temperate environments, open vegetation, and island floras have the highest representation of wind-pollinated plants (Whitehead 1968; Regal 1982). The physical and aerodynamic requirements for successful wind pollination may explain these ecological and geographical correlates. However, it is unclear whether these associations are robust to phylogenetic considerations and whether wind pollination is more likely to originate under these conditions or is simply easier to maintain.

One of the more widely recognized features of wind pollination is the higher frequency of unisexual flowers in wind-pollinated species than in animal-pollinated species (Bawa 1980; Renner and Ricklefs 1995). Several hypotheses for this have been proposed, including a reduction in shared fixed costs between female and male flowers (Lloyd 1982), a more linear male gain curve (Charnov et al. 1976), and limiting self-fertilization because of the unavoidable geitonogamy that seems likely in wind-pollinated plants (Lloyd and Webb 1986; Charlesworth 1993). Some of these explanations are specific to dioecy, while others can be extended to dicliny (unisexual flowers) in general. These hypotheses imply that wind pollination precedes the evolution of unisexual flowers. This particular order of transition is supported in the Poaceae (Malcomber and Kellogg 2006) and *Fraxinus* (Wallander 2001); however, the opposite order appears to occur in *Leucadendron* (Midgley 1987; Hattingh and Giliomee 1989), *Thalictrum* (Kaplan and Mulcahy 1971) and possibly in *Schiedea* (Weller et al. 1998). Therefore, establishing the relative frequency of the two polarities that lead to a correlation between dicliny and anemophily is critical to understanding the functional basis of the correlation.

An association between wind pollination and low ovule number was reported by Linder (1998) in his comparative analysis of the evolution of anemophily. Wind-pollinated flowers have been presumed to produce only single ovules because of the small chance of multiple pollen grains landing on each stigma (Pohl 1929; Dowding 1987). Unlike in animal-pollinated plants, where pollen grains arrive in clumps, the pollen of wind-dispersed species is usually transported as single units, so that the chance of capturing each pollen grain is an independent event. These arguments imply that a reduction in ovule number occurs after the evolution of wind pollination, a sequence supported by the results of Linder (1998).

A suite of morphological traits is commonly associated with wind pollination and constitutes the anemophilous syndrome (Faegri and van der Pijl 1979). Many of the traits can be explained by functional arguments and the aerodynamic requirements for wind pollination. However, it is unclear whether these traits facilitate the evolution of wind pollination or evolve after the origin of wind pollination in particular lineages. In general, wind-pollinated plants have small flowers with highly reduced or no perianth parts. Nectaries are usually absent or nonfunctional (Faegri and van der Pijl 1979). The presence of complex, large, showy flowers may act as a constraint to the evolution of wind pollination, and it appears that nectaries are typically lost after the evolution of wind pollination (Linder 1998). Plants that already have small simple flowers may be more suitable for transporting and capturing some portion of their pollen by wind because the stigmas and anthers are likely to be exposed to air currents (Culley et al.

2002). Plants with a mixed pollination strategy involving both animal and wind pollination (ambophily) tend to have small unshowy flowers (e.g., *Piper* spp.: De Figueiredo and Sazima 2000; *Salix* spp.: Tamura and Kudo 2000; Karrenberg et al. 2002). Similarly, plants that use pollen as a reward (rather than nectar) may produce more pollen and be at a selective advantage if conditions change to favor wind pollination. However, it is not always clear whether wind pollination evolves more frequently in lineages that have small inconspicuous flowers and no nectar or whether these traits are lost after the evolution of wind pollination because of energetic reasons associated with a loss of function.

Here, we use a recent molecular phylogeny of the angiosperms (Soltis et al. 2000) to investigate evidence for correlations between anemophily and specific morphological and ecological traits and to evaluate the evolutionary pathways leading to these associations. Specifically, we investigated whether associations occur between wind pollination and (1) biogeographical and ecological traits, including temperate distributions and open habitat types; (2) sexual traits, including unisexual flowers (dicliny) and plants (dioecy) and low ovule numbers; and (3) a reduction in floral characteristics, including flower size and showiness. In cases where we detected associations, we then investigated the order of transition between the traits using tests of contingent evolution to assess the evolutionary pathways involved. This information enabled us to evaluate several potential hypotheses that have been proposed to explain the evolution of wind pollination from animal pollination.

Methods

Character Coding and Phylogenetic Data

We use the "B series" tree and branch lengths of the molecular phylogeny of angiosperms by Soltis et al. (2000). This tree is based on 567 taxa and three gene sequences (18s rDNA, *rbc*L, and *atp*B). For each species included in the phylogeny, we assigned states for the following 10 characters: pollination mode (animal or wind), floral sexuality (hermaphroditic or unisexual flowers), sexual system (cosexual or dioecious populations), ovules (one or more than one), flower size (small [<1 cm] or medium to large [>1 cm]), flower showiness (showy or plain [green, white, yellow-green]), number of flowers per inflorescence (few [<5] or many [>5]), nectar (present or absent), habitat type (open or closed), and geographical distribution (temperate or tropical).

We obtained information on these character states from Kubitzki (1993, 1998a, 1998b, 2003, 2004), Mabberley (1997), Hutchinson (1964), the database generously provided by Jana Vamosi (used in Vamosi et al. 2003), a variety of online floras, and extensive literature surveys (list of all sources available from the first author). All states were coded as binary characters. The Soltis et al. (2000) study depicts the relationships of genera, although the original phylogeny was estimated using representative species for each genus. We used these species for our character coding. When information was unavailable for a particular species, we used the most common state (>50%) for the members of the genus. We obtained information for all 560 angiosperm species in the phylogeny, of which 68 are wind pollinated.

Testing for Correlated Evolution and Directionality

To test for correlated evolution between wind pollination and alternative character states, we used BayesTraits (Pagel and Meade 2006). We implemented the BayesDiscrete module, which investigates correlated evolution between pairs of discrete binary traits. The program fits continuous-time Markov models to the discrete character data and allows the trait to change states over infinitesimally small intervals of time. The model estimates transition rates and the likelihood associated with different states at each node of the tree and calculates transition probabilities across all possible character states at each node, eliminating the need to assign ancestral states. We used the branch scaling parameter κ, suggested by Pagel (1994), which adjusted the weight of branch lengths in the model and allowed it to take its maximum likelihood value. In all cases, $\kappa < 1$, which reduces the length of longer branches more than shorter ones.

BayesDiscrete tests for correlated evolution in two binary traits by comparing the fit (log likelihood) of two of these continuous-time Markov models. The first is a model in which two traits (e.g., wind pollination and nectar) evolve independently on the tree. This creates two rate coefficients per trait. The other model allows the traits to evolve in a correlated fashion such that the rate of change in one trait depends on the background state of the other. The dependent model has four states, one for each combination of the two binary traits (0, 0; 0, 1; 1, 0; 1, 1; see fig. 1). To determine whether wind pollination is correlated with a trait, we compared the likelihood estimate of the independent model ($L(I)$) to the likelihood estimate of the dependent model ($L(D)$). Because likelihood ratios approximate a χ^2 distribution, support for correlated evolution is indicated when $L(D)$ is significantly greater than $L(I)$, which can be tested by comparing $-2[L(D) - L(I)]$ to a χ^2 distribution with 4 degrees of freedom.

We determined the statistical significance of each of the estimated parameters in the dependent model by restricting individual transition parameters to zero and recalculating the likelihood ratio of the model. We then compared the restricted seven-parameter model to the unrestricted dependent model with a 1-df χ^2 test. A significant likelihood ratio indicates that the transition rate is significantly different from zero. Finally, we tested specific hypotheses about contingent evolution by restricting two of the rates to being equal (e.g., unisexual flowers evolve equally in animal- and wind-pollinated lineages: $q_{12} = q_{34}$). This seven-parameter restricted model can be compared to the full dependent model using the likelihood ratio test, $-2[L(D) - L(q_{12} = q_{34})]$ with a 1-df χ^2 test. A significant likelihood ratio indicates that the parameters are significantly different from one another, demonstrating that the state of trait X influences the direction of evolution of trait Y (Pagel 1994). Because we performed multiple tests, we adjusted the α level using Bonferroni correction.

Results

Traits Correlated with Wind Pollination

We first examined whether traits were associated with pollination mode regardless of phylogenetic considerations. We found that all of the floral characteristics we investigated

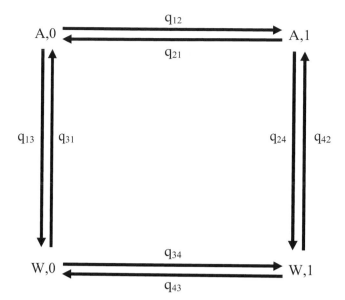

Trait	States
Pollination mode	A=animal, W=wind
Geographical distribution	0=tropical, 1=temperate
Habitat type	0=closed, 1=open
Floral sexuality	0=hermaphrodite, 1=unisexual
Sexual system	0=cosexual, 1=dioecious
Ovules	0=more than one, 1=one
Number of flowers	0=few, 1=several to many
Flower size	0=medium to large, 1=small
Flower showiness	0=showy, 1=plain
Nectar	0=present, 1=absent

Fig. 1 Rate parameters for the eight possible transitions between pollination mode and the binary characters under a model of dependent evolution. The code for pollination mode and states for the nine characters are provided above. The values of 0 and 1 do not necessarily indicate hypotheses about ancestral conditions.

were highly correlated with pollination mode while none of the ecological traits were correlated with pollination mode (table 1). When considering phylogenetic relationships, the results of the maximum likelihood analyses and the likelihood ratio tests indicated that wind pollination evolves in a correlated fashion with habitat type, floral sexuality, sexual system, ovule number, flower size, flower showiness, and nectar presence or absence (table 2). Wind pollination and geographical distribution and number of flowers per inflorescence evolve independently (table 2). To investigate these associations further and to determine the direction and order of transitions underlying correlated evolution, we tested specific hypotheses about each association.

Ecological Traits

Our results indicated that pollination mode and habitat type (closed habitats or open habitats) do not evolve independently. A model of correlated evolution fit the data better,

Table 1

Distribution of Species among the Pollination Modes and States of the Nine Characters Used in This Study

Trait and state	Pollination mode		
	Animal	Wind	χ^{2a}
Geographical distribution:			
Tropical	293	33	2.98
Temperate	199	35	
Habitat type:			
Closed	238	26	2.47
Open	254	42	
Floral sexuality:			
Hermaphrodite	394	18	88.31***
Unisexual	98	50	
Sexual system:			
Cosexual	440	38	53.80***
Dioecious	52	30	
Number of ovules:			
More than one	385	33	27.89***
One	107	35	
Number of flowers:			
Few	143	11	4.98*
Several to many	349	57	
Flower size:			
Medium to large	177	4	24.73***
Small	315	64	
Flower showiness:			
Showy	325	11	61.94***
Plain	167	57	
Nectar:			
Present	398	7	148.76***
Absent	94	61	

[a] χ^2 tests determine whether there is an association between pollination mode and the distribution of species among the two states of each trait.

* $P < 0.05$.

*** $P < 0.0001$.

and the transition rates predicted that wind pollination and open habitats are positively associated (table 2). Furthermore, wind pollination was more often lost in closed habitats (table 2). We were unable to reject the hypothesis that geographical distribution (tropical or temperate) and pollination mode evolved independently.

Floral Traits

Pollination mode and flower size (small or large) do not evolve independently. The transition rates we obtained predict that wind pollination and small flowers are positively correlated (table 2). Although none of the contingent-changes tests were significant, the transition rates indicate that wind pollination evolves more often in small-flowered lineages and once established is seldom lost. Similarly, flower showiness (showy or plain) and pollination vector do not evolve independently. The model of correlated evolution fit the data significantly better than the model of independent evolution, and transition rates predicted that plain flowers and wind pollination are positively associated. In contrast, we cannot reject the hypothesis that pollination mode and flower number (few or many) evolve independently. Finally, nectar (presence

or absence) and pollination mode evolve in a dependent way. Transition rates predict that an absence of nectar and wind pollination are positively correlated. The correlation arises because nectar is lost significantly more often in wind-pollinated clades (table 2).

Sexual Traits

Pollination mode and both floral sexuality and sexual system do not evolve independently. In both cases, a model of correlated evolution fit the data better than a model of independent evolution (table 2). Individual transition rates predicted that wind pollination and unisexual flowers are positively correlated. Furthermore, the contingent-change tests reveal that wind pollination evolves more often in clades with unisexual flowers (usually in species with either monoecious or dioecious sexual systems; table 2). To assess whether the presence of dioecious species is driving this association, we removed all dioecious species from the data set and reran the analysis. Again, we found that the model of correlated evolution fit the data better and that the transition rates predicted that wind pollination and monoecy are associated (data not shown). None of the contingent-changes tests were significant, possibly due to a lack of power, although the trend suggested that wind pollination evolves more often in monoecious lineages (data not shown). With respect to the correlation between dioecy and pollination mode, individual transition rates predicted that wind pollination and dioecy are positively associated. Furthermore, the contingent-change tests indicated that wind pollination evolves more often in dioecious clades (table 2).

Our analysis indicated that ovular condition (single ovule vs. multiple ovules) and pollination mode evolved in a dependent way. The transition rates revealed that single ovules and wind pollination are positively correlated (table 2). Furthermore, the contingent-change tests indicated that single ovules evolve from multiple ovules more often in wind-pollinated clades (table 2).

Discussion

The results of our phylogenetic analysis indicate that wind pollination evolves in a correlated way with open habitats, unisexual flowers, dioecy, the uniovulate condition, small plain flowers, and a lack of nectar. In contrast, wind pollination and geographical distribution and the number of flowers per inflorescence evolve independently. Although several of these associations have been examined previously (e.g., Regal 1982; Linder 1998; Vamosi et al. 2003), our study is the first to investigate correlations across the angiosperm phylogeny. Furthermore, in several cases we were able to detect contingent evolution and identify the most common background on which the evolution of a trait occurs. We now discuss potential adaptive explanations for the associations that our comparative analyses have revealed and comment on some of the limitations of using large-scale phylogenies for this type of analysis. Finally, we discuss unresolved questions and propose several future avenues of research that might be profitably pursued.

Ecological Traits

We found no evidence for correlated evolution between geographical distribution and wind pollination. This result may

Table 2

Likelihood Ratio Values for Tests of Correlated Evolution between Pollination Modes and Ecological and Morphological Traits for 560 Species Using the BayesTraits Program and the Phylogeny of Soltis et al. (2000)

Trait	Likelihood ratio of dependent vs. independent model	Trait (state 1) evolves more often in wind-pollinated clades	Wind pollination evolves more often in clades with trait (state 1)	Wind pollination lost more often in clades with opposite trait (state 0)	Opposite trait (state 0) evolves more often in animal-pollinated clades
Geographical distribution	6.92	1.06	2.43	.10	2.34
Habitat	14.85**	.53	2.50	7.37*	.06
Floral sexuality	58.32***	.27	13.63***	.78	6.15
Sexual system	53.35***	.62	19.30***	7.92*	9.62**
Ovules	21.43***	6.79*	.13	1.50	1.47
Number of flowers	8.20	2.32	.00	2.34	2.22
Flower size	34.83***	.82	1.04	5.55	2.03
Flower showiness	55.63***	.34	.42	5.69	2.93
Nectar	132.14***	15.44***	6.09	10.31**	1.26

Note. Likelihood ratio values are for tests of four hypotheses, which are not mutually exclusive, about contingent evolution.
* $P < 0.05$.
** $P < 0.01$.
*** $P < 0.0001$.

at first appear to challenge many observations of the higher frequency of wind-pollinated species in temperate regions (e.g., Regal 1982; but see Bawa et al. 1985; Bullock 1994). However, in our study we were specifically interested in the correlated evolution of wind pollination, not in the frequency of wind pollination in contrasting geographical regions. For example, the high abundance of anemophily in temperate regions found by Regal (1982) may largely reflect the predominance of Poales (grasses, sedges, rushes, etc.) and Fagales (beeches, oaks, etc.) in many ecosystems. In our data set, each of these groups was represented by a single evolutionary transition. Although there is a greater abundance of wind-pollinated versus animal-pollinated plants in many temperate ecosystems, our analysis provides no evidence that ecological conditions in temperate regions preferentially select for the evolution of wind pollination from animal pollination, or that wind-pollinated plants have migrated from tropical to temperate regions because of more favorable environments.

The aerodynamic requirements for wind pollination occur in habitats with open vegetation that allow for moderate wind speeds (Whitehead 1983; Niklas 1985; Dowding 1987). Although the finding of independent evolution between pollination mode and geographical region (temperate or tropical) may seem to contradict this, we did not limit our coding of tropical regions to tropical forests but included species from other tropical ecosystems including savannas and grasslands. However, we specifically tested the effect of vegetation structure by looking for correlated evolution with open versus closed habitats because plant size and density are likely to influence the efficacy of pollen dispersal in wind-pollinated plants. We found that open habitats and wind pollination evolve in a dependent way and are positively associated. In addition, wind pollination was lost more often when it occurred in closed habitats.

These results suggest that there may be constraints on the origin of wind pollination in closed habitats and that wind-pollinated plants are more likely to persist in open habitats.

Support for this idea is indicated by the frequency with which plants in closed forests in tropical regions use a mixture of wind and insect pollination (e.g., Piperaceae: De Figueiredo and Sazima 2000; Arecaceae: Uhl and Moore 1977; Listabarth 1993; Berry and Gorchov 2004). Additionally, although we did not test for the effect of altitude (due to insufficient data), increasing altitudinal gradients are often coupled with a decrease in vegetative cover. Several studies have shown that altitudinal gradients, which impact both the pollinator community and habitat type, can select for wind pollination (Berry and Calvo 1989; Gomez and Zamora 1996). Our findings suggest that most wind-pollinated species are likely to be limited in distribution by the structure of the surrounding vegetation, including the seasonal phase of canopy cover.

Floral Traits

Not unexpectedly, we found strong evidence that nectar and pollination mode evolve in a correlated manner. Our results indicate that nectar is lost more often in wind-pollinated clades, as one might expect on energetic grounds. However, our results also suggest that the presence of nectar in animal-pollinated species does not act as a constraint to the evolution of wind pollination. It has been proposed that wind pollination evolves more readily in nectarless lineages in which pollen is used as reward for pollinators (e.g., *Thalictrum*: Kaplan and Mulcahy 1971). Although this may be true in particular cases, we found no evidence that nectar generally limits possibilities for transitions to anemophily. In contrast, although the pattern was less clear, it appears that large showy flowers may constrain the evolution of wind pollination. We found that wind pollination evolves more often in taxa with small unshowy flowers. Linder (1998) proposed that wind pollination evolved more often in animal-pollinated groups with poorly developed or undifferentiated perianths, such as the magnoliids, caryophyllids, and rosids, and rarely in groups with zygomorphic flowers. Our data support this proposal. Finally, we found no evidence for correlated evolution between the number of flowers per inflorescence and wind pollination. Weller et al. (2006) reported that the number of flowers, per se, in *Schiedea* was less important for wind pollination in comparison with inflorescence condensation, a composite measure of the number of flowers and the length of the inflorescence. Hence, inflorescence architecture probably plays a more important role in the evolution of wind-pollinated plants than the number of flowers produced within an inflorescence.

A strong association between wind pollination and reduced floral morphology is apparent throughout the literature. However, it is unclear whether this association arises as an adaptation to wind pollination or because large complex flowers constrain the evolution of wind pollination. Our results suggest that large flowers act as a constraint to the evolution of wind pollination, perhaps because they limit the exposure of anthers and stigmas to the airstream. However, there is no a priori expectation for why colorful flowers should also constrain the evolution of wind pollination. Therefore, it is possible that these associations arise because wind pollination evolves more often in species that are pollinated by generalist insects, including flies and small pollen-collecting bees, which are often associated with plants that have small white or pale-colored unshowy flowers. This is supported by the observation that most ambophilous taxa are pollinated by generalist pollinators (e.g., *Salix*: Peeters and Totland 1999; Tamura and Kudo 2000; *Piper*: De Figueiredo and Sazima 2000; *Linanthus*: Goodwillie 1999; *Thalictrum*: Kaplan and Mulcahy 1971; *Schiedea*: Weller et al. 1998, 2006). These patterns suggest that the evolution of wind pollination occurs in lineages with reduced floral morphology. Subsequent selection against attractive structures would then intensify the correlation with wind pollination.

Sexual Traits

Our finding that wind pollination is strongly correlated with dicliny and dioecy is in accord with several previous studies (e.g., Bawa 1980; Charlesworth 1993; Renner and Ricklefs 1995; Vamosi et al. 2003). There is no comprehensive mechanistic explanation for the association between wind pollination and unisexual flowers, although some conceptual arguments are compelling (see Charlesworth 1993). In hermaphrodite animal-pollinated plants, female and male functions usually share the costs of floral display and pollinator rewards (Lloyd 1982). Pollinators perform two services in one visit, delivering pollen to the stigma and picking up pollen from anthers. However, in wind-pollinated plants, the removal and capture of pollen are independent events, and different structural requirements are necessary for optimal pollen dispersal and pollen capture (Niklas 1985; Friedman and Harder 2004). Indeed, spatial interference between female and male structures in a flower may be directly disadvantageous and explain the high incidence of dichogamy and herkogamy in wind-pollinated species (Lloyd and Webb 1986; Webb and Lloyd 1986). Nonetheless, geitonogamous selfing may be inevitable for wind-pollinated plants, resulting in strong selection for dioecy as a mechanism of inbreeding avoidance. Benefits of sexual segregation, such as flexibility for altering male and female investment and differential positioning of

flowers for optimal pollen dispersal versus capture, are likely to be important in the evolution of unisexual flowers. Mechanistic studies examining the benefits of unisexual flowers in wind-pollinated species are necessary for understanding the selective factors responsible for this frequent association.

An important finding of our study concerns the order of transitions involving dicliny and wind pollination. We found that wind pollination evolves more often after the establishment of dicliny in unrelated lineages. This pattern was evident for species with unisexual flowers and also for those that were purely dioecious. This is the first time this pattern has been identified in a large-scale comparative analysis and is important because case studies of particular taxa provide conflicting scenarios. For example, it has been reported that dicliny precedes wind pollination in *Leucadendron* (Proteaceae: Midgley 1987) and *Thalictrum* (Ranunculaceae: Kaplan and Mulcahy 1971), whereas the opposite order apparently occurs in the Poaceae (e.g., *Buchloe, Distichlis, Scleropogon, Spinifex*: Connor 1979 and references therein) and in *Fraxinus* (Oleaceae: Wallander 2001), where wind pollination has originated at least three times, with dioecy evolving from androdioecy after wind pollination on at least three occasions. There are likely to be different selective factors favoring unisexual flowers depending on the ecological conditions, but our finding that wind pollination evolves more often in diclinous lineages suggests a common functional basis for this association.

Here we outline an evolutionary scenario for why wind pollination evolves more often in diclinous lineages. Dioecy is correlated with small unshowy flowers (Vamosi et al. 2003) that are usually pollinated by generalist pollinators (Charlesworth 1993; Bawa 1994). The floral morphology of these plants may make the evolution of wind pollination an especially feasible option because pollen dispersal and capture are not impeded by large or complex perianths. If pollinators become scarce or ineffective, reducing fertility as a result of pollen limitation, diclinous species may evolve wind pollination to ensure more effective pollen dispersal between plants (reviewed in Culley et al. 2002). These same ecological conditions commonly promote the evolution of selfing as a mechanism of reproductive assurance (reviewed in Eckert et al. 2006). However, the presence of unisexual flowers would in most lineages preclude the evolution of selfing through autonomous self-pollination (but see Ågren and Schemske 1993). According to this hypothesis, insufficient pollinator service resulting in pollen limitation could elicit two quite different evolutionary transitions in pollination systems, depending on the floral condition of ancestral populations. In populations with hermaphroditic flowers, autonomous self-pollination would relieve pollen limitation, resulting in the evolution of selfing. In contrast, in populations with unisexual flowers, wind pollination may serve the same role by increasing the proficiency of cross-pollen dispersal. Thus, similar ecological conditions and selective agents could result in two very different evolutionary outcomes due to contrasting ancestral traits.

Several studies provide support for a scenario in which wind pollination evolves in dioecious lineages as a response to pollinator scarcity. Weller et al. (1998) report that some diclinous species of *Schiedea* (Caryophyllaceae) suffer from pollinator limitation, with the evolution of wind pollination a common outcome. In *Leucadendron*, the genus is exclusively dioecious, and the vast majority of species is animal pollinated. However, there are at least four independent transitions from animal pollination to wind pollination (Midgley 1987; Hattingh and Giliomee 1989; Barker et al. 2004). *Leucadendron* is endemic to the fynbos shrublands of the Cape Floristic Region of South Africa, where competition for pollinators may be intense and pollen limitation of seed set is commonplace (Steiner 1988; Johnson and Bond 1997). Whether transitions from animal pollination to wind pollination in *Leucadendron* are driven by pollen limitation and the requirements of more effective cross pollination is not known. Our finding that wind pollination evolves more often in lineages with unisexual flowers suggests that wind pollination evolves to relieve pollinator limitation (but see Cox 1991 for alternative explanations), particularly because a reversion to perfect flowers and autonomous self-pollination would be highly unlikely.

A similar argument might suggest that wind pollination should evolve more frequently in self-incompatible lineages following pollinator loss. However, this transition may be rare because self-incompatibility can break down quite readily in some taxa (see Igic et al. 2008) and certainly occurs more easily than transitions from dioecy to hermaphroditism (Bull and Charnov 1985). In self-incompatible *Linanthus parviflorus*, the evolution of wind pollination offers reproductive assurance against unreliable pollinators (Goodwillie 1999). In other species of *Linanthus*, the breakdown of self-incompatibility and the evolution of self-fertilization provide reproductive assurance (Goodwillie 2001). It would be interesting to determine the ecological and life-history contexts in which these contrasting outcomes occur. The transition to wind pollination in self-incompatible lineages has also been reported in the genus *Espeletia* (Asteraceae: Berry and Calvo 1989). In species where the evolution of selfing is prevented due to strong incompatibility and limited genetic variation for self-compatibility, there may be selection for wind pollination when pollinators are scarce.

We found strong evidence for a reduction in ovule number to one after the evolution of wind pollination. This suggests that the uniovulate condition is an adaptation to wind pollination. Further evidence for this transition is in the Poales, where there have been repeated reductions from multiple ovules per carpel to solitary ovules (Linder 1998; Linder and Rudall 2005). The most common explanation for this association is that wind-pollinated plants are unlikely to capture sufficient pollen grains to fertilize many ovules. However, experimental studies involving the measurement of pollen loads of naturally pollinated taxa of Poaceae, Restionaceae, Rosaceae, and Proteaceae (Honig et al. 1992; Linder and Midgley 1996; Friedman and Harder 2004) found amounts of pollen on stigmas that far exceeded ovule number (mean pollen grains per ovule range from 3 to 100). It is therefore unlikely that pollen limitation alone is responsible for selection of decreased ovule number in wind-pollinated species.

Several morphological and aerodynamic features of wind-pollinated plants may favor an optimal strategy of few ovules per flower. The relatively low cost of producing flowers in wind-pollinated plants may favor a packaging strategy with few ovules per flower and more flowers per plant. The model developed by Burd (1995) provides support for this idea by showing that higher floral costs generally favor more ovules per flower, although his model only considered animal-pollinated plants. Also, because wind pollination is a stochastic process where plants may capture pollen from a variety of potential mates, uniovulate carpels may be a mechanism to increase pollen-tube competition. Finally, by producing more flowers with fewer ovules, the spatial separation of flowers may increase the volume of air sampled by a plant and the probability of capturing pollen grains. A more mechanistic understanding of the functional relation between wind pollination and ovule number would be informative

Caveats and Future Research

There are other traits that may be correlated with wind pollination that would be interesting to investigate. For some of these traits, we attempted to include them in our analysis but were unable to compile sufficient data for all taxa. Most noticeable are those traits associated with pollen. Copious pollen production and smooth, dry, small pollen grains are commonly cited as attributes of anemophilous pollen (Faegri and van der Pijl 1979; Whitehead 1983; Proctor et al. 1996). Linder (2000) reported a correlation between pollen aperture type and wind pollination across angiosperm families, and he also proposed that circular apertures may be linked to speciation. Other attributes that could yield informative results include altitude, life form, and characteristics of stigmas.

Some of the traits associated with wind pollination may not be independent and could complicate some of the correlations we report. We are not aware of analytical methods to investigate correlated evolution for more than two traits simultaneously in a phylogenetic context. The effects of a third trait could best be examined by looking at correlated evolution within higher-level phylogenies that are variable in only the trait of interest. However, there are few clades that have the repeated evolution of wind pollination and adequate variation in ecological and morphological traits (but see Wallander 2001). In an effort to understand the associations between traits, we examined all pairwise intercorrelations between the traits in this study that evolve in a correlated way with wind pollination (21 different pairwise associations; results not shown). We found that many of the morphological traits evolve in a correlated way with one another but that habitat type did not evolve in a correlated way with any trait except the presence of nectar. Unfortunately, where we detected significant intercorrelations between traits (e.g., the positive association between unisexual flowers and single ovules, $P < 0.001$) we cannot determine whether this association is driving correlations with wind pollination or whether the association with wind pollination is causing the intercorrelation. Until multivariate methods become available, it is impossible to tease apart the relative contributions of different associations to the overall positive correlation of a trait with wind pollination.

A potential bias in the methodology using transition rates may occur when the character states at the tips of the phylogeny are unequal and are poor indicators of the stationary frequencies. This issue has been recently addressed in the context of ecological specialization of insects (Nosil and Mooers 2005), where it was shown that there can be false detection of higher transition rates to the more common state. Currently implemented methods of character state reconstruction assume that the rate of character change is the same over the entire phylogeny, which is unlikely to be the case, particularly if the trait itself can influence speciation and/or extinction. However, because we found about equal occurrences of traits evolving before or after wind pollination (table 2) this problem is unlikely to be a major factor influencing our results.

A serious concern for comparative studies of reproductive transitions is the frequent use in the literature of floral characters to infer pollination mode. Here, we were looking for correlations between morphological traits and wind pollination. For some of the taxa included in our study, it is likely that their assignment as wind-pollinated species in the literature was based purely on morphological characters associated with the "anemophilous syndrome" (Faegri and van der Pijl 1979). Clearly, it is then circular to test for correlations with wind pollination. Definitively determining pollination mode in some groups (e.g., large tropical trees) can be time consuming. Pollinator exclusion experiments using bags are commonly employed, but this method is inherently flawed because it changes the aerodynamic environment around inflorescences. Studies have shown that the structure of the stigma and/or inflorescence can have significant consequences for pollen flow (Niklas 1987; Niklas and Buchmann 1987; Linder and Midgley 1996; Friedman and Harder 2005), and so bagging experiments may inadvertently limit pollen dispersal by wind. Unfortunately, without detailed field observations and experiments, it is difficult to know with certainty whether a plant is insect pollinated or wind pollinated or both.

Our study included only a few species per family and only one species per genus. As a result, many transitions occurring at higher taxonomic levels will be undetected in our analysis. For example, in the Poaceae, which are entirely wind pollinated, unisexuality has evolved multiple times, and evidence also suggests frequent reversions to hermaphroditism (Connor 1981; Malcomber and Kellogg 2006). However, neither of the species included in this study are dioecious (*Oryza sativa* and *Zea mays*), and so these transitions are not accounted for. Similarly, Dodd et al. (1999) report a single origin of wind pollination in the Fagales, a result that we also observed in our study. However, a phylogeny of the Fagaceae indicates that there may be multiple origins of wind pollination (Manos et al. 2001). Applying comparative approaches similar to those used here to well-resolved phylogenies of particular angiosperm clades should further distinguish traits that facilitate the evolution of wind pollination from those that are direct adaptations to wind pollination.

There are many fundamental questions about wind pollination that remain unresolved. For example, the comparative study by Dodd et al. (1999) showed that transitions between biotic and abiotic pollination are strongly asymmetric, so shifts from biotic to abiotic pollination happen much more frequently and are also correlated with a net decrease in speciation rate. However, there are important exceptions to this, including the Poaceae, which is highly species rich with over 10,000 species (Doust and Kellogg 2002), and the Fagales, which is a species-rich group of wind-pollinated trees but may in fact have relatively low rates of diversification (Magallón and Sanderson 2001). Our study indicates that wind pollination and geographical distribution do not evolve in a correlated way, although there may be much greater abundance of wind-pollinated plants in temperate regions (Regal 1982). It is possible that wind-pollinated lineages in temperate regions are more likely to persist and undergo speciation, a hypothesis we were unable to test with this data set.

The transition from animal pollination to wind pollination remains a central problem in plant evolutionary biology. Using phylogenetic evidence, our study demonstrated correlated evolution between wind pollination and a range of traits, including open habitats, unisexual flowers, dioecy, uniovules, small plain flowers, and a lack of nectar. For five of these traits (open habitats, unisexual flowers, dioecy, uniovules, and lack of nectar), we found evidence of contingent evolution, allowing us to make predictions about the mechanisms responsible for the associations. Our study raises novel predictions about the causes of correlations between wind pollination and unisexual flowers and between wind pollination and a reduction in ovule number. These ideas would benefit from further exploration. Studies that target specific groups and test mechanistic hypotheses are essential for understanding the functional basis for the evolution and maintenance of wind pollination.

Acknowledgments

We are grateful to Emily Darling for help in compiling the data, Jana Vamosi for statistical advice and for sharing her data set, Pam Soltis for the tree files used by Soltis et al. (2000), and Mario Vallejo-Marín for discussion and advice. This research was supported by the Natural Sciences and Engineering Research Council of Canada, through a Canada Graduate Scholarship (J. Friedman), and by funding from a Discovery Grant and the Canada Research Chairs Program (S. C. H. Barrett).

Literature Cited

Ackerman JD 2000 Abiotic pollen and pollination: ecological, functional, and evolutionary perspectives. Plant Syst Evol 222:167–185.

Ågren J, DW Schemske 1993 Outcrossing rate and inbreeding depression in two annual monoecious herbs, *Begonia hirsuta* and *B. semiovata*. Evolution 47:125–135.

Barker NP, A Vanderpoorten, CM Morton, JP Rourke 2004 Phylogeny, biogeography, and the evolution of life-history traits in *Leucadendron* (Proteaceae). Mol Phylogenet Evol 33:845–860.

Bawa KS 1980 Evolution of dioecy in flowering plants. Annu Rev Ecol Syst 11:15–39.

——— 1994 Pollinators of tropical dioecious angiosperms: a reassessment? no, not yet. Am J Bot 81:456–460.

Bawa KS, SH Bullock, DR Perry, RE Coville, MH Grayum 1985 Reproductive biology of tropical lowland rain-forest trees. 2. Pollination systems. Am J Bot 72:346–356.

Berry EJ, DL Gorchov 2004 Reproductive biology of the dioecious understorey palm *Chamaedorea radicalis* in a Mexican cloud forest: pollination vector, flowering phenology and female fecundity. J Trop Ecol 20:369–376.

Berry PE, RN Calvo 1989 Wind pollination, self-incompatibility, and altitudinal shifts in pollination systems in the high Andean genus *Espeletia* (Asteraceae). Am J Bot 76:1602–1614.

Bull JJ, EL Charnov 1985 On irreversible evolution. Evolution 39: 1149–1155.

Bullock SH 1994 Wind pollination of Neotropical dioecious trees. Biotropica 26:172–179.

Burd M 1995 Ovule packaging in stochastic pollination and fertilization environments. Evolution 49:100–109.

Charlesworth D 1993 Why are unisexual flowers associated with wind pollination and unspecialized pollinators? Am Nat 141:481–490.

Charnov EL, JM Smith, JJ Bull 1976 Why be an hermaphrodite? Nature 263:125–126.

Connor HE 1979 Breeding systems in the grasses: a survey. N Z J Bot 17:547–574.

——— 1981 Evolution of the reproductive systems in the Gramineae. Ann Mo Bot Gard 68:48–74.

Cox PA 1991 Abiotic pollination: an evolutionary escape for animal-pollinated angiosperms. Philos Trans R Soc B 333:217–224.

Culley TM, SG Weller, AK Sakai 2002 The evolution of wind pollination in angiosperms. Trends Ecol Evol 17:361–369.

De Figueiredo RA, M Sazima 2000 Pollination biology of Piperaceae species in southeastern Brazil. Ann Bot 85:455–460.

Dodd ME, J Silvertown, MW Chase 1999 Phylogenetic analysis of trait evolution and species diversity variation among angiosperm families. Evolution 53:732–744.

Donoghue MJ 1989 Phylogenies and the analysis of evolutionary sequences, with examples from seed plants. Evolution 43:1137–1156.

Doust AN, EA Kellogg 2002 Inflorescence diversification in the panicoid "bristle grass" clade (Paniceae, Poaceae): evidence from molecular phylogenies and developmental morphology. Am J Bot 89:1203–1222.

Dowding P 1987 Wind pollination mechanisms and aerobiology. Int Rev Cytol 107:421–437.

Eckert CG, KE Samis, S Dart 2006 Reproductive assurance and the evolution of uniparental reproduction in flowering plants. Pages 183–203 in LD Harder, SCH Barrett, eds. Ecology and evolution of flowers. Oxford University Press, Oxford.

Faegri K, L van der Pijl 1979 The principles of pollination ecology. 3rd rev ed. Pergamon, Oxford.

Felsenstein J 1985 Phylogenies and the comparative method. Am Nat 125:1–15.

Freckleton RP, PH Harvey, M Pagel 2002 Phylogenetic analysis and comparative data: a test and review of evidence. Am Nat 160:712–726.

Friedman J, LD Harder 2004 Inflorescence architecture and wind pollination in six grass species. Funct Ecol 18:851–860.

——— 2005 Functional associations of floret and inflorescence traits among grass species. Am J Bot 92:1862–1870.

Gomez JM, R Zamora 1996 Wind pollination in high-mountain populations of *Hormathophylla spinosa* (Cruciferae). Am J Bot 83: 580–585.

Goodwillie C 1999 Wind pollination and reproductive assurance in *Linanthus parviflorus* (Polemoniaceae), a self-incompatible annual. Am J Bot 86:948–954.

——— 2001 Pollen limitation and the evolution of self-compatibility in *Linanthus* (Polemoniaceae). Int J Plant Sci 162:1283–1292.

Harvey PH, MD Pagel 1991 The comparative method in evolutionary biology. Oxford University Press, Oxford.

Hattingh V, JH Giliomee 1989 Pollination of certain *Leucadendron* species (Proteaceae). S Afr J Bot 55:387–393.

Honig MA, HP Linder, WJ Bond 1992 Efficacy of wind pollination: pollen load size and natural microgametophyte populations in wind-pollinated *Staberoha banksii* (Restionaceae). Am J Bot 79:443–448.

Hutchinson J 1964 The genera of flowering plants. Vols 1, 2. Clarendon, Oxford.

Igic B, R Lande, JR Kohn 2008 Loss of self-incompatibility and its evolutionary consequences. Int J Plant Sci 169:93–104.

Johnson SD, WJ Bond 1997 Evidence for widespread pollen limitation of fruiting success in Cape wildflowers. Oecologia 109: 530–534.

Kaplan SM, DL Mulcahy 1971 Mode of pollination and floral sexuality in *Thalictrum*. Evolution 25:659–668.

Karrenberg S, J Kollmann, PJ Edwards 2002 Pollen vectors and inflorescence morphology in four species of *Salix*. Plant Syst Evol 235: 181–188.

Kubitzki K 1993 The families and genera of vascular plants. Vol 2. Flowering plants, dicotyledons: magnoliid, hamamelid, and caryophyllid families. Springer, Berlin.

——— 1998a The families and genera of vascular plants. Vol 3. Flowering plants, monocotyledons: Lilianae (except Orchidaceae). Springer, Berlin.

——— 1998b The families and genera of vascular plants. Vol 4. Flowering plants, monocotyledons: Alismatanae and Commelinanae (except Gramineae). Springer, Berlin.

——— 2003 The families and genera of vascular plants. Vol 5. Flowering plants, dicotyledons: Malvales, Capparales, and nonbetalain Caryophyllales. Springer, Berlin.

——— 2004 The families and genera of vascular plants. Vol 6. Flowering plants, dicotyledons: Celastrales, Oxalidales, Rosales, Cornales, Ericales. Springer, Berlin.

Linder HP 1998 Morphology and the evolution of wind pollination. Pages 123–125 in SJ Owens, PJ Rudall, eds. Reproductive biology in systematics, conservation and economic botany. Royal Botanic Gardens, Kew.

——— 2000 Pollen morphology and wind pollination in angiosperms. Pages 73–88 in MM Harley, CM Morton, S Blackmore, eds. Pollen and spores: morphology and biology. Royal Botanic Gardens, Kew.

Linder HP, J Midgley 1996 Anemophilous plants select pollen from their own species from the air. Oecologia 108:85–87.

Linder HP, PJ Rudall 2005 Evolutionary history of Poales. Annu Rev Ecol Evol Syst 36:107–124.

Listabarth C 1993 Insect-induced wind pollination of the palm *Chamaedorea pinnatifrons* and pollination in the related *Wendlandiella* sp. Biodivers Conserv 2:39–50.

Lloyd DG 1982 Selection of combined versus separate sexes in seed plants. Am Nat 120:571–585.

Lloyd DG, CJ Webb 1986 The avoidance of interference between the presentation of pollen and stigmas in angiosperms. I. Dichogamy. N Z J Bot 24:135–162.

Mabberley DJ 1997 The plant book: a portable dictionary of the vascular plants. 2nd ed. Cambridge University Press, Cambridge.

Maddison WP 1990 A method for testing the correlated evolution of two binary characters: are gains or losses concentrated on certain branches of a phylogenetic tree? Evolution 44:539–557.

Magallón S, MJ Sanderson 2001 Absolute diversification rates in angiosperm clades. Evolution 55:1762–1780.

Malcomber ST, EA Kellogg 2006 Evolution of unisexual flowers in grasses (Poaceae) and the putative sex-determination gene, TASSELSEED2 (TS2). New Phytol 170:885–899.

Manos PS, ZK Zhou, CH Cannon 2001 Systematics of Fagaceae: phylogenetic tests of reproductive trait evolution. Int J Plant Sci 162:1361–1379.

Midgley JJ 1987 The derivation, utility and implications of a divergence index for the fynbos genus *Leucadendron* (Proteaceae). Bot J Linn Soc 95:137–152.

Niklas KJ 1985 The aerodynamics of wind pollination. Bot Rev 51: 328–386.

——— 1987 Pollen capture and wind-induced movement of compact and diffuse grass panicles: implications for pollination efficiency. Am J Bot 74:74–89.

Niklas KJ, SL Buchmann 1987 The aerodynamics of pollen capture in two sympatric *Ephedra* species. Evolution 41:104–123.

Nosil P, AO Mooers 2005 Testing hypotheses about ecological specialization using phylogenetic trees. Evolution 59:2256–2263.

Pagel M 1994 Detecting correlated evolution on phylogenies: a general method for the comparative analysis of discrete characters. Proc R Soc B 255:37–45.

——— 1999 The maximum likelihood approach to reconstructing ancestral character states of discrete characters on phylogenies. Syst Biol 48:612–622.

Pagel M, A Meade 2006 Bayesian analysis of correlated evolution of discrete characters by reversible-jump Markov chain Monte Carlo. Am Nat 167:808–825.

Peeters L, O Totland 1999 Wind to insect pollination ratios and floral traits in five alpine *Salix* species. Can J Bot 77:556–563.

Pohl F 1929 Beziehungen zwischen Pollenbeschaffenheit, Bestäubungsart und Fruchtknotenbau. Beih Bot Centrabl 46:247–285.

Proctor M, P Yeo, A Lack 1996 The natural history of pollination. Harper Collins, London.

Regal PJ 1982 Pollination by wind and animals: ecology of geographic patterns. Annu Rev Ecol Syst 13:497–524.

Renner SS, RE Ricklefs 1995 Dioecy and its correlates in the flowering plants. Am J Bot 82:596–606.

Routley MB, RI Bertin, BC Husband 2004 Correlated evolution of dichogamy and self-incompatibility: a phylogenetic perspective. Int J Plant Sci 165:983–993.

Sargent RD, SP Otto 2004 A phylogenetic analysis of pollination mode and the evolution of dichogamy in angiosperms. Evol Ecol Res 6:1183–1199.

Schluter D, T Price, AO Mooers, D Ludwig 1997 Likelihood of ancestor states in adaptive radiation. Evolution 51:1699–1711.

Soltis DE, PS Soltis, MW Chase, ME Mort, TD Albach, M Zanis, V Savolaninen, et al 2000 Angiosperm phylogeny inferred from 18S rDNA, *rbcL*, and *atpB* sequences. Bot J Linn Soc 133:381–461.

Steiner KE 1988 Dioecism and its correlates in the Cape flora of South Africa. Am J Bot 75:1742–1754.

Tamura S, G Kudo 2000 Wind pollination and insect pollination of two temperate willow species, *Salix miyabeana* and *Salix sachalinensis*. Plant Ecol 147:185–192.

Uhl NW, HE Moore 1977 Correlations of inflorescence, flower structure, and floral anatomy with pollination in some palms. Biotropica 9:170–190.

Vamosi JC, SP Otto, SCH Barrett 2003 Phylogenetic analysis of the ecological correlates of dioecy in angiosperms. J Evol Biol 16:1006–1018.

Wallander E 2001 Evolution of wind-pollination in *Fraxinus* (Oleaceae). PhD diss. Göteborg University.

Webb CJ, DG Lloyd 1986 The avoidance of interference between the presentation of pollen and stigmas in angiosperms. 2. Herkogamy. N Z J Bot 24:163–178.

Weller SG, AK Sakai, TM Culley, DR Campbell, AK Dunbar-Wallis 2006 Predicting the pathway to wind pollination: heritabilities and genetic correlations of inflorescence traits associated with wind pollination in *Schiedea salicaria* (Caryophyllaceae). J Evol Biol 19:331–342.

Weller SG, AK Sakai, AE Rankin, A Golonka, B Kutcher, KE Ashby 1998 Dioecy and the evolution of pollination systems in *Schiedea* and *Alsinidendron* (Caryophyllaceae: Alsinoideae) in the Hawaiian Islands. Am J Bot 85:1377–1388.

Whitehead DR 1968 Wind pollination in the angiosperms: evolutionary and environmental considerations. Evolution 23:28–35.

——— 1983 Wind pollination: some ecological and evolutionary perspectives. Pages 97–108 *in* L Real, ed. Pollination biology. Academic Press, Orlando, FL.

Int. J. Plant Sci. 169(1):59–78. 2008.
1058-5893/2008/16901-0006$15.00   DOI: 10.1086/523364

FUNCTION AND EVOLUTION OF AGGREGATED POLLEN IN ANGIOSPERMS

Lawrence D. Harder* and Steven D. Johnson[†]

*Department of Biological Sciences, University of Calgary, Calgary, Alberta T2N 1N4, Canada; and[†]School of Biological and
Conservation Sciences, University of KwaZulu-Natal, Postal Bag X01, Scottsville, Pietermaritzburg 3209, South Africa

The evolution of different forms of pollen aggregation (tetrads, polyads, pollen threads, pollinia) from individual monads is a recurring transition in angiosperm history, having occurred independently at least 39 times. Aggregation should evolve only under special circumstances, because diminishing returns associated with pollen removal and receipt instead favor monads that act largely independently. All forms of aggregation result in sibling pollen grains acting together, but they seem to evolve to ease different limitations on siring success: tetrads may evolve most commonly when pollinators visit infrequently, pollen threads may be most beneficial when ovules become available synchronously, and pollinia greatly increase the probability that a pollen grain removed by a pollinator reaches a conspecific stigma. Once pollen aggregation evolves, its implications for gametophytic competition and the relatedness of seeds within fruits probably influence further reproductive evolution, especially the frequency with which pollen from a single donor sires all seeds in a fruit. This latter effect, rather than improvements in pollination efficiency, probably accounts for the common association of pollen aggregation with low pollen : ovule ratios. The ability of orchid pollinia to reduce diminishing returns during pollination may explain both the floral diversity and the widespread occurrence of deceit pollination in this clade.

Keywords: gametophytic competition, pollen-transfer efficiency, pollinia, pollination, polyad, tetrad, viscin threads.

Introduction

Outcrossing angiosperms produce several orders of magnitude more pollen grains than seeds. Each seed represents prior fertilization by a successful pollen grain, so that large pollen : seed ratios illustrate that the vast majority of pollen grains fail in their primary function. To the extent that siring success differs among pollen-producing plants, this low success rate should impose strong fecundity and sexual selection on pollen characteristics. Not surprisingly, pollen exhibits extensive interspecific variation in size, form, and physiology (Muller 1979), which undoubtedly represents the outcome of such selection. Among the many pollen adaptations, pollen aggregation stands out as a mechanism that enhances the chance that a pollen grain will contribute genes to a seed. Whereas most angiosperms shed pollen as relatively independent monads, various forms of aggregation (tetrads, polyads, viscin threads, pollinia) have evolved repeatedly within angiosperms. Species with aggregated pollen tend to have much lower pollen : ovule ratios than species with monads (Cruden 2000), indicating that individual grains of species with aggregated pollen have a comparatively high chance of contributing to the next generation of sporophytes. Thus, the evolution of pollen aggregation from individual monads represents a significant functional transition in angiosperm reproduction.

How aggregation enhances siring success is unclear for two general reasons. First, aggregation has manifold effects, as it can affect the rate of pollen removal, pollen transport, pollen deposition on individual stigmas, postpollination processes that result in ovule fertilization, and the relatedness of developing seeds. Disentangling the relative importance of aggregation for these processes presents a challenge, given the formidable difficulties of studying pollen dispersal and siring success in natural populations. Furthermore, different forms of pollen aggregation may generate contrasting advantages, so the benefits and costs of aggregation may not be universal to all types of aggregation. The second source of uncertainty arises because the diverse hypotheses for the adaptive benefits of pollen aggregation remain largely unexplored theoretically and untested empirically.

Almost 30 years ago, Willson (1979) proposed that variation in pollen aggregation among plants traits reflects sexual selection through a process akin to sperm competition in animals. Since then, understanding of the function of male floral traits and their evolution has progressed considerably. However, pollen aggregation has not been considered thoroughly in the context of the general theory that has developed, and this important transition in angiosperm evolution has not been reviewed thoroughly.

In this article, we consider the variety of pollen aggregation, its occurrence among angiosperms, its influence on pollen performance, and its associated consequences for reproductive evolution in angiosperms. We begin by reviewing the various types of pollen aggregation, the incidence of transitions from monads to aggregation among angiosperm families, and the association between pollen aggregation and clade diversity. Next, we consider general consequences of sibling pollen grains acting together which both explain the widespread production of distinct, rather than aggregated, grains and

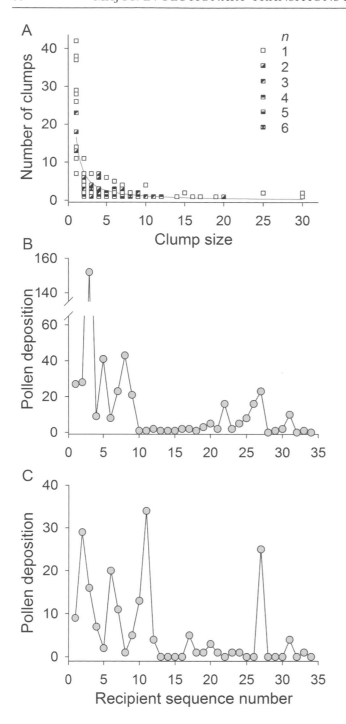

identify specific outcomes that must be realized for aggregation to be selected. Given this conceptual framework, we then consider empirical evidence for the influences of aggregation on pollen fates, including pollen removal, pollen transport, and postpollination success, and their likely roles in the origin and elaboration of aggregation. Finally, we consider several aspects of plant reproduction that may evolve once aggregation is established. This review illustrates that pollen aggregation has extensive and sometimes fundamental consequences for the diversification of angiosperm reproduction.

Types and Phylogenetic Distribution of Aggregated Pollen

Pollenkitt and Tryphine

Since adhesion of pollen both to vector and stigma plays such a central role in angiosperm reproduction, it is surprising that so little is known about the exine coatings which confer adhesive properties on the grains of many species. (Dickinson et al. 2000, p. 302)

During the final stages of pollen production, the tapetal cells lining the inner surface of an anther degenerate, coating the pollen grains with viscous pollenkitt or tryphine (Dickinson et al. 2000; Pacini and Hesse 2005). In most angiosperms, degeneration is complete and produces lipid-rich pollenkitt, whereas less complete degeneration in the Brassicaceae produces chemically more complex tryphine, which contains tapetal organelles (Dickinson et al. 2000). Pollenkitt is produced by almost all animal-pollinated species other than Brassicaceae, but it is absent, or at least much reduced, in many wind-pollinated and buzz-pollinated species (Pacini and Hesse 2005). Many functions have been proposed for pollenkitt and tryphine, including roles during pollen dispersal and pollen-pistil interactions (Dickinson et al. 2000; Pacini and Hesse 2005). However, most of these functions remain to be assessed experimentally, and their possible implications have not been examined theoretically.

Owing to their viscosity, pollenkitt and tryphine cause pollen grains to adhere to anther walls, to each other, and to pollinators (see photographs in Hesse 1980). Consequently, pollen grains of animal-pollinated plants tend not to travel independently, even if they are produced as separate monads. For example, figure 1A illustrates that 86% of *Narcissus assoanus* pollen grains on the tongues of butterfly pollinators were being transported in clumps, rather than as separate grains, with a median clump size of seven grains and a maximum of ca. 150 grains. Pollenkitt and tryphine result in rather haphazard aggregation of pollen grains, with considerable variation in clump size (e.g., fig. 1A). In addition, clump size probably changes during dispersal as the action of pollen vectors breaks large clumps apart and combines individual grains and small clumps (see Lisci et al. 1996). Such aggregation and

Fig. 1 Clumping of individual pollen grains and its effect on pollen dispersal. *A*, Frequency distribution of clump sizes of *Narcissus assoanus* pollen on the proboscides of 16 *Cleopatra* butterflies (*Gonepteryx cleopatra*) collected north of Montpellier, France. Different symbol types indicate the number of butterflies with the same number of clumps of a given size (e.g., two butterflies each had one clump of 20 grains). Five clumps with >30 grains (maximum = 150 grains) were not included. *B*, *C*, Two examples of dispersal of monads from individual donor inflorescences to sequences of recipient flowers by individual bumblebees, illustrating extensive stochasticity in the dispersal process around a generally declining trend. *B*, Dispersal of pollen from a long-styled *Pontederia cordata* inflorescence to short-styled recipient flowers, tracked by morph-specific differences in pollen size (L. D. Harder and S. C. H. Barrett, unpublished data). *C*, Dispersal of *Brassica rapa* pollen from a donor plant with a β-glucuronidase reporter gene to wild-type recipients (N. M. Williams and L. D. Harder, unpublished data).

the considerable variation in clump size undoubtedly contribute to the extensive stochasticity in pollen dispersal by individual pollinators from individual donor inflorescences to successively visited flowers (e.g., fig. 1B, 1C).

Tetrads and Polyads

Lack of separation of pollen grains during pollen production can generate structural aggregations, typically involving four pollen grains (tetrads) but with up to 32 grains in polyads of some African *Acacia* species (fig. 2A; Knox and Kenrick 1983). Some angiosperms, primarily noneudicots, undergo successive microsporogenesis, during which callose is deposited after each meiotic division, producing tetrads for varying shapes (including linear; Furness et al. 2002). In contrast, most eudicots undergo simultaneous microsporogenesis, whereby the two meiotic divisions occur before callose is deposited, to produce the four microspores of each tetrahedral pollen tetrad (Furness et al. 2002). Typically, the tetrads separate to produce monads; however, if this dissociation does not occur, mature pollen grains remain as permanent tetrads. Some species, especially in the Mimosoideae (Fabaceae), produce polyads consisting of multiple tetrads (fig. 2A), which arise because the sporogenous cell divides mitotically before the resulting pollen mother cells undergo meiosis (Knox and Kenrick 1983; Seijo and Solís Neffa 2004). Polyad production can be associated with considerable reduction in pollen production per anther. For example, the anthers of Australian *Acacia* species are divided

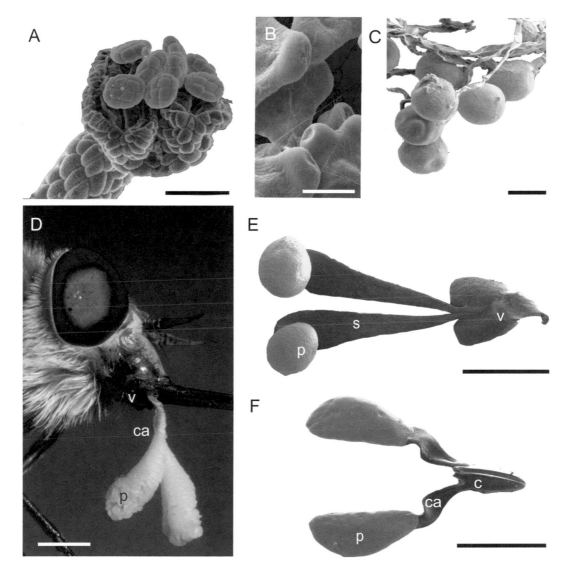

Fig. 2 Representative examples of pollen aggregation in angiosperms. *A*, Polyads consisting of 16 pollen grains in an *Acacia ataxacantha* anther (Mimosoideae [Fabaceae]). *B*, Viscin threads of *Oenothera biennis* (Onagraceae) pollen. *C*, Nonsporopollenin threads entangling *Strelitzia nicolai* (Strelitziaceae) pollen. *D*, Pollinarium of the orchid *Disa harveyana* with two sectile pollinia attached to the proboscis of *Philoliche rostrata* (Tabanidae). *E*, Pollinarium of the orchid *Cyrtorchis arcuata*, with two solid pollinia. *F*, Pollinarium of the milkweed *Pachycarpus grandiflorus*, with two solid pollinia (Asclepiadoideae [Apocynaceae]). False yellow color is added to pollen structures. *c* = corpusculum, *ca* = caudicle, *p* = pollinium, *s* = stipe, *v* = viscidium. Scale bars: *A*, 50 μm; *B*, *C*, 10 μm; *D–F*, 1 mm.

into eight loculi, each of which produces a single polyad, with typically 16 pollen grains per polyad (Knox and Kenrick 1983: also see fig. 2A).

Pollen Threads

The pollen of species in a few families is aggregated by various types of threads that entangle grains. Hesse et al. (2000) distinguished threads composed of sporopollenin (viscin threads; fig. 2B), which are projections from the outer layer of the pollen wall (ectexine), from those composed of other substances, primarily pollenkitt, that adhere to pollen grains (fig. 2C). The production of viscin threads is an integral process of microsporogenesis for species that possess them (see Rowley and Skvarla 2006), so they are produced by every grain in an anther. These threads are flexible but not elastic (Hesse et al. 2000). Only species in the Onagraceae (fig. 2B) and Rhodoreae [Ericaceae] produce viscin threads, and pollen in these clades is also shed as tetrads. In contrast, nonsporopollenin threads develop after pollen grains have been produced and are often not associated with each pollen grain. These threads occur in scattered species or genera in the Annonaceae, Araceae, Aristolochiaceae, Caesalpinioideae [Fabaceae], Heliconiaceae, Hydrocharitaceae, Marcgraviaceae, Passifloraceae, Potamogetonaceae, and Strelitziaceae (fig. 2C; Hesse et al. 2000).

Pollinia and Pollinaria

Pollinia are cohesive masses of many pollen grains that are removed as a unit from anthers (Johnson and Edwards 2000; Verhoeven and Venter 2001). They are the most evolutionarily derived form of pollen aggregation and are known from only two families, Orchidaceae and Apocynaceae. In derived orchids (Epidendroideae and Orchidoideae) and milkweeds (Secamondoideae and Asclepiadoideae), two or more pollinia and accessory structures for attachment to pollinators constitute a pollinarium (fig. 2D, 2E).

Orchids exhibit diverse types of pollen aggregation, from a sticky smear of monads in most Cypripedioideae to various forms of pollinia with monads or tetrads (fig. 3A). The simplest pollinia, found in the Vanilloideae, consist of a loose sticky mass of monads that adheres to the pollinator without a viscidium. Analogous loose pollinia lacking detachable viscidia also occur in some cypripedioid and epidendroid orchids, notably the tribe Neottiea. In some Neottiea, such as *Epipactis* and *Listera*, attachment of loose pollinia to pollinators is aided by a droplet of glue secreted by the rostellum (an elaboration of the stigma). Mealy pollinia with detachable viscidia occur in the Diurideae and Cranchideae. Sectile pollinia, in which distinct subunits of pollen known as massulae are held together by viscin, occur in several tribes, including the Orchideae and Diseae (fig. 2D). For mealy and sectile pollinia, clumps of pollen or wedge-shaped massulae, rather than the whole pollinium, adhere to individual stigmas, so that numerous flowers can be pollinated by a single pollinium (Peakall 1989; Johnson and Nilsson 1999). Hard pollinia (fig. 2E) predominate in the largest orchid clade (Epidendroideae) and are deposited on stigmas as an entire unit. Multiple pollinia can allow a single pollinarium to pollinate more than one flower; however, pollen deposition on stigmas

of epidendroid orchids usually involves simultaneous deposition of all pollinia from a single pollinarium (Nilsson et al. 1992; Alexandersson 1999).

The accessory structures of an orchid pollinarium include a sticky viscidium derived from stigmatic tissue, which attaches the pollinarium to a pollinator, and a connecting structure known as a "caudicle," if it is derived from sporogenous tissue, or a "stipe," if it is derived from stylar or stigmatic tissue (Dressler 1993; Johnson and Edwards 2000). Caudicles and stipes of many orchids bend or twist slowly after a pollinarium is removed from the anther. This reorientation reduces the likelihood of geitonogamous self-pollination, as pollinators have usually moved to a different plant by the time pollinia are positioned so that they can contact stigmas (Johnson et al. 2004; Peter and Johnson 2006). Caudicles, in particular, are highly elastic, which makes the attached pollinaria resistant to loss by pollinator grooming.

The Apocynaceae also exhibit various forms of pollen aggregation, although with less diversity than in the orchids (fig. 3B). Most Periplocoideae produce free tetrads, but some genera in this subfamily possess pollinia deposited onto spoon-shaped translators, which adhere to pollinators by a sticky viscidium similar to that of orchids. Flowers of Asclepiadoideae and Secamonoideae produce five pollinaria, each with two or four hard pollinia covered in a hard outer wall, giving the entire mass a smooth bony appearance (fig. 2F). Each pollinium connects to a clasping corpusculum, which attaches the pollinarium to a pollinator via a translator (sometimes termed a "caudicle"; fig. 2F). Both corpusculum and translator are derived from stigmatic secretions (Verhoeven and Venter 2001). Curiously, the pollinia attached to a single corpusculum are derived from adjacent halves of different anthers. Sideways torsion is usually required for pollinia to be inserted into the stigmatic grooves, which are orientated vertically at 90° relative to the anthers. Thus, as with orchid pollinia, many milkweed pollinia undergo gradual reconfiguration after removal from a flower (e.g., Wyatt 1976), and variation in the timing of this process is also likely to reduce geitonogamous self-pollination.

Phylogenetic Distribution of Pollen Aggregates

Angiosperm history has involved recurring experimentation with pollen aggregation, as the transition from monads to various types of aggregation has occurred repeatedly. Most angiosperms, especially animal-pollinated species, have mechanisms that aggregate pollen, so the evolution of derived forms of aggregation (tetrads, polyads, viscin threads, and pollinia) must be considered in the context of ancestral populations of plants with monads that adhered to each other to varying degrees owing to the presence of pollenkitt. Forty-two of the 457 families recognized by the APG (2003) include species with tetrads, polyads, pollen threads, or pollinia, and these families are scattered throughout the angiosperms (table 1). Maximum parsimony analysis (Mesquite, ver. 1.12; Maddison and Maddison 2006) of the occurrence of pollen aggregation based on the phylogenic tree of Davies et al. (2004) suggests that aggregated pollen has evolved independently at least 39 times. This evolution typically occurred within families, as 34 of the families with aggregated pollen also include species with monads. In addition, the evolution of aggregated pollen usually resulted in

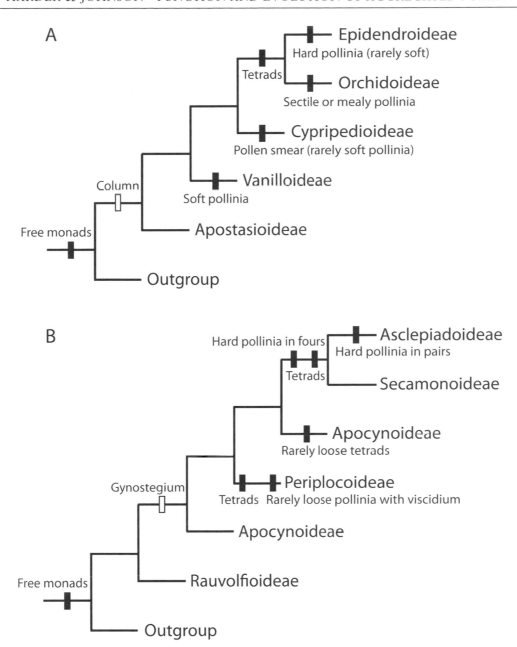

Fig. 3 Phylogenetic hypotheses for transitions in the evolution of pollen aggregation in (*A*) the Orchidaceae and (*B*) the Apocynaceae. Phylogenetic trees of subfamilies and character mapping (based on maximum parsimony) are adapted from Kocyan et al. (2004; Orchidaceae) and Livshultz et al. (2007; Apocynaceae). Filled bars indicate transitions of pollen traits, whereas open bars indicate the fusion of the androecium and gynoecium. In *B*, branches representing the Apocynoideae simplify the relationships identified by Livshultz et al. (2007), who proposed that this paraphyletic group comprises at least seven clades.

the production of tetrads (39 families). Thus, most origins of aggregated pollen probably involved the relatively easy retention of an intermediate stage in pollen production, with little additional evolutionary innovation (table 1). In contrast, the rare evolution of viscin threads (Onagraceae and Rhodoreae [Ericaceae]) and pollinia (Orchidaceae and Apocynaceae) suggests strong evolutionary constraints that have seldom been overcome. Aggregated pollen occurs primarily in animal-pollinated taxa, although it is also found in the water-pollinated Hydrocharitaceae (including Najadaceae) and four wind-pollinated clades

(Thurniaceae + Cyperaceae + Juncaceae, Myrothamnaceae, Scheuchzeriaceae, and Typhaceae).

Although apparently rare, reversions from tetrads to monads are known (Walker and Doyle 1975; Freudenstein 1999; Doyle et al. 2004). For example, the pseudomonad of Cyperaceae represents a tetrad in which three grains do not develop (Walker and Doyle 1975). In contrast, the true monads of *Orthilia* (Pyroloideae, Ericaceae) seem to represent a case in which the terminal stage of pollen development has reevolved (Freudenstein 1999). Such examples suggest that, at least for tetrads, the advantages of

Table 1

Taxonomic Distribution of Different Types of Pollen Aggregation among Angiosperm Families Recognized by APG (2003)

Aggregation type	Family
Dyads	Ericaceae (Styphelioideae), Podostemaceae, Scheuchzeriaceae
Tetrads	Actinidiaceae, Agavaceae, Anisophylleaceae, Annonaceae, Apocynaceae (Periplocoideae), Araceae, Berberidaceae, Bromeliaceae, Celastraceae, Clusiaceae, Cornaceae, Cyperaceae, Droseraceae, Ericaceae, Escalloniaceae, Fabaceae, Gentianaceae, Goodeniaceae, Gunneraceae, Hydrostachyaceae, Juncaceae, Lactoridaceae, Meliaceae, Monimiaceae, Muntingiaceae, Nepenthaceae, Onagraceae, Orchidaceae, Pedaliaccae, Philydraceae, Rubiaceae, Sarcolaenaceae, Solanaceae, Tamaricaceae, Thurniaceae, Torricelliaceae, Typhaceae, Velloziaceae, Winteraceae
Polyads	Annonaceae, Celastraceae, Fabaceae, Hydrocharitaceae
Viscin threads	Ericaceae (Rhodoreae), Onagraceae
Pollinia	Apocynaceae (Secamonoideae + Asclepiadoideae), Orchidaceae

Note. Based on data from Watson and Dallwitz (2006).

pollen aggregation depend on the prevailing conditions for pollination and/or mating, rather than being universal.

Pollen evolution in both the Orchidaceae and Apocynaceae has proceeded from monads through increasing aggregation, culminating in alternate forms of pollinaria (fig. 3, filled bars). In both clades, the evolution of pollinia was preceded by the fusion of the androecium and the gynoecium, producing the column (orchids) or gymnostegium (Apocynaceae: fig. 3, open bars), which initially allowed stigmatic fluid to participate in adhering pollen to pollinators. This fusion and the resulting functional integration may represent an essential transition that facilitated the subsequent evolution of pollinia.

Within orchids, tetrads and pollinia originated at different times (fig. 3A), suggesting that they serve different functions. Species in the basal Apostasioideae have free pollen monads, are nectarless, and are pollinated by pollen-collecting bees. The vanilloid orchids, most of which have soft pollinia of monads, are generally regarded as a sister clade to the remainder of the orchids (e.g., Kocyan et al. 2004), although in some phylogenies, this position is occupied by the Cypripedioideae (e.g., Cameron et al. 1999). The Cypripedioideae mostly deposit a sticky smear of monads onto flower visitors, so the soft pollinia in some derived taxa in this subfamily, notably *Phragmidium longifolium*, presumably evolved independently of soft pollinia in other orchid clades. The Orchidoideae, which dominate the orchid floras of temperate regions, tend to have either mealy or sectile pollinia. Within the Epidendroideae, the largest orchid clade, the number of pollinia per pollinarium seems to have generally been reduced from eight to two (Cameron et al. 1999), but examples of increases in pollinium number are also known (e.g., Whitten et al. 2000). Although most Epidendroideae produce hard pollinia, species in some, probably derived, tribes, such as the Neottieae (*Cephalan-*

thera, Epipactis, and *Listera*), produce soft pollinia (Dressler 1993).

The evolution of pollen aggregation within the Apocynaceae is not fully resolved, because of uncertainty about the affinities of the Periplocoideae. Some analyses proposed that the Periplocoideae constitute a sister clade to the Asclepiadoideae-Secamonoideae (e.g., Fishbein 2001), but the most recent analyses suggest that the Periplocoideae is a monophyletic clade within the paraphyletic Apocynoideae (Livshultz et al. 2007; fig. 3B). Early-branching lineages in the Apocynoideae have independent monads, whereas most Periplocoideae produce tetrads (Verhoeven and Venter 2001). Tetrads in *Apocynum*, which in Livshultz et al.'s (2007) phylogenetic tree is in the sister clade to Asclepiadoideae-Secamonoideae, are clearly not homologous with those of other species in the family (Nilsson et al. 1993) and are thus considered to be independently derived (fig. 3B). Rudimentary pollinia appear to have evolved at least three times within the Periplocoideae (Ionta and Judd 2007). Pollinaria with hard pollinia connected to a clasping corpusculum appear to have originated once in the common ancestor of the Asclepiadoideae and Secamonoideae. Reconstructions of likely ancestors by Livshultz et al. (2007) support Fishbein's (2001) suggestion that pollinaria with four pollinia (as found in the Secamonoideae) represent the ancestral condition in the Asclepiadoideae-Secamonoideae clade, whereas pollinaria with two pollinia (all Asclepiadoideae) are derived (fig. 3B).

On average, families with aggregated pollen contain more species (mean = 1200 species, lower SE = 286 species, upper SE = 376 species, based on ln-transformed data) than families with only monads (mean = 304 species, lower SE = 24 species, upper SE = 26 species; fig. 4A). This association is also evident in comparisons of sister clades that contain only species with monads and those that include species with aggregated pollen (fig. 4B). Thus, aggregated pollen may facilitate species diversification. Alternatively, given that not all species within families with aggregated pollen exhibit this trait, large families may simply provide more opportunities for the evolution of aggregation (or any other trait).

Diminishing Returns in Pollen Function

Aggregation of pollen necessarily causes pollen grains to be removed together, to travel together, to be deposited together on individual stigmas, and to increase the chance of seeds in a fruit being full siblings. Each of these consequences of aggregation has distinct implications for plant reproduction, so pollen aggregation will be selected only if the cumulative effects of aggregation improve reproductive performance. If success during any of these events depends on the number of sibling pollen grains that participate, the resulting "density dependence" will influence whether pollen aggregation is advantageous. Therefore, before considering the circumstances that may permit and facilitate the evolution of pollen aggregation, we review two reproductive processes for which outcomes depend on the number of sibling pollen grains involved: pollen export and competition for ovule fertilization. Both processes can involve diminishing returns, which arise when an investment of 1 unit of a "resource" produces an outcome of magnitude

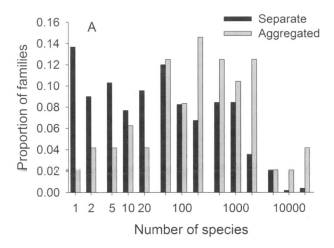

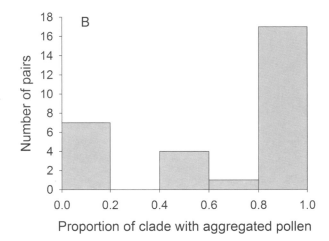

Fig. 4 Associations of species richness with separate versus aggregated pollen, based on data from Watson and Dallwitz (2006). *A*, Relative frequency distributions of species richness for families that include only species with separate pollen grains or some species with aggregated pollen. The two types of families differ significantly in species richness (likelihood ratio test, $G_1 = 33.61$, $P < 0.001$; generalized linear model, with negative binomial distribution and log-link function). *B*, Proportions of the total number of species in clades represented by subclades (usually a family) that include species with aggregated pollen. The complementary sister subclade in each clade includes only species with separate pollen grains. Sister subclades were based on Davies et al. (2004). A one-tailed Wilcoxon signed-rank test based on these proportions found higher species richness in subclades with aggregated pollen ($P < 0.05$).

r, whereas the outcome from a single investment of $n > 1$ units is less than nr. In such situations, division of R resources into multiple smaller investments produces a greater total outcome than fewer, larger subdivisions.

Pollen Export

Pollen dispersal by individual plants is probably often subject to diminishing returns on increased pollen removal per pollinator (Harder and Thomson 1989; Harder and Barrett 1996). At least three nonexclusive processes can create this re-

lation. First, bees groom during most flights between flowers (Harder 1990), which dislodges pollen from pollinators' bodies and thus excludes it from subsequent dispersal (Thomson 1986). Because both the chance of a bee grooming and grooming intensity increase with the amount of pollen removed from the last-visited flower (Harder 1990), removal of many grains from a flower decreases the chance of an individual grain reaching a conspecific stigma (Harder and Wilson 1997). Second, layering of pollen on the bodies of pollinators that groom infrequently, such as hummingbirds and Lepidoptera, can cause diminishing returns, because the rate of pollen burial (and its exclusion from deposition on stigmas) increases with the amount of pollen that is removed from each flower and occupies the outer pollen layer (Harder and Wilson 1997). Diminishing returns caused by layering can be particularly severe if the accumulation of pollen eventually causes all of the pollen carried by a pollinator to fall off (see Johnson et al. 2005). The third process, self-pollination between flowers (geitonogamy), depends on the number of flowers that a pollinator visits on a plant. The more flowers with functional stigmas that a pollinator visits after picking up pollen from the same plant, the smaller the proportion of removed pollen that leaves the plant for potential export (Harder and Barrett 1995, 1996). The diminishing returns caused by this pollen discounting are aggravated if a pollinator removes much pollen from each flower (Harder and Barrett 1996). Geitonogamy is probably a widespread cause of diminishing returns during pollen dispersal, as it occurs commonly and can be the main mode of self-pollination, regardless of the type of pollinator (reviewed by Harder et al. [2004]; see also Harder and Johnson 2005; Johnson et al. 2005). If any or all of these three processes are active, each pollinator will export a smaller proportion of the pollen it removes from a plant if it removes many grains than if it removes few grains.

The general consequences of diminishing returns on increased pollen removal can be appreciated by considering a simple model of pollen dispersal (for a full analysis, see Harder and Wilson 1994; LeBuhn and Holsinger 1998). Suppose a plant exposes P pollen grains during a brief period, all of which are removed by n pollinators. The amount of pollen exported by pollinator i (e_i) increases with pollen removal (p_i) according to

$$e_i = ap_i^b, \qquad (1)$$

where a depicts the chance of a pollen grain reaching a stigma, regardless of the number of grains with which it is removed, and b represents removal-dependent effects on dispersal. For equation (1), $b = 1$ indicates a linear relation between removal and export, whereas $0 < b < 1$ represents diminishing returns (fig. 5A). Two studies of this relation for species with monads reported $b \approx 0.3$ (Harder and Thomson 1989; M. B. Routley and L. D. Harder, unpublished data). For simplicity, suppose that each of the n pollinators removes P/n pollen grains, so total pollen export is

$$E = \sum_{i=1}^{n} e_i = na\left(\frac{P}{n}\right)^b = aP^b n^{1-b} \qquad (2)$$

(fig. 5B). In equation (2), n^{1-b} represents the ratio of total export by n pollinators to that realized if a single pollinator

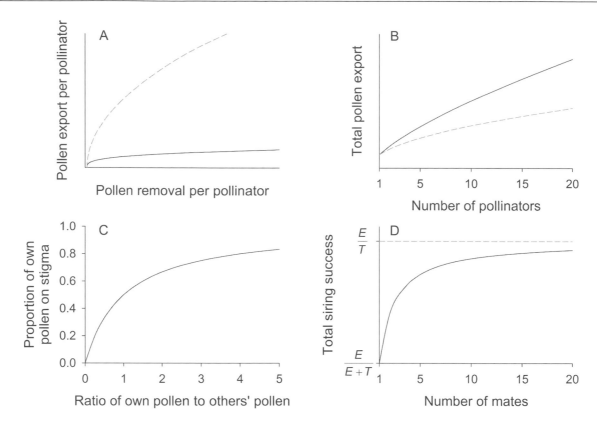

Fig. 5 Theoretical examples of diminishing returns for a plant's paternal success. *A* and *B* depict the effects of pollen removal per pollinator on pollen export per pollinator (eq. [1]) and total export (eq. [2]), respectively. Solid curves: $a = 1$, $b = 0.3$; dashed curves: $a = 1$, $b = 0.5$. *C* and *D* illustrate the consequences of increased deposition of a focal donor's pollen on individual stigmas for the donor's share of pollen tubes (eq. [3]) and total siring success on all stigmas (eq. [4]), respectively.

removed all of a plant's pollen. This model exposes three predictions about animal-pollinated plants when pollen export is subject to diminishing returns (i.e., $0 < b < 1$). First, P^b in equation (2) proposes that total pollen export exhibits diminishing returns with increased pollen production. This relation favors the evolution of hermaphroditism and also controls the evolutionary allocation of reproductive investment in pollen versus ovules (Charnov 1982; Zhang 2006; see "Pollen : Ovule Ratios"). Second, total pollen export increases (in a decelerating manner) with the number of pollinators attracted to a plant (n^{1-b}) if they all share in pollen removal (fig. 5*B*). Therefore, selection for increased siring success should increase pollinator attraction, within constraints imposed by the costs of attraction. The diverse and elaborate signaling and rewarding traits that distinguish animal-pollinated angiosperms from abiotically pollinated species are consistent with this expectation. Third, diminishing returns should favor restricted pollen removal by individual pollinators, which probably explains the diverse mechanisms for pollen packaging and dispensing exhibited by angiosperms (Harder and Thomson 1989; Thomson et al. 2000). The optimal restriction of pollen removal depends on the mean and variation in pollinator availability and on time-dependent aspects of pollen viability and ovule availability (Harder and Wilson 1994). Thus, the diminishing returns associated with pollen dispersal probably influence many aspects of floral evolution.

Gametophytic Competition for Ovule Fertilization

Once pollen reaches a stigma, it can again be subject to diminishing returns during competition to fertilize ovules, as another simple model illustrates. Suppose a plant contributes t_j pollen tubes to the pistil of recipient flower j, which compete with a total of T pollen tubes from other plants. If the $T + t_j$ pollen grains exceed the number needed to fertilize all ovules and all pollen tubes have an equal chance of fertilizing ovules, the proportion of ovules fertilized by the plant of interest in flower j will be

$$f_j = \frac{t_j}{T + t_j}. \tag{3}$$

If the number of pollen tubes from other plants is fixed, then two aspects of competition affect the siring success of the focal donor in this pistil. First, an increase in a donor's contribution of pollen to a stigma, t_j, enhances its share of pollen tubes compared to those from other donors, so its siring success increases in a decelerating manner with increases in its contribution of pollen tubes to flower j (fig. 5*C*). Second, an increase in t_j also intensifies competition between the focal donor's own pollen grains, reducing the chance of any specific pollen grain fertilizing an ovule (also see Queller 1984). This local mate competition creates diminishing returns for increases in a plant's contributions of pollen to individual stigmas, as long as pollen import is sufficient to maximize ovule fertilization.

The diminishing returns associated with pollen tube competition promote dispersal of pollen to many recipients, rather than a few. For example, suppose a focal donor exports a total of E pollen grains to m recipient flowers, so $t_j = E/m$. Therefore, this donor's total siring success,

$$S \propto \sum_{j}^{m} f_j = \sum_{j}^{m} \frac{E/m}{T + E/m} = \frac{mE}{E + mT}, \qquad (4)$$

increases from a minimum of $E/(E + T)$ when a donor plant mates with only one recipient to an asymptote of E/T if it mates with many recipients (fig. 5D).

General Implications

The preceding models demonstrate that diminishing returns select for subdivision, rather than aggregation. In particular, diminishing returns select for both restricted pollen removal per pollinator to diversify pollen dispersal among many pollinators (eq. [2]) and diversification in the number of mates to which pollen is dispersed (eq. [4]). The occurrence of aggregated pollen contradicts both of these expectations. Therefore, aggregated pollen must evolve in special circumstances with either no diminishing returns or countervailing influences. With these expectations in mind, we now consider what circumstances could be responsible for the origin and maintenance of aggregated pollen in angiosperms.

Conditions Promoting Pollen Aggregation

As a male character, pollen aggregation can evolve only if it enhances siring success. Therefore, we now examine specific conditions under which pollen aggregation may enhance siring; these are summarized in table 2A.

Pollen Removal

Aggregated pollen must generally increase the number of pollen grains removed by individual pollinators. In many monandrous orchids, both pollinaria share a viscidium (e.g., *Cyrtorchis arcuata*; fig. 2E) or have two closely adjacent viscidia on the rostellum, such that a single visit usually results in removal of all of a flower's pollen. According to the preceding overview of the implications of diminishing returns during pollen dispersal, permissive pollen removal is expected if ecological circumstances do not allow plants to benefit from restricting pollen removal by individual pollinators. We now consider three such circumstances.

Despite diminishing returns during pollen dispersal, pollen removal per pollinator should not be restricted when flowers are visited infrequently (Harder and Thomson 1989; Harder and Wilson 1994), either because pollinators are rare in the environment or because the species is less attractive than other coflowering species. The extent to which infrequent pollinator visits prompted the evolution of pollen aggregation is unclear. Orchids are notoriously pollen limited (Darwin 1877; Tremblay et al. 2005), and their flowers commonly wilt with pollinia remaining in anthers (Harder 2000); however, as we describe below ("Deceit Pollination"), pollen removal failure in orchids may often be an indirect consequence of reproductive strategies associated with pollen aggregation. Some of the families listed in table 1 include species that experience pollen limitation because they either are unrewarding (e.g., Podophyllaceae [=Berberidaceae]; Laverty 1992) or occupy environments with few or unpredictable pollinators (e.g., Ericaceae; Kudo and Suzuki 2002). More generally, in his review of pollen limitation, Burd (1994) included data for species from 14 of the families listed in table 1, and pollen limitation was detected in 12 of these families. For the two families in which pollen limitation was not observed, the single species studied in the Agavaceae and at least one of the three species in the Rubiaceae do not produce pollen aggregates (pollen state of the two other

Table 2

Aspects of Reproduction That Could Influence the Evolution of Pollen Aggregation

Reproductive phase, condition/consequence	Benefit of pollen aggregation
A. Conditions promoting aggregation:	
Pollen removal:	
Infrequent pollinators	Increased chance of dispersal; access to many vacant stigmas
Brief pollen viability	Increased chance of dispersal while viable
Synchronous ovule availability	Increased access to ovules
Pollen transport:	
Low transfer efficiency	Increased access to stigmas
Stigmas susceptible to usurpation	Priority access to ovules
Pollen tube growth and fertilization:	
Many ovules per ovary	Increased access to ovules, given successful dispersal
B. Accentuating consequences of aggregation:	
Pollen transport:	
Weak diminishing returns during transport	Enhanced transfer efficiency; total export becomes independent of number of pollinators
Lower variance in export to individual stigmas	Reduced local mate competition for fertilization
Pollen mixtures on stigmas involve lower male diversity	Reduced intermale competition for fertilization
Seed development:	
Developing seeds are mostly full siblings	Reduced embryo competition; opportunity for altruistic suicide

Rubiaceae is undetermined). This association is admittedly crude and warrants more detailed analysis, but it is consistent with pollen aggregation serving as one adaptive response to insufficient pollinator visitation.

Pollen removal should also not be restricted if a species' pollen has brief viability (Harder and Wilson 1994). In general, binucleate pollen, in which the germinative cell has not divided into sperm when the pollen is shed, is less active metabolically during dispersal and remains viable longer than trinucleate pollen (Hoekstra and Bruinsma 1975), so aggregated pollen might be favored in trinucleate clades. However, data from Watson and Dallwitz (2006; based largely on Brewbaker 1967) do not support this expectation. In particular, the incidence of families with aggregated pollen does not differ between families with only binucleate pollen (10.1% of 238 families) and those with only trinucleate pollen (8.7% of 69 families; Fisher's exact test, $P = 0.822$). Furthermore, orchid and milkweed pollen remains viable for several days (e.g., Morse 1987; Proctor 1998; Luyt and Johnson 2001), even after removal from anthers. Thus, limited pollen viability is probably not a common influence on the evolution of pollen aggregation.

If ovules become available relatively synchronously within a population, permissive pollen removal by a flower's first few visitors increases siring chances, despite diminishing returns during dispersal (Harder and Wilson 1994). Although such temporal patterns are not widespread, they may have played a role in the evolution of viscin threads in the Onagraceae. Anyone who has observed the rapid anthesis in populations of moth-pollinated *Oenothera* species followed by the brief period (ca. 1 h) of crepuscular activity by hawk moths can appreciate the advantage of permissive pollen removal in these species. However, these genera occur in a derived clade within the Onagraceae (Levin et al. 2003) and so need not represent conditions during the evolution of viscin threads, which is a synapomorphy for this family (Levin et al. 2003). Nevertheless, *Ludwigia*, the earliest-branching extant genus in this family (Levin et al. 2003), also exhibits relatively synchronous anthesis. For example, Gimenes et al. (1996) reported that the one-day flowers of *Ludwigia elegans* open between 0800 and 0900 hours, with anther dehiscence 30–60 min later. Specialized bees begin visiting once ambient temperature exceeds 18°C, and they cease pollen collection before 1200 hours. Such temporal limitation of mating opportunities should strongly favor rapid pollen removal to facilitate prompt access to newly available ovules. Pollen aggregation, such as viscin threads, provides one mechanism for rapid dispersal. However, many species with aggregated pollen do not have synchronous anthesis (notably orchids and milkweeds), so this mating circumstance does not provide a general explanation for the evolution of pollen aggregation.

Pollen Transport

Aggregated pollen could be beneficial if it enhances the probability that a removed pollen grain reaches a conspecific stigma (pollen-transfer efficiency). Pollen-transfer efficiency can be estimated for a plant population as the ratio of the average number of pollen grains on stigmas to the average pollen removal from flowers after pollination is complete (removal =

production − remaining). For hermaphroditic plants, this measure does not distinguish between self- and cross pollination and so tends to overestimate pollen export between plants. Nevertheless, comparison of estimates of pollen-transfer efficiency for 16 species with monads, 54 orchid species, and 28 asclepiad species reveals that pollinia enhance average pollination efficiency by up to two orders of magnitude (fig. 6B; comparison of species with monads and those with pollinia, $F_{1,92} = 92.60$, $P < 0.001$). We found estimates of pollen removal and deposition for only one species with tetrads, *Drosera tracyi* (Droseraceae; Wilson 1995), which had somewhat elevated transfer efficiency, compared to species with granular pollen, but lower

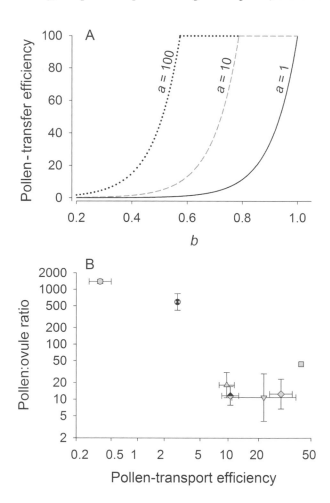

Fig. 6 Influences on the percentage of pollen removed from flowers that reaches conspecific stigmas (pollen-transfer efficiency) and its relation to pollen : ovule ratios. *A*, Theoretical effect of the diminishing-return parameter, *b*, on pollen-transfer efficiency, ap^{b-1}, based on equation (1), for three values of the scaling parameter, *a*. Total pollen removal = 50,000 grains. *B*, Relation of mean (±SE) pollen : ovule ratio to mean (±SE) pollen-transfer efficiency for species that present their pollen as monads (gray circle) or tetrads and polyads (half-filled circle); orchids with smear (*Cypripedium*: square), solid (gray diamond), mealy (down-pointing triangle), and massulate pollinia (up-pointing triangle); and asclepiads with pollinia (half-filled diamond). Based on published records and unpublished observations. Data for pollen-transfer efficiency and pollen : ovule ratios were collected from different sources (see appendix) and generally do not involve the same species. Note logit-scaled abscissa and ln-scaled ordinate.

efficiency than species with pollinia (fig. 6B; $P > 0.05$ in both cases). In contrast, for *Leucothoe racemosa* (Ericaceae), another species with tetrads, only 0.88% of pollen production was found on stigmas (a conservative measure of efficiency; L. D. Harder, unpublished data), which is similar to the pollen-transfer efficiency for species with monads (fig. 6B). Thus, whether tetrads enhance pollen-transfer efficiency remains to be clarified by the collection of more data. We found no measurements of pollen-transfer efficiency for species with viscin threads, but 7.1% of the pollen produced by *Chamerion angustifolium* (Onagraceae) was found on stigmas (L. Grinevitch and L. D. Harder, unpublished data), which exceeds the average efficiency for species with monads by an order of magnitude (fig. 6B).

Pollen aggregation could increase pollen transport either without affecting diminishing returns (e.g., larger *a* in eq. [1]; fig. 6A) or by alleviating diminishing returns (e.g., larger *b* in eq. [1]; fig. 6A). These nonexclusive alternatives are difficult to assess with existing information, because the relation of pollen export to removal by individual pollinators has been measured so seldom. Removal-independent improvements in pollen-transfer efficiency may underlie the evolution of aggregated pollen in some species with underwater pollination. Ackerman (1995) demonstrated that many such species produce either linear pollen monads (e.g., *Zostera*, Zosteraceae) or linear pollen aggregates (e.g., *Thalassia* and *Halophila*, Hydrocharitaceae). The linear form of pollen dispersal units induces tumbling in a moving current, which increases the probability of contact with a nearby stigma. Unfortunately, pollen-transfer efficiency has not been measured for such species, so whether aggregated pollen improves efficiency for submerged pollination remains a matter of conjecture.

The greatly enhanced pollination efficiency of orchids and milkweeds undoubtedly results because they glue their pollen, or pollinaria, to pollinators (fig. 2D), thereby reducing transport losses and relaxing diminishing returns during export. For milkweeds and derived orchids, this function is served by accessory structures of pollinaria (corpusculum and viscidium, respectively; fig. 2D), which evolved after pollen aggregation. However, the high pollen-transfer efficiency for the smear pollinium of *Cypripedium acaule* in figure 6B (O'Connell and Johnston 1998), which represents a basal clade in the Orchidaceae (see Freudenstein et al. 2004), compared to that of species with more independent monads, indicates that high efficiency probably predated the evolution of these elaborations. Both the stickiness of a *C. acaule* pollen mass and its transport on the thoracic dorsum of bees greatly reduce losses associated with bee grooming. As grooming is an important cause of diminishing returns for bee-pollinated species (see "Pollen Export"), the *Cypripedium* pollinium probably improves pollen-transfer efficiency by alleviating this dependence of export probability on the number of pollen grains removed per pollinator (e.g., larger *b* in eq. [1]; fig. 6A). The remarkable improvement in pollination efficiency associated with such an innovation would result in strong selection for this particular type of pollen aggregation.

Fertilization Opportunities

It is probably because of the variable number of ovules per ovary that different types of pollen dispersing units

are present in angiosperms. (Pacini and Hesse 2005, pp. 399–400)

Ovaries with multiple ovules provide more siring opportunities when multiple pollen grains from a donor reach a stigma, so it is not surprising that most families listed in table 1 have multiovulate ovaries. However, in contrast to Pacini and Hesse's assertion (and similar claims by Darwin [1877]), multiovulate ovaries per se are neither a necessary nor a sufficient condition for the evolution of pollen aggregation. For example, Kress (1981) found little difference in mean or median ovule number between genera in the Annonaceae with pollen monads and those with tetrads. Indeed, in the Annonaceae, tetrads have evolved in several uniovulate clades (e.g., *Pseudoxandra*, *Fusaea*), and uniovulate ovaries may have evolved from multiovulate ancestors in clades with aggregated pollen (Doyle and Le Thomas 1997). Furthermore, aggregation has not evolved in most clades with profuse ovule production, such as the Cucurbitaceae and Phyrmaceae. Therefore, although multiovulate ovaries undoubtedly facilitate the evolution of pollen aggregation, some other advantage is probably needed for its establishment.

Consequences of Aggregation That Promote Elaboration

The evolution of pollen aggregation precipitates several reproductive consequences that could favor additional evolution and further improve siring ability. These consequences largely involve postpollination processes, which determine fertilization success and the relatedness of seeds within individual fruits (table 2B).

Success of Male Gametophytes in Pistils

Pollen aggregation affects the dispersion of pollen grains among recipient stigmas, especially the frequency with which a specific donor plant is the sole contributor of pollen to individual stigmas, which in turn influences the extent and nature of competition between pollen tubes for access to ovules (fig. 5C). For a given pollen production, aggregation necessarily reduces the number of plants to which a plant can export pollen. The pollen removed by a single pollinator or gust of wind can reach only the subset of potential recipient plants in the population located along the vector's subsequent path. By reducing the number of "visits" required to remove a plant's pollen, aggregation restricts the number of recipient subsets that can be accessed. In addition, for animal-pollinated species, aggregation modifies the distribution of pollen from a donor plant on a single pollinator among the potential recipient flowers. For species with monads, a visit to a recipient flower usually removes a fraction of the donor pollen on the pollinator (Morris et al. 1994), so donor pollen is carried over to be delivered to multiple recipients (e.g., fig. 1B, 1C). Aggregation restricts the opportunity for pollen carryover (see Galen and Plowright 1985; Peakall 1989; Peakall and Beattie 1991; Nilsson et al. 1992; Johnson and Nilsson 1999; Johnson et al. 2005), especially for species with solid pollinia (e.g., Nilsson et al. 1992). This restriction can be compounded in species subject to geitonogamy, because self-pollination discounts the pollen available to be carried over for export to other plants.

Despite restricting the number of plants that receive a plant's pollen, pollen aggregation should generally reduce variation in the pollen contributed to each recipient, which may ease local mate competition among a donor's pollen grains. For species with monads, dispersal of pollen from a specific donor plant by a single pollinator creates a general decline in export to successive recipient flowers (Morris et al. 1994). To appreciate the implications of this pattern for siring success, consider figure 1B, which illustrates the dispersal of monad pollen from a *Pontederia cordata* inflorescence by a single pollinator. Several of the initial recipient flowers in this sequence received >40 donor pollen grains, with one flower receiving 152 grains. However, these large pollen contributions to individual stigmas largely represent lost siring opportunities, because *P. cordata* flowers contain only one ovule, so that each of these successfully dispersed pollen grains has a small chance of fertilizing an ovule. In contrast, production of pollinia, and perhaps viscin threads, probably moderates the variance in pollen contributed to individual stigmas, thereby reducing siring opportunities lost to local mate competition. This advantage of aggregation may also apply to some polyads (see *Acacia* examples described below) but not to pollen tetrads, which likely exhibit dispersal patterns similar to those observed for monads.

Restricted pollen carryover should also tend to reduce competition for fertilization between pollen tubes from different donors. Owing to carryover, stigmas of species with monads generally receive pollen from several donors during a single pollinator visit (e.g., Karron et al. 2006), resulting in multiple paternity among seeds in individual fruits of multiovulate species (see Campbell 1998; Mitchell et al. 2005). In contrast, if aggregation limits pollen export to fewer recipient stigmas, then recipient stigmas must also import pollen from fewer donors, on average. For example, a milkweed flower has five stigmatic chambers and so could receive pollinia from up to five donors, but Queller (1984) found that 59% of *Asclepias exaltata* flowers and 82% of *Asclepias viridiflora* flowers received only one pollinarium, Gold and Shore (1995) found no *Asclepias syriaca* fruits sired by more than one male, and Broyles and Wyatt (1990) found only a few *A. exaltata* fruits with multiple paternity. Similarly, Trapnell and Hamrick (2006) found no evidence of within-fruit multiple paternity in 15 populations of an epiphytic orchid, *Laelia rubescens*. Indeed, stigmas of epidendroid orchids rarely receive pollinia from more than a single pollinarium (Nilsson et al. 1992; Dressler 1993; Alexandersson 1999).

Aggregated pollen may further reduce gametophytic competition among pollen donors if pollen receipt shortens floral longevity, thereby reducing the chance of subsequent receipt of pollen from other donors. Lankinen et al. (2006) proposed that pollination-induced wilting, which occurs throughout orchids and is also known from other clades with aggregated pollen (see van Doorn 1997), is caused by pollen-borne chemicals and represents a male strategy to limit gametophytic competition. The benefits of such a strategy generally increase with the number of pollen grains from a single donor that arrive simultaneously on a stigma and so should be particularly favored for species with aggregated pollen (Lankinen et al. 2006). However, if mate choice is possible and sufficiently advantageous and pollen arrives relatively frequently, selection should favor female counterstrategies.

How might local mate and intermale competition affect selection on pollen aggregation? In general, while ovules remain unfertilized, a specific donor's siring success should increase roughly linearly with its pollen export to a stigma, regardless of the number of other donors contributing pollen. In contrast, if pistils receive enough pollen to fertilize most ovules, increased export to individual stigmas imposes diminishing returns (fig. 5C), so pollen should instead be dispersed to more recipients (fig. 5D). Thus, postpollination benefits of pollen aggregation should be greatest in two nonexclusive types of species: those in which insufficient pollen receipt commonly limits seed production during a flowering season and those in which only single donors typically contribute pollen to individual stigmas. As discussed above, both of these characteristics are expected outcomes of the dispersal of aggregated pollen, so postpollination benefits of aggregation may often reinforce selection on dispersal characteristics. Such synergy between pollination and postpollination outcomes would increase the selective advantage of pollen aggregation.

The preceding overview of the postpollination consequences of pollen aggregation focused on the male perspective, but the pattern of pollen dispersal also depends on stigmatic characteristics. In particular, if each stigma removes much of the pollen carried by a pollinator, then little pollen will be carried over to other recipients, and each pollinator will deliver pollen from few donors. For example, the rostella of many orchids scrape individual pollinia from pollinators (Nazarov and Gerlach 1997), and the mechanism of insertion of pollen into a stigmatic chamber of a milkweed flower allows receipt of only one pollinium per stigmatic chamber, so that a pollinarium can contribute pollen to a maximum of only two (Asclepiodeae) or four (Secamonoideae) stigmas (Kunze 1991). Thus, limited mate diversity in such species may involve both the mechanism of pollen deposition and pollen aggregation. Alternatively, small stigmas may limit mate diversity, despite the possibility of extensive pollen carryover, if they cannot capture many pollen grains. For example, the stigmas of many acacias have a terminal cup that is large enough to accept one polyad comfortably (Kenrick and Knox 1982; Tandon et al. 2001). Data from two *Acacia* species indicate that stigmas received a single polyad more frequently than expected from a random (Poisson) distribution of polyads among stigmas (*Acacia retinodes*: $X_2^2 = 20.78$, $P < 0.001$, Knox and Kenrick 1983; *Acacia senegal*: $X_3^2 = 46.99$, $P < 0.001$, Tandon et al. 2001). Similarly, for *Acacia melanoxylon*, a species with 16-grain polyads, Muona et al. (1991) found that only 8% and 15% of the pods assayed from two populations contained seeds sired by two donors rather than one. These orchid, milkweed, and *Acacia* examples reveal that the postpollination implications of pollen aggregation must be considered in the context of the floral mechanisms responsible for both female and male influences on pollen dispersal.

Seed Development

Pollen aggregation increases the chance that seeds in a multiovulate ovary are full siblings, rather than half-siblings, which influences opportunities for sibling competition, altruism, and parent-offspring conflict (Hamilton 1964; Kress 1981; Uma Shaanker et al. 1988; Lloyd 2000). A maternal plant is equally related to all of its outcrossed offspring and so should invest resources equally among them, unless they differ in quality because of the paternal genetic contribution. In contrast, if the chance of seedling establishment increases with seed size, an individual seed benefits directly from appropriating as much of the maternal resources as possible. However, if maternal resources are limited, the resulting competition reduces the number of siblings that can mature and may waste resources if some less competitive seeds consume some resources but then die. Such waste reduces both the direct fitness of the maternal (and paternal) plant and the indirect (collective) fitness of offspring realized through the transmission of shared alleles (Hamilton 1964; Lloyd 2000). High relatedness among offspring reduces the individual benefits of an offspring's competition with its siblings and maternal parent and so should promote resource sharing among siblings, reducing competitive waste. Consequently, pollen aggregation could enhance total seed production and reduce seed size variance within fruits.

Of course, such selection depends on seed production consequences for entire plants, rather than individual flowers. To the extent that pollen aggregation limits pollen carryover, it should increase mate diversity among flowers on individual plants. Such a negative association of within- and among-fruit mate diversity has been observed for an acacia (Muona et al. 1991) and two milkweed species (Broyles and Wyatt 1990; Gold and Shore 1995). High mate diversity among fruits, but not within fruits, provides opportunities for maternal plants to implement "family planning" by allocating resources preferentially to fruits with genetically superior offspring (e.g., Bookman 1984; Torres et al. 2002). This mechanism of offspring choice would be more economical than preferential allocation among seeds within individual fruits, because the fixed costs of fruit production (e.g., investment in pericarp) are expended primarily on fruits with many viable seeds (see Bookman 1984). However, family planning requires fertilization of more flowers than mature into fruits, whereas pollen limitation is common for many species with aggregated pollen (Burd 1994; Tremblay et al. 2005). Thus, pollen aggregation does not generally provide opportunities for offspring selection, and so this consequence of aggregation will not contribute universally to its further evolutionary elaboration.

The low diversity of mates contributing pollen to individual flowers associated with pollen aggregation also influences the consequences of self-pollination. If most flowers receive pollen from only one donor plant, then self-pollination in species with either self-incompatibility mechanisms (e.g., *Acacia*: Kenrick and Knox 1989; Asclepiadoideae: Wyatt and Lipow 2007) or strong inbreeding depression (e.g., some Onagraceae: Husband and Schemske 1996; some Orchidaceae: Smithson 2006) will greatly depress fruit set. For example, Shore (1993) found that self-pollination accounted for 66% of pollinia on stigmas in a population of *A. syriaca*, a species with notoriously low fruit set (Kephart 1987; Morse 1994).

Such observations led Wyatt and Lipow (2007) to propose "that the evolution of postzygotic self-incompatibility (in the Apocynaceae) created conditions in which compatible cross pollen was wasted whenever it occurred in mixture with incompatible pollen, because such mixed pollen loads would likely cause the entire fruit to abort.... This maladaptive condition may have been the stimulus for the evolution of pollinia" (p. 477). We find this hypothesis unsatisfying for several

reasons. First, as a male character, pollen aggregation must evolve to improve siring success, so that it can evolve in species with strong self-incompatibility or subject to strong inbreeding depression only if it increases pollen export and/or competitive ability in pistils on other plants. Second, although pollinia apparently evolved several times in the Apocynaceae, they have not evolved in the myriad other dicot families with self-incompatibility or inbreeding depression. Third, the pollinia of the Apocynaceae bear many similarities to those of orchids, but orchids mostly lack self-incompatibility systems (Johnson and Edwards 2000), causing Wyatt and Lipow (2007) to invoke different causes for pollinium evolution in these two families. In contrast, a common cause (enhanced pollen-transfer efficiency) provides a more parsimonious explanation, especially given the unusual fusion of the androecium and gynoecium before pollinia evolution in both families (fig. 3), which we interpret as a necessary step in pollinium evolution. Finally, Wyatt and Lipow's hypothesis led them to propose "that the greater efficiency of pollen delivery is an epiphenomenon, a fortunate happenstance of the evolution of pollinia," because "the likelihood of selection for pollination efficiency being so strong and consistent across so many derived taxa of the Apocynaceae ... seems remote," especially "given the wide range of environments, including pollinator diversity, that various Apocynaceae experience" (Wyatt and Lipow 2007, p. 481). In contrast, figure 6*B* illustrates that pollinia improve pollen-transfer efficiency by almost two orders of magnitude for milkweeds and orchids, compared to species with monads, even though the 29 milkweed species represented in this figure occupy different continents and are pollinated by diverse pollinators, including bees, butterflies, wasps, and birds. Together, these arguments favor enhanced pollen-transfer efficiency as the primary influence on pollinium evolution in both milkweeds and orchids.

Consequences of Aggregated Pollen for Floral Evolution

The preceding review indicated that pollen aggregation has diverse direct effects on plant reproduction. Such influences should in turn affect selection on other floral traits that influence reproductive performance, several of which we now consider.

Pollinaria and Floral Diversification

Among clades with aggregated pollen, orchids and milkweeds are unique in that aggregation has not been the final stage in the evolution of pollen dispersal units. Instead, in derived groups, pollinia are combined with unique structures for attachment to pollinators that often can also reorient to reduce self-pollination (fig. 2*D*–2*F*; see "Pollinia and Pollinaria"). In both orchids and milkweeds, these accessory structures are produced in whole or in part by the stigma, so pollen dispersal units comprise both male and female tissues. This sexual cooperation in pollen dispersal is one manifestation of unusual flower structures in both groups, in which the gynoecium and androecium are fused into a common structure (orchid column, milkweed gynostegium). In contrast, the rarity of ancillary elaboration of pollen aggregation in other clades may largely reflect the widespread structural and functional independence of gynoecium and androecium, which serves to reduce sexual interference in most angiosperms (Barrett 2002; secondary pollen presentation [Yeo 1993] being a notable exception).

The precision in pollen exchange with pollinators afforded by the combination of adhesive pollinia or pollinaria and close proximity of anthers and stigma on the column/gynostegium probably plays a key role in the remarkable floral diversification among both the ca. 22,000 species of orchids and the ca. 2400 species of Secamonoideae and Asclepiadoideae. Together, these traits greatly increase the precision of pollen placement on pollinators. For example, Maad and Nilsson (2004) described the pollination of two sister species of *Platanthera* that differed in the position of attachment of pollinaria on moth pollinators: proboscis versus eye. Such precision allows these sympatric species to use the same pollinators with limited hybridization (see Kephart and Theiss 2004 for a milkweed example). Similarly, because of precise pollen placement, simple changes in traits that affect where pollinators contact the column/gynostegium, such as the length of the nectar spur, can facilitate evolutionary transitions between different pollinators (e.g., Steiner 1989; Johnson 1997; Johnson and Steiner 1997; Johnson et al. 1998). Thus, pollen aggregation in the orchids and milkweeds may directly facilitate floral diversification and speciation and so represent a key innovation.

Deceit Pollination

Deceit pollination, in which pollinators are not rewarded for their services, is rare among angiosperms as a whole but occurs in ca. 8000 orchid species, which is roughly one-third of the family (Dressler 1990). We propose that this high incidence of deceit pollination among orchids evolved because of the relaxed diminishing returns during pollen dispersal resulting from reduced pollen loss during transport and limited geitonogamy in species with pollinarium reorientation (see "Pollen Transport"). As equation (2) indicates, relaxed diminishing returns during pollen dispersal (i.e., as b approaches 1) cause total pollen export to become increasingly independent of whether a few or many pollinators disperse a plant's pollen. Given the costs of flower maintenance (Ashman and Schoen 1996) and pollinator attraction (showy corollas, rewards), plants subject to weak or absent diminishing returns on pollen removal should allow the first few pollinators that visit to remove all of their pollen. Because such plants do not need to be as attractive as those subject to strong diminishing returns, deceit pollination becomes a practical option. Of course, deceit pollination involves risks, because pollinators learn to avoid unrewarding plants (Smithson and MacNair 1997). As a result, deceitful species generally experience low pollinator visitation, as pollen is left in the anthers of deceitful orchid species when flowers wilt more frequently than in those of rewarding orchids, on average (Harder 2000). However, lack of rewards also reduces the number of flowers visited by individual pollinators, which reduces geitonogamy (Johnson et al. 2004; Jersakova and Johnson 2006), further easing diminishing returns and reinforcing the evolution of deceit.

Unlike orchids, all milkweeds studied to date provide nectar and so do not engage in purely deceitful pollination, although some species use misinformation to attract or trap flies (e.g., Meve and Liede 1994; Masinde 2004). Indeed,

some milkweeds produce much more nectar than species with monads and equivalent flower size and number per inflorescence (Harder and Barrett 1992). Presumably, the excessive nectar production of some milkweeds and the trap flowers of *Ceropegia* (Masinde 2004) are necessary to retain pollinators on flowers for long periods to increase the chance of pollinarium removal and/or pollinium insertion from the complex flowers (see Kunze 1991). Long visits likely also caused the high frequency of self-pollination observed by Shore (1993) in an *Asclepias syriaca* population. As self-pollination in this species requires insect activity, the incidence of self-pollination likely involves geitonogamy and so should impose strong diminishing returns. Thus, the high transfer efficiency for milkweeds (fig. 6B) probably reflects improved aspects of dispersal that arise independently of the number of pollen grains removed per pollinator (e.g., increased a in fig. 6A).

Pollen : Ovule Ratios

If the Orchideae had elaborated as much pollen as is produced by other plants, relatively (sic) to the number of seeds which they yield, they would have had to produce a most extravagant amount, and this would have caused great exhaustion. Such exhaustion is avoided by pollen not being produced in any great superfluity owing to the many special contrivances for its safe transportal from plant to plant, and for placing it securely on the stigma. (Darwin 1877, pp. 288–289)

Outcrossing species with monads typically produce more that 1000 pollen grains per ovule (Cruden 2000), which greatly exceeds the mean pollen : ovule ratios of 570 for 17 genera with tetrads or polyads, 166 for eight genera with viscin threads, and 10 for 23 genera with pollinia (based on ln-transformed data from Erbar and Langlotz 2005; fig. 6B). Darwin's intuition that species with aggregated pollen produce relatively few pollen grains per ovule because their pollination systems deliver pollen efficiently to fertilize available ovules is a recurring theme in the pollen aggregation literature (Cruden 1977; Cruden and Jensen 1979; Kenrick and Knox 1982; Nazarov and Gerlach 1997; Erbar and Langlotz 2005), even though no published study of pollen : ovule ratios has also reported pollen-transfer efficiency. Nevertheless, the summary of available data in figure 6B illustrates the anticipated negative association between pollen : ovule ratio and pollination efficiency ($r = 0.83$, df = 5, $P < 0.025$).

The association depicted in figure 6B seems at odds with theoretical analyses of pollen : ovule ratios. Charnov (1982) rejected Cruden's (1977) suggestion "that P/O's reflect the likelihood of sufficient pollen grains reaching each stigma to result in maximum seed set" (p. 32) as a proposal that pollen grain production is governed by ovule production. Instead, Charnov correctly noted that pollen grains and ovules are alternate means of contributing genes to offspring for hermaphrodites, and so selection should act on allocation of reproductive resources in the production of both pollen and ovules. Theory that incorporates Charnov's perspective predicts that the optimal pollen : ovule ratio equalizes the rate of change in maternal and paternal fitness contributions with increasing investment (marginal fitness; Charnov 1982; Lloyd

1984). This result led Lloyd (1984) to conclude that "it is not the efficiency of pollination as such that controls gender allocations. It is the way in which pollinator actions affect the shape of the paternal fitness curve (with increased paternal investment) that is important" (p. 299). How can this conclusion be reconciled with figure 6B?

We proposed above that, at least in orchids, pollen aggregation improves pollen-transfer efficiency by alleviating the diminishing returns associated with pollen dispersal, which should strongly influence "the shape of the paternal fitness curve." However, the resulting more linear relation of paternal fitness to allocation in pollen should increase relative investment in pollen relative to ovules, whereas figure 6B illustrates that pollen : ovule ratio declines with increased efficiency. Thus, the explanation for low pollen : ovule ratios in plants with aggregated pollen must lie elsewhere.

Instead, resolution of the apparent conflict between theory and observation may be found in Lloyd's (1984) conclusion that "[a]n upper limit on paternal fitness offers the most promise of explaining the observed deviations (in sex allocation) emphasizing maternal expenditure" (p. 298). Lloyd identified restriction in the number of fertilizations resulting from individual pollen-removing visits as one cause of such an upper limit. As discussed above, pollen aggregation both reduces the number of visits required to remove all of a flower's pollen and can result in strong local mate competition if more sibling pollen grains reach individual stigmas than are needed to fertilize all available ovules. In this case, individuals that allocate more reproductive resources to ovule production and less to pollen production should contribute more genes to offspring, resulting in selection for a lower pollen : ovule ratio. Such selection should be particularly intense if individual donor plants are the sole contributors of the pollen received by individual stigmas, which seems to occur commonly for species with aggregated pollen (see also Wyatt et al. 2000). In turn, relatively exclusive access to individual stigmas should select for correlation between the typical number of pollen grains deposited on stigmas and ovule number. According to this interpretation, the association of low pollen : ovule ratios with high pollination efficiency in many species with polyads or solid pollinia is not causal but instead evolves as a correlated consequence of aggregation for low mate diversity within fruits.

Conclusion

The preceding overview suggests that the different types of pollen aggregation are not alternate solutions to a single problem but instead may originate to mitigate contrasting limitations on siring ability. In particular, tetrads may be especially advantageous for species subject to infrequent pollinator visits, viscin threads promote rapid removal when ovules become available synchronously (at least for Onagraceae), and aggregation enhances pollen-transfer efficiency for species with submerged pollination, or viscin threads, or pollinia that limit grooming losses. These alternate influences on aggregation may act together, as some species have two means of aggregating pollen (e.g., viscin threads and tetrads in *Rhododendron*), underscoring the conclusion that aggregation probably serves several functions. Consequently, pollen aggregation cannot be considered a single recurring trait with different degrees of elaboration. Indeed, even the pollinia of orchids and milkweeds seem not to be functionally comparable. Given the relative frequency of transitions from monads to different types of pollen aggregation, adaptations that enhance pollen-transfer efficiency seem to originate less often than those that contend with infrequent pollination or limited ovule availability.

Once aggregation evolves, it may precipitate further evolution in reproductive traits, restriction of mate diversity within fruits, increased relative allocation in female function, and pollination by deceit. These far-reaching effects arise because pollen aggregation influences all phases of reproduction from pollen removal to seed production. As a result, the transition from monads to aggregated pollen can influence subsequent phenotypic, functional, and phylogenetic diversification. However, these changes must occur in the context of overall reproductive function, and so they are realized to differing extents among the lineages in which pollen aggregation has originated.

Acknowledgments

We thank Spencer Barrett for the opportunity to contribute to this volume. In addition, we thank the following individuals for their contributions: the staff of the Centre for Electron Microscopy at the University of KwaZulu-Natal assisted with the pollen images; N. Hobbhahn translated Hesse (1980); J. D. Thomson and K. Goodell provided unpublished data on pollen-transfer efficiency; G. Ionta, T. Livshultz, and R. Wyatt provided copies of unpublished manuscripts; and M. W. Chase and M. Fishbein contributed valuable comments on the manuscript. The Natural Sciences and Engineering Research Council of Canada (L. D. Harder) and the National Research Foundation of South Africa (S. D. Johnson) funded this research.

Appendix

Sources of the Observations of Pollen-Transfer Efficiency and Pollen : Ovule Ratios Summarized in Figure 6*B*

Pollen-Transfer Efficiency

Table A1

Sources for Measurements of the Percentage of Pollen Removed from Flowers That Was Found on Conspecific Stigmas

Aggregation type	Sources
Monads	Ornduff 1970*a*, 1970*b*, 1971; Webb and Bawa 1983; Snow and Roubik 1987; Galen and Stanton 1989; Harder and Thomson 1989; Young and Stanton 1990; Wilson and Thomson 1991; van der Meulen 1992 (two species); Aizen and Raffaele 1996; Freitas and Paxton 1998; Harder 2000 (three species); Hiei and Suzuki 2001; Thomson and Goodell 2001 (two species)
Tetrads	Wilson 1995
Pollinia:	
Asclepiad	Wyatt 1976; Bertin and Willson 1980 (two species); Kunze and Liede 1991; Pleasants 1991; Liede 1994 (three species); Broyles and Wyatt 1995; Lipow and Wyatt 1998; Pauw 1998; Vieira and Shepherd 2002 (seven species); Ollerton et al. 2003 (nine species); Tanaka et al. 2006; S. D. Johnson and L. D. Harder, unpublished data[a]
Orchid massulate	Harder 2000 (seven species); S. D. Johnson and L. D. Harder, unpublished data[b]
Orchid mealy	Harder 2000 (two species)
Orchid smear	O'Connell and Johnston 1998
Orchid solid	Boyden 1982; Nilsson et al. 1986; Ackerman and Montalvo 1990; Christensen 1992; Pettersson and Nilsson 1993; Ackerman et al. 1994; Bartareau 1995; Singer and Cocucci 1997; Alexandersson 1999; Harder 2000; S. D. Johnson and L. D. Harder, unpublished data[c]

Note. In analyses, the observation for *Calypso bulbosa* was the average of results from Boyden (1982), Alexandersson (1999), and Harder (2000).

[a] *Gomphocarpus fruticosus*, 15.2%.

[b] *Anacamptis pyramidalis*, 7.6%; *Brownleea macroceras*, 12.1%; *Coeloglossum viride*, 7.7%; *Dactylhoriza incarnata*, 17.6%; *Disa cephalotes*, 7.3%; *Disa chrysostachy*, 6.8%; *Disa cooperi*, 6.8%; *Disa ferruginea*, 8.9%; *Disa graminifolia*, 11.9%; *Disa hircicornis*, 6.8%; *Disa pulchra*, 16.9%; *Disa uniflora*, 6.5%; *Disa versicolor*, 11.3%; *Gymnadenia conopsea*, 16.6%; *Ophrys sphegodes*, 14.3%; *Orchis militaris*, 8.7%; *Orchis morio*, 8.1%; *Orchis ustulata*, 1.6%; *Platanthera bifolia*, 4.6%; *Platanthera chlorantha*, 9.3%; *Satyrium bicorne*, 6.7%; *Satyrium coriifolium*, 9.0%; *Satyrium erectum*, 17.6%; *Satyrium hallackii*, 6.9%; *Satyrium longicauda*, 12.7%; *Satyrium longicolle*, 8.3%; *Satyrium membranaceum*, 10.8%; *Satyrium microrynchum*, 6.0%.

[c] *Eulophia cucullata*, 20.5%; *Eulophia parviflora*, 30.3%; *Eulophia welwitschii*, 52.6%; *Eulophia zeyheri*, 11.8%; *Mystacidium capense*, 56.2%; *Mystacidium gracile*, 35.2%; *Mystacidium venosum*, 42.8%; *Rangaeris muscicola*, 23.4%.

Pollen : Ovule Ratios

Information on pollen : ovule ratios was drawn from three published sources. The estimate for species with pollen monads is the average (based on log-transformed data) for xenogamous and facultative xenogamous species reported in table 1 of Cruden (1977). Most observations for species with aggregated pollen were extracted from Erbar and Langlotz's (2005) compilation, as summarized in table A2.

We used means for species for which Erbar and Langlotz (2005) presented multiple observations. Finally, we included observations for *Cypripedium calceolus* (orchid smear pollinium), *Calypso bulbosa*, and *Corallorhiza striata* (orchid solid pollinia) from Lukasiewicz (1999) and observations for *Earina aestivalis*, *Earina autumnalis*, *Earina mucronata*, and *Winika cunninghamii* (orchid solid pollinia) from Lehnebach and Robertson (2004).

Table A2

Species for Which Pollen : Ovule Ratios Were Obtained from Erbar and Langlotz (2005)

Aggregation type, family	Species
Tetrads:	
Annonaceae	Three *Asimina* spp.
Apocynaceae = Asclepiadaceae	*Periploca aphylla*
Clusiaceae	*Kielmeyera coriacea*
Ericaceae	*Andromeda polifolia*, *Calluna vulgaris*, *Moneses uniflora*, *Pernettya rigida*, three *Pyrola* spp., four *Vaccinium* spp.
Lactoridaceae	*Lactoris fernandeziana*
Winteraceae	*Pseudowintera colorata*

Table A2

(*Continued*)

Aggregation type, family	Species
Polyads:	
Ericaceae (=Pyrolaceae)	*Chimaphila umbellata*
Fabaceae	Four *Calliandra* spp., seven *Inga* spp., *Mimosa bimucronata*
Pollinia:	
Asclepiad:	
Apocynaceae (=Asclepiadaceae)	28 *Asclepias* spp., two *Calotropis* spp., *Ceropegia woodii*, two *Cynanchum* spp., *Hoya carnosa*, *Huemia* sp., five *Matelea* spp., *Oxypetalum caerulea*, *Pergularia daemia*, two *Sarcostemma* spp., *Vincetoxicum officinale*
Orchid massulate:	
Orchidaceae	*Comperia comperiana*, five *Dactylorhiza* spp., *Goodyera repens*, three *Ophrys* spp., six *Orchis* spp., two *Platanthera* spp., *Steveniella satyroides*
Orchid mealy:	
Orchidaceae	*Isotria* spp., *Listera ovata* (Orchidaceae)
Orchid solid:	
Orchidaceae	*Coryanthes senghasiana* (Orchidaceae)
Viscin threads:	
Ericaceae	Eight *Rhododendron* spp.
Onagraceae	*Calyophus serralatus*, two *Camissonia* spp., *Circaea canadensis*, seven *Clarkia* spp., 14 *Epilobium* spp., *Gaura drummondii*, *Oenothera biennis*

Literature Cited

Ackerman JD 1995 Convergence of filiform pollen morphologies in seagrasses: functional mechanisms. Evol Ecol 9:139–153.

Ackerman JD, AM Montalvo 1990 Short- and long-term limitations to fruit production in a tropical orchid. Ecology 71:263–272.

Ackerman JD, JA Rodríguez Robles, EJ Meléndez 1994 A meager nectar offering by an epiphytic orchid is better than nothing. Biotropica 26: 44–49.

Aizen MA, E Raffaele 1996 Nectar production and pollination in *Alstroemeria aurea*: responses to level and pattern of flowering shoot defoliation. Oikos 76:312–322.

Alexandersson R 1999 Reproductive biology of the deceptive orchid *Calypso bulbosa*. PhD thesis. University of Umeå.

APG (Angiosperm Phylogeny Group) 2003 An update of the Angiosperm Phylogeny Group classification for the orders and families of flowering plants. Bot J Linn Soc 141:399–436.

Ashman T-L, DJ Schoen 1996 Floral longevity: fitness consequences and resource costs. Pages 112–139 *in* DG Lloyd, SCH Barrett, eds. Floral biology: studies on floral evolution in animal-pollinated plants. Chapman & Hall, New York.

Barrett SCH 2002 Sexual interference of the floral kind. Heredity 88: 154–159.

Bartareau T 1995 Pollination limitation, costs of capsule production and the capsule-to-flower ratio in *Dendrobium monophyllum* F. Muell. (Orchidaceae). Aust J Ecol 20:257–265.

Bertin RI, MF Willson 1980 Effectiveness of diurnal and nocturnal pollination of two milkweeds. Can J Bot 58:1744–1746.

Bookman SS 1984 Evidence for selective fruit production in *Asclepias*. Evolution 38:72–86.

Boyden TC 1982 The pollination biology of *Calypso bulbosa* var. *americana* (Orchidaceae): initial deception of bumblebee visitors. Oecologia 55:178–184.

Brewbaker JL 1967 The distribution and phylogenetic significance of binucleate and trinucleate pollen grains in the angiosperms. Am J Bot 54:1069–1083.

Broyles SB, R Wyatt 1990 Paternity analysis in a natural population of *Asclepias exaltata*: multiple paternity, functional gender, and the pollen donation hypothesis. Evolution 44:1454–1468.

——— 1995 A reexamination of the pollen-donation hypothesis in an experimental population of *Asclepias exaltata*. Evolution 49: 89–99.

Burd M 1994 Bateman's Principle and plant reproduction: the role of pollen limitation in fruit and seed set. Bot Rev 60:83–139.

Cameron KM, MW Chase, WM Whitten, PJ Kores, DC Jarrell, VA Albert, T Yukawa, HG Hills, DH Goldman 1999 A phylogenetic analysis of the Orchidaceae: evidence from *rbcL* nucleotide sequences. Am J Bot 86:208–224.

Campbell DR 1998 Multiple paternity in fruits of *Ipomopsis aggregata* (Polemoniaceae). Am J Bot 85:1022–1027.

Charnov EL 1982 The theory of sex allocation. Princeton University Press, Princeton, NJ.

Christensen D 1992 Notes on the reproductive biology of *Stelis argenta* Lindl. (Orchidaceae: Pleurothallidinae) in eastern Ecuador. Lindleyana 7:28–33.

Cruden RW 1977 Pollen-ovule ratios: a conservative indicator of breeding systems in flowering plants. Evolution 31:32–46.

——— 2000 Pollen grains: why so many? Plant Syst Evol 222: 143–165.

Cruden RW, KG Jensen 1979 Viscin threads, pollination efficiency and low pollen-ovule ratios. Am J Bot 66:875–879.

Darwin CR 1877 The various contrivances by which orchids are fertilised by insects. 2nd ed. J. Murray, London.

Davies TJ, TG Barraclough, MW Chase, PS Soltis, DE Soltis, V Savolainen 2004 Darwin's abominable mystery: insights from a supertree of the angiosperms. Proc Natl Acad Sci USA 101:1904–1909.

Dickinson HG, CJ Elleman, J Doughty 2000 Pollen coatings: chimaeric genetics and new functions. Sex Plant Reprod 12:302–309.

Doyle JA, A Le Thomas 1997 Significance of palynology for phylogeny of Annonaceae: experiments with removal of pollen characters. Plant Syst Evol 206:133–159.

Doyle JA, H Sauquet, T Scharaschkin, A Le Thomas 2004 Phylogeny, molecular and fossil dating, and biogeographic history of Annonaceae and Myristicaceae (Magnoliales). Int J Plant Sci 165(suppl):S55–S67.

Dressler RL 1990 The orchids: natural history and classification. Harvard University Press, Cambridge, MA.

—— 1993 Phylogeny and classification of the orchid family. Timber, Portland, OR.

Erbar C, M Langlotz 2005 Pollen to ovule ratios: standard or variation—a compilation. Bot Jahrb Syst 126:71–132.

Fishbein M 2001 Evolutionary innovation and diversification in the flowers of Asclepiadaceae. Ann Mo Bot Gard 88:603–623.

Freitas BM, RJ Paxton 1998 A comparison of two pollinators: the introduced honey bee Apis mellifera and an indigenous bee Centris tarsata on cashew Anacardium occidentale in its native range of NE Brazil. J Appl Ecol 35:109–121.

Freudenstein JV 1999 Relationships and character transformation in Pyroloideae (Ericaceae) based on ITS sequences, morphology, and development. Syst Bot 24:398–408.

Freudenstein JV, C van den Berg, DH Goldman, PJ Kores, M Molvray, MW Chase 2004 An expanded plastid DNA phylogeny of Orchidaceae and analysis of jackknife branch support strategy. Am J Bot 91:149–157.

Furness CA, PJ Rudall, BF Sampson 2002 Evolution of microsporogenesis in angiosperms. Int J Plant Sci 163:235–260.

Galen C, RC Plowright 1985 The effects of nectar level and flower development on pollen carry-over in inflorescences of fireweed (Epilobium angustifolium) (Onagraceae). Can J Bot 63:488–491.

Galen C, ML Stanton 1989 Bumble bee pollination and floral morphology: factors influencing pollen dispersal in the alpine sky pilot, Polemonium viscosum (Polemoniaceae). Am J Bot 76:419–426.

Gimenes M, AA Benedito-Silva, MD Marques 1996 Circadian rhythms of pollen and nectar collection by bees on the flowers of Ludwigia elegans (Onagraceae). Biol Rhythm Res 27:281–290.

Gold JJ, JS Shore 1995 Multiple paternity in Asclepias syriaca using a paired-fruit analysis. Can J Bot 73:1212–1216.

Hamilton WD 1964 The genetical evolution of social behaviour. II. J Theor Biol 7:17–52.

Harder LD 1990 Behavioral responses by bumble bees to variation in pollen availability. Oecologia 85:41–47.

—— 2000 Pollen dispersal and the floral diversity of monocotyledons. Pages 243–257 in KL Wilson, D Morrison, eds. Monocots: systematics and evolution. CSIRO, Melbourne.

Harder LD, SCH Barrett 1992 The energy cost of bee pollination for Pontederia cordata. Funct Ecol 6:226–233.

—— 1995 Mating cost of large floral displays in hermaphrodite plants. Nature 373:512–515.

—— 1996 Pollen dispersal and mating patterns in animal-pollinated plants. Pages 140–190 in DG Lloyd, SCH Barrett, eds. Floral biology: studies on floral evolution in animal-pollinated plants. Chapman & Hall, New York.

Harder LD, SD Johnson 2005 Adaptive plasticity of floral display size in animal-pollinated plants. Proc R Soc B 272:2651–2657.

Harder LD, CY Jordan, WE Gross, MB Routley 2004 Beyond floricentrism: the pollination function of inflorescences. Plant Species Biol 19:137–148.

Harder LD, JD Thomson 1989 Evolutionary options for maximizing pollen dispersal of animal-pollinated plants. Am Nat 133:323–344.

Harder LD, WG Wilson 1994 Floral evolution and male reproductive success: optimal dispensing schedules for pollen dispersal by animal-pollinated plants. Evol Ecol 8:542–559.

—— 1997 Theoretical perspectives on pollination. Acta Hortic 437:83–101.

Hesse M 1980 Zur Frage der Anheftung des Pollens an blütenbe-

suchende Insekten mittels Pollenkitt und Viscinfäden. Plant Syst Evol 133:135–148.

Hesse M, S Vogel S, H Halbritter H 2000 Thread-forming structures in angiosperm anthers: their diverse role in pollination ecology. Plant Syst Evol 222:281–292.

Hiei K, K Suzuki 2001 Visitation frequency of Melampyrum roseum var. japonicum (Scrophulariaceae) by three bumblebee species and its relation to pollination efficiency. Can J Bot 79:1167–1174.

Hoekstra FA, J Bruinsma 1975 Respiration and vitality of binucleate and trinucleate pollen. Physiol Plant 34:221–225.

Husband BC, DW Schemske 1996 Evolution of the magnitude and timing of inbreeding depression in plants. Evolution 50:54–70.

Ionta GM, WS Judd 2007 Phylogenetic relationships in Periplocoideae (Apocynaceae s.l.) and insights into the origin of pollinia. Ann Mo Bot Gard 94:360–375.

Jersakova J, SD Johnson 2006 Lack of floral nectar reduces self-pollination in a fly-pollinated orchid. Oecologia 147:60–68.

Johnson SD 1997 Pollination ecotypes of Satyrium hallackii (Orchidaceae) in South Africa. Bot J Linn Soc 123:225–235.

Johnson SD, TJ Edwards 2000 The structure and function of orchid pollinaria. Plant Syst Evol 222:243–269.

Johnson SD, HP Linder, KE Steiner 1998 Phylogeny and radiation of pollination systems in Disa (Orchidaceae). Am J Bot 85:402–411.

Johnson SD, PR Neal, LD Harder 2005 Pollen fates and the limits on male reproductive success in an orchid population. Biol J Linn Soc 86:175–190.

Johnson SD, LA Nilsson 1999 Pollen carryover, geitonogamy, and the evolution of deceptive pollination systems in orchids. Ecology 80:2607–2619.

Johnson SD, CI Peter, J Ågren 2004 The effects of nectar addition on pollen removal and geitonogamy in the non-rewarding orchid Anacamptis morio. Proc R Soc B 271:803–809.

Johnson SD, KE Steiner 1997 Long-tongued fly pollination and evolution of floral spur length in the Disa draconis complex (Orchidaceae). Evolution 51:45–53.

Karron JD, RJ Mitchell, JM Bell 2006 Multiple pollinator visits to Mimulus ringens (Phrymaceae) flowers increase mate number and seed set within fruits Am J Bot 93:1306–1312.

Kenrick J, RB Knox 1982 Function of the polyad in reproduction of Acacia. Ann Bot 50:721–727.

—— 1989 Quantitative analysis of self-incompatibility in trees in seven species of Acacia. J Hered 80:240–245.

Kephart SR 1987 Phenological variation in flowering and fruiting of Asclepias. Am Midl Nat 118:64–76.

Kephart SR, K Theiss 2004 Pollinator-mediated isolation in sympatric milkweeds (Asclepias): do floral morphology and insect behavior influence species boundaries? New Phytol 161:265–277.

Knox RB, J Kenrick 1983 Polyad function in relation to the breeding system of Acacia. Pages 411–417 in DL Mulcahy, E Ottaviano, eds. Pollen: biology and implications for plant breeding. Elsevier, Amsterdam.

Kocyan A, YL Qiu, PK Endress, E Conti 2004 A phylogenetic analysis of Apostasioideae (Orchidaceae) based on ITS, trnL-F and matK sequences. Plant Syst Evol 247:203–213.

Kress WJ 1981 Sibling competition and evolution of pollen unit, ovule number, and pollen vector in angiosperms. Syst Bot 6:101–112.

Kudo G, S Suzuki 2002 Relationships between flowering phenology and fruit-set of dwarf shrubs in alpine fellfields in northern Japan: a comparison with a subarctic heathland in northern Sweden. Arct Antarct Alp Res 34:185–190.

Kunze H 1991 Structure and function in asclepiad pollination. Plant Syst Evol 176:227–253.

Kunze H, S Liede 1991 Observations on pollination in Sarcostemma (Asclepiadaceae). Plant Syst Evol 178:95–105.

Lankinen Å, B Hellriegel, G Bernasconi 2006 Sexual conflict over floral receptivity. Evolution 60:2454–2465.

Laverty TM 1992 Plant interactions for pollinator visits: a test of the magnet species effect. Oecologia 89:502–508.

LeBuhn G, K Holsinger 1998 A sensitivity analysis of pollen-dispensing schedules. Evol Ecol 12:111–121.

Lehnebach CA, AW Robertson 2004 Pollination ecology of four epiphytic orchids of New Zealand. Ann Bot 93:773–781.

Levin RA, WL Wagner, PC Hoch, M Nepokroeff, JC Pires, EA Zimmer, KJ Sytsma 2003 Family-level relationships of Onagraceae based on chloroplast rbcL and ndhF data. Am J Bot 90.107 115.

Liede S 1994 Some observations on pollination in Mexican Asclepiadaceae. Madroño 41:266–276.

Lipow SR, R Wyatt 1998 Reproductive biology and breeding system of Gonolobus suberosus (Asclepiadaceae). J Torrey Bot Soc 125: 183–193.

Lisci M, G Cardinali, E Pacini 1996 Pollen dispersal and role of pollenkitt in Mercurialis annua L. (Euphorbiaceae). Flora 191:385–391.

Livshultz T, DJ Middleton, ME Endress, JK Williams 2007 Phylogeny of Apocynoideae and the APSA clade (Apocynaceae s.l.). Ann Mo Bot Gard 94:324–359.

Lloyd DG 1984 Gender allocations in outcrossing cosexual plants. Pages 277–300 in R Dirzo, J Sarukhán, eds. Perspectives on plant population ecology. Sinauer, Sunderland, MA.

——— 2000 The selection of social actions in families. II. Parental investment. Evol Ecol Res 2:15–28.

Lukasiewicz MJ 1999 Maternal investment, pollination efficiency and pollen : ovule ratios in Alberta orchids. MSc thesis. University of Calgary.

Luyt R, SD Johnson 2001 Hawkmoth pollination of the African epiphytic orchid Mystacidium venosum, with special reference to flower and pollen longevity. Plant Syst Evol 228:49–62.

Maad J, LA Nilsson 2004 On the mechanism of floral shifts in speciation: gained pollination efficiency from tongue- to eye-attachment of pollinia in Platanthera (Orchidaceae). Biol J Linn Soc 83:481–495.

Maddison WP, DR Maddison 2006 Mesquite: a modular system for evolutionary analysis, version 1.12. http://www.mesquiteproject.org.

Masinde PS 2004 Trap-flower fly pollination in East African Ceropegia L. (Apocynaceae). Int J Trop Insect Sci 24:55–72.

Meve U, S Liede 1994 Floral biology and pollination in stapeliads: new results and a literature review. Plant Syst Evol 192:99–116.

Mitchell RJ, JD Karron, KG Holmquist, JM Bell 2005 Patterns of multiple paternity in fruits of Mimulus ringens (Phrymaceae). Am J Bot 92:885–890.

Morris WF, MV Price, NM Waser, JD Thomson, BA Thomson, DA Stratton 1994 Systematic increase in pollen carryover and its consequences for geitonogamy in plant populations. Oikos 71:431–440.

Morse DH 1987 Roles of pollen and ovary age in follicle production of the common milkweed Asclepias syriaca. Am J Bot 74:851–856.

——— 1994 The role of self-pollen in the female reproductive success of common milkweed (Asclepias syriaca: Asclepiadaceae). Am J Bot 81:322–330.

Muona O, GF Moran, JC Bell 1991 Hierarchical patterns of correlated mating in Acacia melanoxylon. Genetics 127:619–626.

Muller J 1979 Form and function in angiosperm pollen. Ann Mo Bot Gard 66:593–632.

Nazarov VV, G Gerlach 1997 The potential seed productivity of orchid flowers and peculiarities of their pollination systems. Lindleyana 12: 188–204.

Nilsson LA, L Jonsson, L Rason, E Randrianjohany 1986 The pollination of Cymbidiella flabellata (Orchidaceae) in Madagascar: a system operated by sphecid wasps. Nord J Bot 6:411–422.

Nilsson LA, E Rabakonandrianina, B Pettersson 1992 Exact tracking of pollen transfer and mating in plants. Nature 360:666–668.

Nilsson S, ME Endress, E Grafstrom 1993 On the relationship of Apocynaceae and Periplocaceae. Grana 2(suppl):3–20.

O'Connell LM, MO Johnston 1998 Male and female pollination success in a deceptive orchid, a selection study. Ecology 79:1246–1260.

Ollerton J, SD Johnson, L Cranmer, S Kellie 2003 The pollination ecology of an assemblage of grassland asclepiads in South Africa. Ann Bot 92:807–834.

Ornduff R 1970a Heteromorphic incompatibility in Jepsonia malvifolia. Bull Torrey Bot Club 97:258–261.

——— 1970b Incompatibility and the pollen economy of Jepsonia parryi. Am J Bot 57:1036–1041.

——— 1971 The reproductive system of Jepsonia heterandra. Evolution 25:300–311.

Pacini E, Hesse M 2005 Pollenkitt: its composition, forms and functions. Flora 200:399–415.

Pauw A 1998 Pollen transfer on bird's tongues. Nature 394:731–732.

Peakall R 1989 A new technique for monitoring pollen flow in orchids. Oecologia 79:361–365.

Peakall R, AJ Beattie 1991 The genetic consequences of worker ant pollination in a self-compatible, clonal orchid. Evolution 45:1837–1848.

Peter CI, Johnson SD 2006 Doing the twist: a test of Darwin's cross-pollination hypothesis for pollination reconfiguration. Biol Lett 2: 65–68.

Pettersson B, LA Nilsson 1993 Floral variation and deceit pollination in Polystachya rosea (Orchidaceae) on an inselberg in Madagascar. Opera Bot 121:237–245.

Pleasants JM 1991 Evidence for short-distance dispersal of pollinia in Asclepias syriaca L. Funct Ecol 5:75–82.

Proctor HC 1998 Effect of pollen age on fruit set, fruit weight, and seed set in three orchid species. Can J Bot 76:420–427.

Queller DC 1984 Pollen-ovule ratios and hermaphrodite sexual allocation strategies. Evolution 38:1148–1151.

Rowley JR, JJ Skvarla 2006 Pollen development in Epilobium (Onagraceae): late microspore stages (a review). Rev Palaeobot Palynol 140: 91–112.

Seijo G, VG Solís Neffa 2004 The cytological origin of the polyads and their significance in the reproductive biology of Mimosa bimucronata. Bot J Linn Soc 144:343–349.

Shore JS 1993 Pollination genetics of the common milkweed, Asclepias syriaca L. Heredity 70:101–108.

Singer RB, AA Cocucci 1997 Pollination of Pteroglossaspis ruwenzoriensis (Rendle) Rolfe (Orchidaceae) by beetles in Argentina. Bot Acta 110:338–342.

Smithson A 2006 Pollinator limitation and inbreeding depression in orchid species with and without nectar rewards. New Phytol 169: 419–430.

Smithson A, MR MacNair 1997 Negative frequency-dependent selection by pollinators on artificial flowers without rewards. Evolution 51:715–723.

Snow AA, DW Roubik 1987 Pollen deposition and removal by bees visiting two tree species in Panama. Biotropica 19:57–63.

Steiner KE 1989 The pollination of Disperis (Orchidaceae) by oil-collecting bees in southern Africa. Lindleyana 4:164–183.

Tanaka H, T Hatano, N Kaneko, S Kawachino, O Kitamura, Y Suzuki, T Tada, Y Yao 2006 Andromonoecious sex expression of flowers and pollinia delivery by insects in a Japanese milkweed Metaplexis japonica (Asclepiadaceae), with special reference to its floral morphology. Plant Species Biol 21:193–199.

Tandon R, KR Shivanna, HYM Ram 2001 Pollination biology and breeding system of Acacia senegal. Bot J Linn Soc 135:251–262.

Thomson JD 1986 Pollen transport and deposition by bumble bees in *Erythronium*: influences of floral nectar and bee grooming. J Ecol 74:329–341.

Thomson JD, K Goodell 2001 Pollen removal and deposition by honeybee and bumblebee visitors to apple and almond flowers. J Appl Ecol 38:1032–1044.

Thomson JD, P Wilson, M Valenzuela, M Malzone 2000 Pollen presentation and pollination syndromes, with special reference to *Penstemon*. Plant Species Biol 15:11–29.

Torres C, MC Eynard, MA Aizen, L Galetto 2002 Selective fruit maturation and seedling performance in *Acacia caven* (Fabaceae). Int J Plant Sci 163:809–813.

Trapnell DW, JL Hamrick 2006 Floral display and mating patterns within populations of the Neotropical epiphytic orchid, *Laelia rubescens* (Orchidaceae). Am J Bot 93:1010–1017.

Tremblay RL, JD Ackerman, JK Zimmerman, RN Calvo 2005 Variation in sexual reproduction in orchids and its evolutionary consequences: a spasmodic journey to diversification. Biol J Linn Soc 84:1–54.

Uma Shaanker R, KN Ganeshaiah, KS Bawa 1988 Parent-offspring conflict, sibling rivalry, and brood size patterns in plants. Annu Rev Ecol Syst 19:177–205.

van der Meulen MB 1992 Limits to reproduction by *Aconitum delphinifolium* and *Delphinium glaucum*. MSc thesis. University of Calgary.

van Doorn WG 1997 Effects of pollination on floral attraction and longevity. J Exp Bot 48:1615–1622.

Verhoeven RL, HJT Venter 2001 Pollen morphology of the Periplocoideae, Secamonoideae, and Asclepiadoideae (Apocynaceae). Ann Mo Bot Gard 88:569–582.

Vieira MF, GJ Shepherd 2002 Removal and insertion of pollinia in flowers of *Oxypetalum* (Asclepiadaceae) in southeastern Brazil. Rev Biol Trop 50:37–43.

Walker JW, JA Doyle 1975 The bases of angiosperm phylogeny: palynology. Ann Mo Bot Gard 62:664–723.

Watson L, MJ Dallwitz 2006 The families of flowering plants: descriptions, illustrations, identification, and information retrieval. http://www.delta-intkey.com/angio.

Webb CJ, KS Bawa 1983 Pollen dispersal by hummingbirds and butterflies: a comparative study of two lowland tropical plants. Evolution 37:1258–1270.

Whitten WM, NH Williams, MW Chase 2000 Subtribal and generic relationships of Maxillarieae (Orchidaceae) with emphasis on Stanhopeinae: combined molecular evidence. Am J Bot 87:1842–1856.

Willson MF 1979 Sexual selection in plants. Am Nat 113:777–790.

Wilson P 1995 Variation in the intensity of pollination in *Drosera tracyi*: selection is strongest when resources are intermediate. Evol Ecol 9:382–396.

Wilson P, JD Thomson 1991 Heterogeneity among floral visitors leads to discordance between removal and deposition of pollen. Ecology 72:1503–1507.

Wyatt R 1976 Pollination and fruit-set in *Asclepias*: a reappraisal. Am J Bot 63:845–851.

Wyatt R, SB Broyles, SR Lipow 2000 Pollen-ovule ratios in milkweeds (Asclepiadaceae): an exception that probes the rule. Syst Bot 25:171–180.

Wyatt R, SR Lipow 2007 A new explanation for the evolution of pollinia and loss of carpel fusion in *Asclepias* and the Apocynaceae s.l. Ann Mo Bot Gard 94:474–484.

Yeo PF 1993 Secondary pollen presentation. Springer, New York.

Young HJ, ML Stanton 1990 Influences of floral variation on pollen removal and seed production in wild radish. Ecology 71:536–547.

Zhang D-Y 2006 Evolutionarily stable reproductive investment and sex allocation in plants. Pages 41–60 *in* LD Harder, SCH Barrett, eds. Ecology and evolution of flowers. Oxford University Press, Oxford.

Int. J. Plant Sci. 169(1):79–92. 2008.
1058-5893/2008/16901-0007$15.00   DOI: 10.1086/523354

ORIGIN OF THE FITTEST AND SURVIVAL OF THE FITTEST: RELATING FEMALE GAMETOPHYTE DEVELOPMENT TO ENDOSPERM GENETICS

William E. Friedman,* Eric N. Madrid,* and Joseph H. Williams†

*Department of Ecology and Evolutionary Biology, University of Colorado, Boulder, Colorado 80309, U.S.A.; and †Department of Ecology and Evolutionary Biology, University of Tennessee, Knoxville, Tennessee 37996, U.S.A.

For more than a century, most biologists have viewed the structural diversity of angiosperm female gametophytes as trivial variants of the reproductive process. However, analysis of variation among angiosperm female gametophytes from an evolutionary developmental perspective can provide new insights into patterns of reproductive innovation and evolution among flowering plants. The key is to link the developmental and structural diversity of angiosperm female gametophytes to evolutionary innovations (perhaps even adaptations) associated with endosperm genetics and ploidy. Selection has been hypothesized to favor endosperms with higher ploidy, higher heterozygosity, higher maternal-to-paternal genome ratios, and reduced opportunity for genetic (interparental and/or parent-offspring) conflict. We evaluate these hypotheses for the seven basic genetic types of endosperm known among flowering plants and interpret their relative importance when mating system is considered. We demonstrate that variation in female gametophyte developmental patterns represents the source material that ultimately creates variation in endosperm genetics. Evolutionary transitions in female gametophyte development are therefore a function of selection directly acting on the resultant phenotypes of endosperms. Thus, the relation between variation in female gametophyte development and variation in endosperm genetic constitution should be seen as one between the origin of structural novelties (origin of the fittest) and its downstream consequences on the relative fitness (survival of the fittest) of these novelties, as expressed in the biology of endosperm.

Keywords: female gametophyte, endosperm, fertilization, inclusive fitness theory, development, modularity, evolution.

Introduction

For more than a century, the intricate details of female gametophyte (embryo sac) development in flowering plants (monosporic vs. bisporic vs. tetrasporic; Polygonum-type vs. Allium-type vs. Fritillaria-type vs. Oenothera-type and so on) have been the bane of plant biologists (fig. 1). Much of this justifiably negative reaction to understanding the diversity of female gametophyte development and structure can be traced to a highly typological approach to the categorization of embryo sacs that has dominated flowering-plant embryological work. At the end of the day, most plant biologists respond to the details of embryology by accepting that there are well-established differences among angiosperm female gametophytes and rejecting that these differences are interesting.

Analysis of the diversity of angiosperm female gametophytes from an evolutionary developmental (evo-devo) perspective (as opposed to a strictly typological one) can be biologically meaningful. The key is to understand female gametophyte diversity (i.e., differences in basic structure) based on general principles of developmental biology and to link this diversity to critical aspects of flowering-plant reproductive biology, such as the fertilization process, and evolution-

ary innovations (perhaps even adaptations) associated with endosperm genetics and ploidy. Our goal, in presenting a developmentally based model of angiosperm female gametophyte evolution, is to reintegrate this important organismic generation into the broader reproductive biology of flowering plants and attempt to move the field of plant embryology beyond its historically typological roots.

In this article, we first examine correlations between patterns of female gametophyte development and the highly variable genetic constructs of endosperm among flowering plants. We show that the known diversity of female gametophyte development and mature structure among angiosperms is tightly linked to and constrained by the biology of endosperm, specifically, endosperm ploidy, maternal-to-paternal genomic ratios, degree of heterozygosity, and genetic relatedness to its compatriot embryo relative to its relatedness to other embryos on the maternal sporophyte.

We argue that the female gametophytes of all angiosperms are fundamentally modular entities (sensu Friedman and Williams 2003); in other words, they are composed of iteratively expressed units. We then go on to demonstrate that variation in mature angiosperm female gametophyte structure is the result of three basic types of developmental modifications or themes: (1) relative timing of the establishment of female gametophyte modules (during or after megasporogenesis), (2) early ontogenetic events that determine the number of developmental modules initiated (one, two, or four),

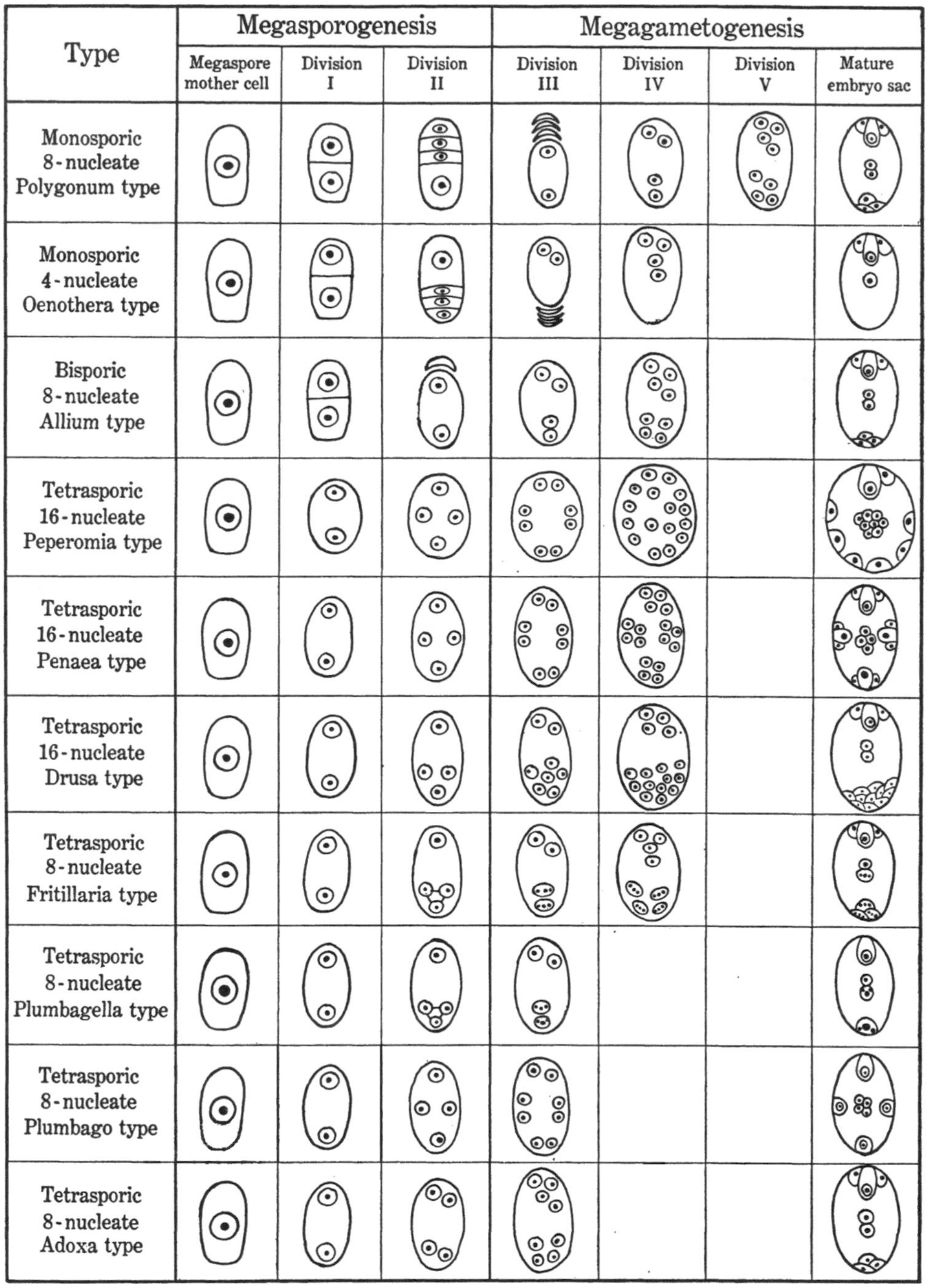

Type	Megasporogenesis			Megagametogenesis			
	Megaspore mother cell	Division I	Division II	Division III	Division IV	Division V	Mature embryo sac
Monosporic 8-nucleate Polygonum type							
Monosporic 4-nucleate Oenothera type							
Bisporic 8-nucleate Allium type							
Tetrasporic 16-nucleate Peperomia type							
Tetrasporic 16-nucleate Penaea type							
Tetrasporic 16-nucleate Drusa type							
Tetrasporic 8-nucleate Fritillaria type							
Tetrasporic 8-nucleate Plumbagella type							
Tetrasporic 8-nucleate Plumbago type							
Tetrasporic 8-nucleate Adoxa type							

Fig. 1 The famous diagram showing the basic types of female gametophyte development and structural diversity from Maheshwari (1950). Maheshwari was not the first to publish such a diagrammatic representation of angiosperm female gametophyte diversity (see, e.g., Chiarugi 1927; Schnarf 1929), nor was he the last (see, e.g., Gifford and Foster 1987; Haig 1990). This diagram and its intellectual descendants show the basic variation in megasporogenesis, meiotic and mitotic division patterns, and mature structure in angiosperms. Various portions of the diagram, in modified form, are used to examine the developmental evolution of angiosperm female gametophytes throughout this article.

and (3) ontogenetic events that result in developmental deviations from the basic (and plesiomorphic) female gametophyte developmental module. In so doing, we hope to illuminate the connection between the origin of structural variation in the angiosperm female gametophyte and the selective forces that have allowed novel endosperm genetic constructs to persist.

Connecting the Female Gametophyte Central Cell to Endosperm Genetics

Angiosperm female gametophytes are ontogenetically and structurally highly variable (fig. 1). They may be monosporic, bisporic, or tetrasporic in origin. The number of synergids in the adult gametophyte ranges from zero to three (Maheshwari 1950; Friedman 2006a, 2006b), while antipodal number varies widely (zero to many). The number of nuclei contributed to the central cell ranges from one (Yoshida 1962; Galati 1985; Williams and Friedman 2002) to 14 (Johnson 1914; Maheshwari 1950). For the purposes of our analysis, variation in ontogeny and adult female gametophyte structure is biologically significant only if it has a downstream effect on the genetic constitution of endosperm.

Endosperm in flowering plants is initiated by the fertilization of the central cell of the female gametophyte during the process of double fertilization. Consequently, there is a direct relation between the genetic constitution of endosperm and the antecedent genetic constitution of the central cell (Palser 1975). Among the innumerable variants of angiosperm female gametophyte development and structure (see Haig 1990 and Johri et al. 1992 for excellent discussions), there appear to be just seven basic types of central cell genetic constitution and hence only seven genetic constructs or types of endosperm (table 1). For the purpose of cross-referencing the genetic constitutions of central cells and endosperms to the embryological literature, we have also circumscribed the classical names of embryo sac types that produce the seven genetic kinds of central cells and endosperms (table 1).

Monosporic diploid endosperms are produced by female gametophytes initiated by a single megaspore that mature a central cell with a single haploid nucleus. The phylogenetically widespread and common monosporic triploid endosperm is typically formed from Polygonum-type female gametophytes but may also be initiated upon fertilization in the recently discovered Amborella-type (nine-nucleate, eight-celled female gametophyte with three synergids; Friedman 2006a, 2006b). Monosporic triploid endosperms are derived from the fertilization of a central cell that contains two polar nuclei that are genetically identical to each other and to the egg nucleus.

All bisporic angiosperm female gametophytes produce bisporic triploid endosperms. The two haploid nuclei of the central cell are derived from the two megaspores from meiosis II of one of the dyads. As such, their coefficient of relatedness (r) is defined as $r = 1 - q$, where q is the frequency of second-division segregation (Bulmer 1986; Haig 1986).

Tetrasporic triploid endosperms are initiated from a central cell with two polar nuclei that are lineal descendants of each dyad from meiosis I. As such, their coefficient of relatedness is defined as $r = q/2$, assuming that the egg is the mitotic sister nucleus of the micropylar polar nucleus, as expected (Bulmer 1986; Haig 1986). Tetrasporic pentaploid endosperms arise from a number of different types of female gametophytes. These female gametophytes are always initiated by four megaspore nuclei within a coenomegaspore, and the central cell contains a lineal descendant of each of the megaspores (this is assumed to be the case in those embryo sacs that form restitution nuclei).

Tetrasporic nonaploid (9N) and decapentaploid (15N) endosperms are confined to the genus Peperomia (Piperaceae; Johnson 1900, 1914). Tetrasporic nonaploid endosperms are formed by the fertilization of a central cell with four pairs of nuclei that are each mitotically descended from the original four megaspores that initiate embryo sac development. Tetrasporic decapentaploid endosperms are derived from a central cell that contains two nuclei genetically identical to the egg cell and three sets of four nuclei derived from the three megaspores not associated with the production of the egg (fig. 1; table 1).

The Basics of Endosperm Genetics: Heterozygosity, Ploidy, and Relatedness

There is good reason to carefully analyze the genetic constructs of the various entities found within a flowering-plant seed (female gametophyte, embryo, endosperm, and maternal sporophyte). Nearly three-quarters of a century ago, a rich, but sporadic, theoretical literature began to examine the genetic and evolutionary implications of a sexually formed, genetically biparental embryo-nourishing tissue, as found only in angiosperms. Beginning with the pioneering papers of Brink and Cooper (1940, 1947), three basic theories have been developed to explain the variable "consequences" of the

Table 1

The Seven Basic Genetic Constructs of Endosperms, the Mature Female Gametophyte Types That Produce These Endosperms, and Central Cell Genetic Constitution

Endosperm construct	Correlated female gametophytes	Central cell nuclear contents[a]
Monosporic 2N	Oenothera-type, Nuphar-type	M1
Monosporic 3N	Polygonum-type, Amborella-type	2(M1)
Bisporic 3N	Allium-type, Endymion-type, Drusa-type[b]	M1, M2
Tetrasporic 3N	Adoxa-type, Drusa-type[b]	M1, M3, or M4
Tetrasporic 5N	Penaea-type, Plumbago-type Fritillaria-type, Plumbagella-type	M1, M2, M3, M4
Tetrasporic 9N	Peperomia-type	2(M1), 2(M2), 2(M3), 2(M4)
Tetrasporic 15N	Peperomia-type	2(M1), 4(M2), 4(M3), 4(M4)

[a] M1: female nucleus genetically identical to the egg nucleus; M2: female nucleus sister to M1 and derived from meiosis II of dyad I; M3: female nucleus derived from meiosis II of dyad II; M4: female nucleus sister to M3 and derived from meiosis II of dyad II.

[b] For the Drusa-type female gametophyte, the chalazal polar nucleus may be derived either from the same meiotic dyad as the egg nucleus is derived from or from the other dyad. In the first case, the genetics of endosperm are the same as for bisporic 3N; in the second case, the genetics are the same as for tetrasporic 3N.

evolutionary origin and diversification of a sexually formed embryo-nourishing tissue, namely, endosperm.

Endosperm Heterozygosity

Brink and Cooper (1940, 1947) were the first of many (Stebbins 1976; Tiffney 1981; Takhtajan 1991; Donoghue and Scheiner 1992) to suggest that heterosis (essentially heterozygosity) creates in endosperm a more vigorous embryo-nourishing tissue than the haploid (hence hemizygous) embryo-nourishing female gametophytes of nonflowering plants. This theory (hereafter the "heterozygosity hypothesis") leads to the prediction that further increases in the heterozygosity of bisexual endosperm should also be advantageous (Brink and Cooper 1940, 1947; Stebbins 1976; Tiffney 1981; Takhtajan 1991; Donoghue and Scheiner 1992).

Endosperm heterozygosity, defined as the average probability of having two or more different alleles per locus, is expected to increase as the probability of including both maternal alleles in the endosperm increases. This probability depends on the pattern of megasporogenesis and whether mitotic derivatives of one, two, or all four meiotic products (megaspore nuclei) are incorporated into the central cell that initiates an endosperm upon fertilization (table 1).

The egg cell and central cell of all monosporic female gametophytes are composed of lineal mitotic descendants of a single megaspore, such that endosperm heterozygosity is the same as embryo heterozygosity. The two polar nuclei of bisporic female gametophytes are the lineal mitotic descendants of two megaspores derived from the same dyad; hence, both maternal alleles will be represented at endosperm loci

that have been affected by crossing over between nonsister chromatids (table 1). In contrast, the two polar nuclei of the central cell of the tetrasporic Adoxa-type female gametophyte are derived from separate dyads; hence, both maternal alleles will be represented in endosperm for loci that have not been affected by crossing over between nonsister chromatids. In both of these cases, the probability that the two polar nuclei represent one or both maternal alleles depends on the second-division segregation rate: with no crossing over, alleles are always identical by descent in bisporic female gametophytes and are always nonidentical in tetrasporic Adoxa-type female gametophytes, whereas with maximum recombination, both maternal alleles will be represented two-thirds of the time in both cases (Fincham 1994). Among other tetrasporic female gametophytes (those that produce pentaploid or higher endosperms), the central cell contains lineal mitotic descendants of all four megaspores. Therefore, the central cell always includes both alleles, and before fertilization, its heterozygosity is equivalent to that of the maternal genotype (table 1).

We have calculated expected endosperm heterozygosity for monosporic, bisporic, and tetrasporic female gametophytes (table 2; appendix). Our results indicate that in a randomly mating population, heterozygosity at any single locus in an endosperm derived from a monosporic female gametophyte (H) is less than or equal to that of an endosperm formed from a bisporic female gametophyte ($H + qH/2$), where q is the frequency of second-division segregation. Heterozygosity of triploid endosperms derived from bisporic female gametophytes is less than or equal to heterozygosity of triploid endosperms derived from tetrasporic female gametophytes ($3H/2 - qH/4$), which in turn is always less than or equal to the heterozygosity

Table 2

Seven Basic Genetic Constructs of Endosperms among Flowering Plants

Endosperm type	M : P	H	r_{ms-ent}	r_{ent-ce} : r_{ent-oe} Outcross, unrelated fathers	r_{ent-ce} : r_{ent-oe} Outcross, same father	r_{ent-ce} : r_{ent-oe} Self-fertilization
Monosporic 2N	1 : 1	H	1/2	4	2	3/2
Monosporic 3N	2 : 1	H	1/2	3	2	3/2
Bisporic 3N[a]	2 : 1	$H + qH/2$	1/2 to 1	$3 - q$	$2 - 2q/3$	$3/2 - q/3$
Tetrasporic 3N[a]	2 : 1	$3H/2 - qH/4$	1 to 1/2	$2 + q/2$	$4/3 + q/3$	$7/6 + q/6$
Tetrasporic 5N[b]	4 : 1	$3H/2$	1	3/2	6/5	11/10
Tetrasporic 9N	8 : 1	$3H/2$	1	5/4	10/9	19/18
Tetrasporic 15N	14 : 1	$3H/2$	1	1	29/30	29/30
Perisperm[c]	2 : 0	H_0	1	1	1	1

Note. Shown are ploidy, maternal-to-paternal genomic ratios (M : P), heterozygosity, and ratios of coefficients of relatedness of endosperm to its own embryo versus other embryos under conditions of outcrossing to unrelated pollen donors, outcrossing to the same pollen donor, and self-fertilization. H = expected heterozygosity of embryo-nourishing tissue, expressed as a function of theoretical expected heterozygosity in a randomly mating population; H_0 represents observed heterozygosity of sporophytes (as represented by perisperm); for derivations, see appendix. r_{ms-ent} = relatedness of maternal sporophyte to its embryo-nourishing tissue (ent), assuming nonimprinted gene expression with additive effects (see Queller 1989); r_{ent-ce} : r_{ent-oe} = ratio of relatedness of ent to its compatriot embryo (ce) and that to another embryo (oe) on the same maternal sporophyte, assuming nonimprinted gene expression with additive effects (see Queller 1989). q = frequency of second-division segregation.

[a] For the Drusa-type female gametophyte, the chalazal polar nucleus may be derived either from the same meiotic dyad as the egg nucleus is derived from or from the other dyad. In the first case, the genetics of endosperm are the same as for bisporic 3N; in the second case, the genetics are the same as for tetrasporic 3N.

[b] This includes pentaploid endosperms derived from female gametophytes with restitution nuclei (e.g., Plumbagella-type and Fritillaria-type).

[c] For comparative purposes, we have included calculations for the maternal sporophyte-derived tissue, perisperm.

of all other endosperms derived from tetrasporic female gametophytes ($3H/2$). Thus, according to the tenets of the heterozygosity (heterosis) hypothesis, endosperms derived from tetrasporic female gametophytes should be favored over those derived from bisporic female gametophytes, and endosperms derived from both tetrasporic and bisporic female gametophytes should be favored over endosperms derived from monosporic female gametophytes (but see Haig 1986 for a discussion of potential evolutionarily unstable genetic conflict among megaspores in female gametophytes that are bisporic or tetrasporic).

Endosperm Ploidy

A second theory concerned with endosperm origin and evolution suggests that higher levels of ploidy should benefit the embryo-nourishing function of endosperm when, for example, pentaploid endosperms are compared with triploid endosperms or any endosperm is compared with the haploid female gametophytes of nonflowering plants (Stebbins 1974, 1976). The gist of this argument is that higher ploidy will enable higher rates of gene transcription in support of the active physiological role of endosperm. There is certainly widespread evidence to support the concept that higher ploidy levels (e.g., through endopolyploidy or endoreduplication) are correlated with physiologically active regions of the plant body (D'Amato 1984; Galbraith et al. 1991). Thus, the prediction of this theory (hereafter the "ploidy hypothesis") is that evolutionary transitions to higher endosperm ploidy should be favored.

The vast majority of flowering-plant species produce female gametophytes with diploid central cells that contain two haploid polar nuclei or the product of their fusion, the secondary nucleus. However, the central cell does vary among different lineages of angiosperms, from haploid to decatetraploid (14N). Thus, endosperm ploidy among flowering plants ranges from diploid to decapentaploid (in *Peperomia*; Johnson 1914). Pentaploid endosperms derived from the fertilization of tetraploid central cells can be found in diverse angiosperm clades (e.g., *Penaea*, *Plumbago*, *Plumbagella*, and *Fritillaria*; Stephens 1909; Maheshwari 1950; Haig 1990; Johri et al. 1992). Although we have not performed an explicit phylogenetically based comparative analysis of endosperm ploidy, it is interesting to note that two of the most ancient lineages of angiosperms (Nymphaeales and Austrobaileyales) have recently been shown to produce diploid endosperm derived from the fertilization of a haploid central cell (Williams and Friedman 2002, 2004; Friedman and Williams 2003; Friedman et al. 2003). Given the straightforward predictions of the ploidy hypothesis, it would be well worth investigating whether evolutionary transitions to female gametophyte forms that produce higher-ploidy endosperms are largely (NB, but not always, e.g., Onagraceae; Ishikawa 1918; Tobe and Raven 1986) irreversible.

Endosperm Relatedness

A third body of literature that has been developed to help understand the unique genetic constitution of endosperm derives from inclusive-fitness theory. These analyses suggest that the original integration of a paternal genome into the embryo-nourishing tissue of flowering plants had a number of profound effects on how maternal sporophytic resources might be allocated to seeds and their constituent embryos.

Inclusive-fitness analyses of angiosperm reproduction (Charnov 1979; Cook 1981; Westoby and Rice 1982; Queller 1983, 1989, 1994; Willson and Burley 1983; Law and Cannings 1984; Bulmer 1986; Haig 1986; Haig and Westoby 1989a, 1989b; Friedman 1995; Härdling and Nilsson 1999, 2001) indicate that the origin and subsequent evolution of a heterozygous and polyploid endosperm can be viewed as the outcome of (1) conflict between male and female parents over the investment of nutrients in the embryo-nourishing tissues of seeds of a single maternal sporophyte (interparental or intersexual conflict) and/or (2) conflict among sibling embryos for resources from the maternal sporophyte (kin or parent-offspring conflict).

Both interparental conflict and kin conflict hypotheses (hereafter "conflict hypotheses") assume that resources available for the production of seeds by a maternal sporophyte are limiting and that, as a consequence, a subset of embryos/seeds on a given plant will abort or be underprovisioned. Central to these ideas is the supposition that changes in the relatedness of the embryo-nourishing tissue to its own embryo, the maternal and paternal sporophytes, and other embryos and embryo-nourishing tissues on a single maternal sporophyte can affect the relative "aggressiveness" or "selfishness" of an embryo-nourishing tissue to procure nutrients on behalf of its own embryo.

A maternal sporophyte is equally related to all of her progeny (assuming that the maternal sporophyte is equally related, or unrelated, to the paternal sporophytes). Therefore, her fitness is maximized when the subset of embryos that are themselves most fit are successfully reared, while less-fit embryos are aborted. Charnov (1979) was the first to recognize that the evolutionary origin of a second fertilization event and its insertion of a paternal genome into the embryo-nourishing tissue had the effect of increasing the relatedness of the embryo-nourishing tissue to its own (same seed) embryo, when compared with its relatedness to other embryos, as well as decreasing its relatedness to the maternal sporophyte. Irrespective of whether interparental conflict or parent-offspring conflict drives the system, the maternal sporophyte should favor increases in its genomic and allelic contributions to its proxy embryo-nourishing tissue, the endosperm, to decrease the selfish behavior of this genetically biparental entity. These evolutionary transitions can be accomplished by increasing the number of maternal genomes contributed to the central cell (in essence, more maternal nuclei) and/or by increasing the number of megaspore genomes represented in the central cell through polysporic initiation of the female gametophyte (e.g., transitions from monosporic to bisporic, from monosporic to tetrasporic, or from bisporic to tetrasporic). Any and all changes in the maternal genetic contribution to the central cell will, of course, be manifest in the genetic constitution of the subsequently formed endosperm.

To analyze genetic conflict among parents, offspring, and endosperms on a single maternal sporophyte, Westoby and Rice (1982), Queller (1983, 1989, 1994), Willson and Burley (1983), Bulmer (1986), Haig (1986, 1987), and Haig and Westoby (1988, 1989a, 1989b) pioneered the use of relatedness ratios for flowering plant seeds. Their analyses derive from the basic construct of "Hamilton's rule," which circumscribes the conditions under which natural selection will favor the evolution of altruistic behavior. In essence, the fitness cost (c) of an altruistic behavior by an individual (multiplied by its relatedness to itself, $r_{a,a}$) must be less than the benefit to a relative (b) multiplied by the relatedness of the altruist to the beneficiary ($r_{a,b}$),

$$r_{a,b}b > r_{a,a}c. \tag{1}$$

Thus, the benefit-to-cost ratio must be greater than the inverse of the relatedness ratio for natural selection to favor the expression of an altruistic behavior,

$$\frac{b}{c} > \frac{r_{a,a}}{r_{a,b}}. \tag{2}$$

In the case of endosperm and its nourishing relationship with its compatriot embryo, Hamilton's rule can be extended (Queller 1989, 1994) to take into account that the "actor" in a flowering plant seed is the individual endosperm, whose acquisition of nutrients from the maternal sporophyte affects the fitness of the embryo in its own seed as well as the fitness of embryos in other seeds. Under conditions of resource limitation, additional provisioning of an endosperm for its compatriot embryo (ce) will come at the cost of resources that another endosperm could have acquired from the maternal sporophyte for its embryo (other embryo [oe]). Thus, relatedness (r) in the standard equation (2) becomes the relatedness of the actor (embryo-nourishing tissue, endosperm) to the embryo whose fitness it affects. This leads to

$$\frac{b}{c} > \frac{r_{ent\text{-}ce}}{r_{ent\text{-}oe}}, \tag{3}$$

where $r_{ent\text{-}ce}$ is the relatedness of an embryo-nourishing tissue (ent), endosperm, to its compatriot embryo (within the same seed) and $r_{ent\text{-}oe}$ is the relatedness of the same endosperm to another embryo on the same maternal sporophyte.

A larger relatedness ratio ($r_{ent\text{-}ce}/r_{ent\text{-}oe}$) value is indicative of a higher degree of relatedness of an endosperm to its compatriot embryo relative to its relatedness to other embryos on a maternal sporophyte. The larger the relatedness ratio ($r_{ent\text{-}ce}/r_{ent\text{-}oe}$), the greater the conflict between maternal and paternal interests (interparental conflict) and/or between maternal parent and offspring interests (parent-offspring conflict)—and the more selfish or aggressive an endosperm is predicted to be in garnering resources for its own embryo at a cost to the fitness of other embryos and, ultimately, to the maternal sporophyte. In essence, relatedness ratios ($r_{ent\text{-}ce}/r_{ent\text{-}oe}$) reveal a tipping point (threshold) at which the benefit-to-cost ratio (from the perspective of an individual endosperm and its compatriot embryo) will favor the termination of maternal sporophyte investment into an endosperm (cost to the endosperm and its compatriot embryo), with consequent supplemental investment into another seed or seeds (benefit to related endosperm and its compatriot embryo).

The original relatedness ratio analyses for angiosperm seeds (Westoby and Rice 1982; Queller 1983, 1989, 1994; Willson and Burley 1983; Bulmer 1986; Haig 1986, 1987; Haig and Westoby 1988, 1989a, 1989b) were typically calcu-

lated for triploid endosperms derived from monosporic female gametophytes. We have extended these analyses to all seven types of endosperm genetic constructs found among extant flowering plants (table 2). We have also examined the effects on relatedness ratios of seeds sired by pollen derived from a single paternal sporophyte unrelated to the maternal sporophyte and under conditions of self-fertilization (table 2).

As revealed in table 2, evolutionary transitions to polyspory, as well as increases in the maternal genomic contribution to the central cell (and endosperm), serve to increase the relatedness of the endosperm to the maternal sporophyte (and vice versa) and decrease the ratio of the relatedness of the endosperm (embryo-nourishing tissue) to its own compatriot embryo ($r_{ent\text{-}ce}$) versus its relatedness to other embryos ($r_{ent\text{-}oe}$) on the maternal sporophyte ($r_{ent\text{-}ce} : r_{ent\text{-}oe}$). The effect of such changes in endosperm genetics is to decrease the selfish or aggressive behavior of an individual endosperm and increasingly coalign the resource allocation strategy of an endosperm with that of the maternal sporophyte (decreased parent-offspring and/or interparental conflict). Conflict over resource allocation strategy disappears when the relatedness ratio $r_{ent\text{-}ce} : r_{ent\text{-}oe}$ is 1.0, a condition found only in the tetrasporic decapentaploid endosperm of *Peperomia hispidula*, or when the maternal sporophyte substitutes her own tissue (perisperm) for an endosperm to nourish her offspring. Interestingly, if not enigmatically, *Peperomia* species have a minimally developed endosperm and use perisperm as the major embryo-nourishing tissue within a seed.

Effects of Biparental Inbreeding and Self-Fertilization

The relative magnitude of effects predicted by the heterozygosity and conflict hypotheses changes when populations are less than completely outbreeding (table 2). Our calculations show that the relatedness ratio $r_{ent\text{-}ce} : r_{ent\text{-}oe}$ decreases for any type of endosperm genetic construct when two seeds are sired by pollen from the same paternal sporophyte and further decreases when self-fertilization occurs. Consequently, under conditions of inbreeding, genetic conflict is diminished (lower relatedness ratios), ploidy predictions are unaltered, and the fitness benefits of heterozygosity assume greater importance (as an arbiter of deleterious effects of inbreeding). However, under long-term inbreeding, the benefits of increased heterozygosity and of reduced conflict are small, because there is little allelic variation left in the population. Thus, the advantages of evolutionary transitions to bispory and tetraspory should be greatest in highly outcrossing populations that are undergoing transitions to inbreeding or that typically experience periodic bouts of inbreeding. Bispory and tetraspory provide both a buffer from the effects of inbreeding and an escape from the effects of conflict.

Predictions of Homoplasy

Although the heterozygosity hypothesis, ploidy hypothesis, and conflict hypotheses are essentially independent explanations of the potential fitness consequences of changes in endosperm genetic constructs, each of these theories is coaligned in the predictions it makes about trends that should emerge in the evolutionary history of female gametophyte development and downstream endosperm genetics. Endosperm hybrid vigor (heterozygosity) increases in parallel with transitions from monospory to bispory to tetraspory, just as the relative relatedness of an endosperm to its own embryo versus other embryos decreases.

An important caveat to bear in mind is that the predictions of the heterozygosity hypothesis, ploidy hypothesis, and conflict hypotheses do not operate within a vacuum. Predicted trends need not occur inevitably. However, if patterns of increasing heterozygosity, increasing ploidy, and decreasing relative relatedness of the endosperm to its own embryo versus other embryos are selectively favored, considerable homoplasy should be apparent in endosperm genetic constructs among flowering plants. Hence, we predict that homoplasy should also be apparent in mature female gametophyte structure as well as underlying patterns of development.

Relating Female Gametophyte Evo-Devo to Endosperm Genetics: Heterochrony and the Modular Nature of the Angiosperm Female Gametophyte

Variation in the genetic constitution of the target of the second fertilization event (the central cell) directly affects the genetic constitution of the subsequently formed endosperm. The question is, how do innovations in female gametophyte development create new genetic constructs in the central cell?

Developmental analyses of four-nucleate, four-celled female gametophytes in the ancient angiosperm lineages Nymphaeales and Austrobaileyales led to the insight that the female gametophytes of most, if not all, flowering plants are likely to be fundamentally modular entities (Friedman and Williams 2003, 2004) composed of quartets of nuclei (sensu Favre-Duchartre 1977; Battaglia 1989; Haig 1990). The basic, and plesiomorphic, angiosperm female gametophyte developmental module proceeds through three critical ontogenetic stages: (1) positioning of a single nucleus within a developmentally autonomous cytoplasmic domain of the female gametophyte, (2) two free-nuclear mitoses to yield four nuclei within that domain, and (3) partitioning of three uninucleate cells adjacent to the pole such that the fourth nucleus is confined to the central cell of the female gametophyte (fig. 2; Friedman and Williams 2003). We wish to be explicit as to what we mean by the term "module" in its various biological manifestations. The "morphological" sense of a module makes

reference to a static adult structure (terminal ontogenetic stage). A "developmental" module, in this case, refers to the compartmentalized ontogenetic events within a cytoplasmically autonomous domain of the female gametophyte.

Variation in mature angiosperm female gametophyte structure results from developmental differences in one or more of three basic aspects of ontogeny: (1) early ontogenetic events that determine the number of developmental modules initiated (one, two, or four) in a female gametophyte, (2) relative timing of the establishment of female gametophyte modules (after or during megasporogenesis), and (3) ontogenetic events that result in developmental deviations from the plesiomorphic female gametophyte developmental module.

Evolution of Module Number in Angiosperm Female Gametophytes: Ploidy and Conflict Consequences

The key to evolutionary transitions between angiosperm female gametophytes with different numbers of developmental (and hence, morphological) modules lies in the modification of early developmental events to position nuclei within one, two, or four cytoplasmic and developmentally autonomous domains (Friedman and Williams 2003). In Nymphaeales and Austrobaileyales (fig. 3; Schisandra-type), a single modular domain is established by the functional megaspore nucleus at the micropylar pole of the female gametophyte. The chalazal region of the female gametophyte remains "unfilled" throughout ontogeny. Subsequent cellularization yields a three-celled egg apparatus, while the fourth nucleus is contributed to the central cell (fig. 3; Friedman and Williams 2003).

In angiosperms with a Polygonum-type female gametophyte (fig. 3), the uninucleate functional megaspore divides mitotically to produce two daughter nuclei that migrate to opposite poles (domains). Each of these nuclei initiates an independent developmental module that produces four free nuclei (for a total of eight free nuclei). At the eight-nucleate stage, cytokinesis partitions three nuclei into three cells at each pole, while the remaining free nucleus from each of the two modular quartets is contributed to the common cytoplasm of the central cell. Thus, after cellularization, the Polygonum-type female gametophyte is seven-celled and eight-nucleate. Likewise, early developmental establishment of four cytoplasmic domains

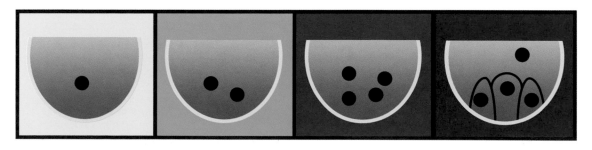

Fig. 2 The plesiomorphic angiosperm developmental module. A single nucleus within a cytoplasmically autonomous region of a female gametophyte undergoes two free-nuclear divisions to yield four free nuclei. Three of these nuclei are partitioned into uninucleate cells, and the fourth nucleus is contributed to the common cytoplasm of the central cell, where it will contribute to the second fertilization event to initiate endosperm. The yellow background indicates the time of module establishment, green indicates a two-nucleate module, blue indicates a four-nucleate module, and red indicates a mature cellularized module.

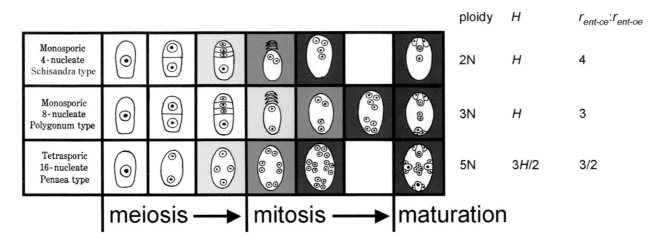

Fig. 3 An example of how changes in module number in an angiosperm female gametophyte can affect the mature structure and nuclear contents of the central cell. In these three cases, a plesiomorphic module ontogeny is apparent. In the Schisandra-type (Nuphar-type) female gametophyte, a single developmental module is initiated and yields a three-celled egg apparatus and a uninucleate central cell (hence, diploid endosperm). In the Polygonum-type, two developmental modules are established at opposite poles of the female gametophyte. The result is a three-celled egg apparatus (derived from the micropylar module), a set of three antipodals (derived from the chalazal module), and a binucleate central cell (hence, triploid endosperm) with a nuclear contribution from each of the two modules. In the Penaea-type, four developmental modules are initiated, and a four-nucleate central cell forms (hence, pentaploid endosperm). As previously discussed (Friedman and Williams 2003), initiation of "extra" modules in the female gametophyte has the effect of increasing the ploidy of endosperm. In addition, the ratio of endosperm relatedness to its own embryo compared with other embryos ($r_{ent-ce} : r_{ent-oe}$) is diminished with increased nuclear contributions to the endosperm by the female gametophyte. This ratio is shown for panmictic outcrossing only (calculations for other mating systems can be found in table 2). The yellow background indicates the time of module establishment, green indicates two-nucleate modules, blue indicates four-nucleate modules, and red indicates mature cellularized modules. The micropylar pole is up, and the chalazal pole is down.

can yield a 16-nucleate (e.g., *Gunnera* and *Penaea*; Stephens 1909; Samuels 1912; Arias and Williams forthcoming) structure consisting of four modules (fig. 3; Penaea-type).

We have speculated (Friedman and Williams 2004) that differences in patterns of nuclear positioning and cytoplasmic regionalization at early ontogenetic stages may hold the key to understanding and predicting the number of modules that are initiated in angiosperm female gametophytes. In essence, the number of modules initiated can be viewed as establishing the body plan (bauplan) of the embryo sac.

To demonstrate how changes in module number can directly affect the genetic constitution of endosperm, we have compared the ontogenies of Schisandra-type (Yoshida 1962; Friedman et al. 2003), also referred to as the Nuphar-type (Williams and Friedman 2002; one plesiomorphic developmental module), Polygonum-type (two plesiomorphic developmental modules), and Penaea-type (four plesiomorphic developmental modules) female gametophytes (fig. 3). Since each plesiomorphic embryo sac module (fig. 2) contributes a single haploid polar nucleus to the central cell, changes in the number of modules initiated directly alter the ploidy of the resulting central cell and endosperm. In addition to modified endosperm ploidy, increases in the relative maternal genomic contributions to the central cell and endosperm also drive the critical ratio $r_{ent-ce} : r_{ent-oe}$ (under conditions of outcrossing) from 4, in the case of a diploid endosperm, to 3/2, in the case of a pentaploid endosperm in angiosperms with a Penaea-type female gametophyte. As discussed above, the closer this ratio is to 1, the more closely aligned the behavior of an endosperm should be to that which maximizes the reproductive

fitness of the maternal sporophyte in terms of which progeny should be provisioned and which should be aborted (under resource-limited conditions). Finally heterozygosity in endosperms derived from four-module female gametophytes is higher than that in endosperms derived from two modules. This is a by-product of the correlated tetrasporic condition, which ensures that both maternal alleles at each locus will be incorporated into the endosperm.

Assuming that the predictions of the endosperm ploidy hypothesis, heterozygosity hypothesis, and conflict hypotheses are correct, increases in the number of female gametophyte developmental modules should have been favored over the course of angiosperm evolutionary history. The benefits of higher module number should be manifest in trends toward increased endosperm ploidy, increased maternal-to-paternal genomic ratios, increased relatedness of the maternal sporophyte to the endosperms contained within its seeds, and diminished conflict through decreased ratios of relatedness of endosperm to its compatriot embryo versus other embryos (hence, a less "selfish" endosperm).

Relative Timing of Module Establishment: Heterochronic Effects and Genetic Outcomes

The overwhelming majority of angiosperm species produce female gametophytes that are monosporic and initiate two developmental modules (the Polygonum-type). The establishment of module identity in these female gametophytes occurs after megasporogenesis and the first free-nuclear mitotic division of the embryo sac (fig. 4, Polygonum-type, yellow box).

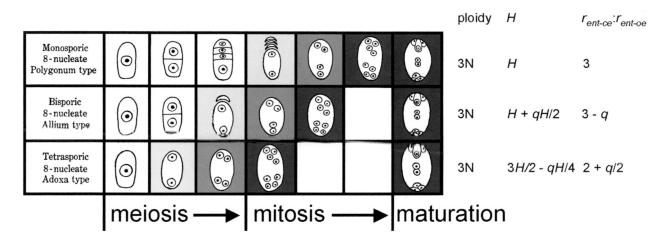

Fig. 4 An example of how heterochronic changes in relative timing of module establishment in an angiosperm female gametophyte can affect the genetic relationships of the nuclear contents of the central cell. In these three cases, a plesiomorphic module ontogeny is apparent, and a two-module, seven-celled, eight-nucleate female gametophyte is formed. The result of pushing module initiation to the end of meiosis II (Allium-type) or meiosis I (Adoxa-type) is to decrease the ratio of endosperm relatedness to its own embryo over that with other embryos ($r_{ent-ce} : r_{ent-oe}$; shown for panmictic outcrossing only; calculations for other breeding systems can be found in table 2) and increase the heterozygosity of endosperm. In these three examples, central cell ploidy and endosperm ploidy are unchanged. The yellow background indicates the time of module establishment, green indicates two-nucleate modules, blue indicates four-nucleate modules, and red indicates mature cellularized modules. The micropylar pole is up, and the chalazal pole is down.

Thus, all nuclei within mature monosporic two-module female gametophytes are derived from a single megaspore (the functional megaspore) and are mitotic relatives; the egg and the two polar nuclei are genetically identical.

In contrast to Polygonum-type female gametophytes, all other types of angiosperm female gametophytes establish developmental modules during or at the completion of megasporogenesis. Monosporic one-module female gametophytes initiate their single developmental module at the end of meiosis II from a cytoplasmic domain that contains the single functional megaspore nucleus (fig. 3, Schisandra-type, yellow box). All bisporic and tetrasporic embryo sacs, with the exception of Adoxa-type female gametophytes, establish the basic number of modules at the end of meiosis II (figs. 4, 5). In Adoxa-type embryo sacs, the two developmental modules are established at the end of meiosis I (fig. 4, yellow box).

Figure 4 shows three ontogenetic paths that result in mature (two-module) seven-celled, eight-nucleate female gametophytes. In essence, these three types of female gametophytes are structurally identical at maturity. What differs in each case is the point of establishment of the two modules (fig. 4, yellow boxes) relative to the process of megasporogenesis and free-nuclear mitotic development of the female gametophyte. As a consequence of heterochronic shifts in module initiation, the genetic relatednesses of the polar nuclei to each other and to the egg cell differ. In the Polygonum-type embryo sac (fig. 4), the polar nuclei are genetically identical. In the Allium-type female gametophyte (fig. 4), the two polar nuclei are derived from meiosis II of a single dyad (and hence are related by $1 - q$, where q is the frequency of second-division segregation; table 1). In the Adoxa-type female gametophyte (fig. 4), the two polar nuclei are derived from separate dyads and are related by $q/2$ (table 1). The relatedness of the polar nuclei to each other and to the egg will be strongly influenced by the second-division segregation rate, which is maximal when genes are far from the centromere.

Overall, changes in the developmental timing of module establishment (fig. 4; table 2) have profound consequences for both levels of endosperm heterozygosity (higher in bisporic than in monosporic and higher in tetrasporic than in bisporic) and ratios of relatedness of an endosperm to its own embryo to relatedness with other embryos on a maternal sporophyte (polyspory drives down the selfishness of an endosperm and diminishes parent-offspring and interparental conflict). Thus, both the heterozygosity hypothesis and the conflict hypotheses predict evolutionary developmental trends toward earlier establishment of modules and the possibility of considerable homoplasy in these transitions among flowering plants.

Effects of Developmental Changes to Plesiomorphic Module Ontogeny

The plesiomorphic angiosperm module is initiated with a single nucleus within a cytoplasmically autonomous zone of the female gametophyte. Two free-nuclear divisions ensue, and maturation of the module is accomplished through the partitioning of three of the four nuclei into uninucleate, parietally positioned cells. The fourth nucleus of a plesiomorphic module is contributed to the common cytoplasm of the central cell (fig. 2). Considerable structural diversity among angiosperm female gametophytes can be attributed to developmental deviations from this plesiomorphic pattern. These modifications fall into three distinct classes: early ontogenetic changes that affect the establishment of modules, heterochronic modifications that lead to the maturation of juvenilized modules, and heterotopic modifications that alter cell fate in the mature module such that additional nuclei are contributed to the central cell (and, ultimately, the endosperm).

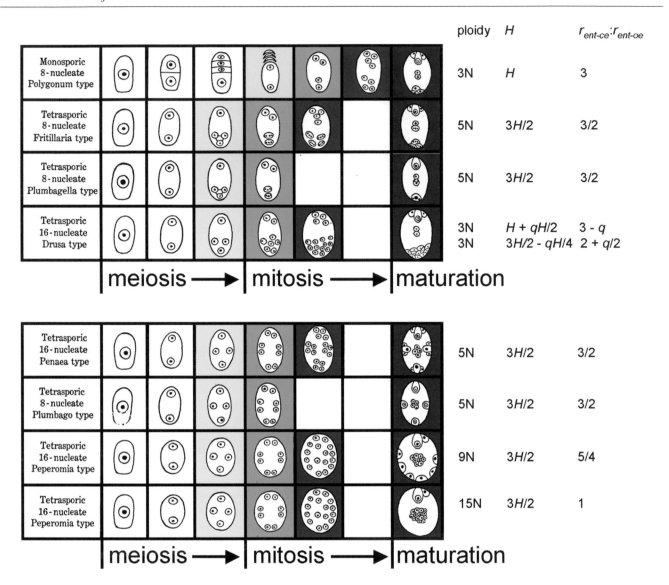

Fig. 5   Examples of how changes in the plesiomorphic module development in an angiosperm female gametophyte can affect the mature structure and nuclear contents of the central cell. In the top panel, the Polygonum-type female gametophyte (with plesiomorphic modules) is compared with other types of two-module female gametophytes with apomorphic module developmental patterns. Both Fritillaria- and Plumbagella-type female gametophytes initiate a chalazal module with three haploid nuclei that produce two triploid restitution nuclei (early ontogenetic modification). In addition, Plumbagella-type female gametophytes cellularize at the two-nucleate stage of module ontogeny (paedomorphosis). The bottom panel shows apomorphic module ontogenies in female gametophytes with four modules. The developmental and genetic consequences of the apomorphic modules are shown for panmictic outcrossing only (calculations for other breeding systems can be found in table 2). The yellow background indicates the time of module establishment, green indicates two-nucleate modules, blue indicates four-nucleate modules, and red indicates mature cellularized modules. The micropylar pole is up, and the chalazal pole is down.

Among the most enigmatic and remarkable female gameto-phyte types are those in which three haploid nuclei enter into a common division that results in the production of two trip-loid nuclei, the so-called restitution nuclei. This pattern of de-velopment is found in Fritillaria- and Plumbagella-type female gametophytes and is almost certainly highly homoplasious among angiosperms (found in Piperaceae, Tamaricaceae, Lilia-ceae, Plumbaginaceae, Asteraceae, and Cornaceae; Maheshwari 1950; Haig 1990; Johri et al. 1992). Female gametophytes that form triploid restitution nuclei initiate two developmental modules from a tetrasporic coenomegaspore (fig. 5, Fritillaria-type and Plumbagella-type, yellow boxes). However, the cha-lazal module contains three nuclei instead of the usual one nucleus (thus, an early modification of module patterning). The transition to the two-nucleate stage of module ontogeny in-volves a mitotic division of the single-nucleate micropylar mod-ule and the formation of two triploid restitution nuclei in the chalazal module. The net result of this apomorphic ontogeny is the production of a triploid polar nucleus in the chalazal mod-ule, increased central cell and endosperm ploidy, and decreased relative relatedness of an endosperm to its own embryo versus other embryos on a maternal sporophyte (diminished conflict).

Heterochronic truncation of module ontogeny can be found in two types of female gametophytes, the Plumbagella-type and the Plumbago-type (fig. 5). In these female gametophytes, each developmental module yields two (Plumbagella-type) or four (Plumbago-type) two-nucleate modules. Cellularization of each two-nucleate module creates a single parietal cell, with the other nucleus being contributed to the central cell. These paedomorphic modules do not alter the ploidy, heterozygosity, or relative relatedness values involved in reproduction relative to their presumed ancestral types (Fritillaria-type in the case of Plumbagella-type, Penaea-type in the case of Plumbago-type).

Ontogenetic modifications late in module development can have a profound effect on the genetics of endosperm. This is most apparent in the Piperaceae, where different species of *Peperomia* have been reported to contribute as many as 14 nuclei to the central cell and produce a decapentaploid endosperm (Johnson 1914). Peperomia-types of female gametophytes are tetrasporic and initiate four developmental modules that yield 16 total nuclei (fig. 5). Rather than partition three nuclei into uninucleate cells and contribute the fourth nucleus to the central cell, in certain *Peperomia* species, these modules alter cell/nucleus position (and hence fate) and contribute two free nuclei per module to the central cell (nonaploid endosperm). In the most extreme case (*Peperomia hispidula*), three of the four developmental modules contribute all four nuclei to the central cell, while the egg-producing module contributes two nuclei to the central cell and produces an egg and a single synergid.

The major consequence of heterotopic changes that alter cell/nucleus position and fate within modules in *Peperomia* is a significant increase in the ploidy of the resultant endosperm, which, in turn, drives down the relative relatedness (selfishness) of an endosperm to its own embryo as compared with its relatedness to other embryos ($r_{\text{ent-ce}} : r_{\text{ent-oe}}$). Altered developmental identity of what were ancestrally somatic female gametophyte nuclei to gametic female gametophyte (central cell) nuclei results in increased ploidy and decreased relatedness ratios but does not change endosperm heterozygosity relative to the presumed ancestral female gametophyte type (Penaea-type). As noted above, endosperm in *Peperomia* plays only a minor role in nourishing the embryo; perisperm (maternal sporophyte tissue) serves as the principal embryonourishing tissue.

Why Are Monosporic Triploid Endosperms So Common among Flowering Plants?

The heterozygosity, ploidy, and conflict hypotheses each predict that evolutionary transitions to higher levels of endosperm heterozygosity and ploidy and diminished levels of genetic conflict should be selectively favored. Our analyses of the underlying female gametophyte developmental patterns that generate changes in central cell genetic constitution demonstrate how such variation in endosperm genetic profiles can evolve. Therefore, bisporic and tetrasporic higher-ploidy endosperms (with their correlated female gametophytes) might

be expected to be common among angiosperms. Thus, it is reasonable to ask why well over 80% (Palser 1975; this is likely to be a conservative estimate) of extant angiosperm species produce a monosporic triploid endosperm with relatively low levels of heterozygosity and ploidy and relatively high levels of interparental and/or parent-offspring conflict. To counter this seeming paradox, it is essential to identify when, in the evolutionary history of a clade, a character evolved, as well as the number of times a type of character transition occurred (homoplasy).

The plesiomorphic condition for the angiosperm female gametophyte is unresolved (Friedman 2006b). Nevertheless, the Polygonum-type female gametophyte evolved no later than the common ancestor of monocots, eudicots, and eumagnoliids (Williams and Friedman 2004), roughly 12 million years after the origin of flowering plants, and may have been present in the common ancestor of all angiosperms (Friedman 2006b). As such, monosporic triploid endosperm retains the benefit of position in that the clade defined by its origin (all angiosperms or all angiosperms except *Amborella*, Nymphaeales, and Austrobaileyales) includes no less than 99% of the quarter-million extant angiosperm species. Explicit examination of character transitions among angiosperms indicates that if a Polygonum-type female gametophyte was present in the common ancestor of all angiosperms, triploid monosporic endosperm evolved only once in the entire history of flowering plants. If the Schisandra-type (Nuphar-type) female gametophyte that yields a monosporic diploid endosperm is plesiomorphic for flowering plants, monosporic triploid endosperm will have evolved only twice (once in the ancestor of *Amborella* and once in the common ancestor of monocots, eudicots, and eumagnoliids). Interestingly, such a homoplastic transition from diploid to triploid endosperms is predicted by the ploidy and conflict hypotheses previously discussed.

Homoplasy has long been viewed as evidence of potential adaptation or the product of natural selection (Wake 1991; Armbruster 1996). While the total number of angiosperm species with bisporic and tetrasporic higher-ploidy endosperms is relatively modest, it is clear that transitions away from the monosporic condition have occurred many times. To date, evolutionary transitions from polyspory to monospory (except for pseudomonosporic cases) or from higher endosperm ploidy to lower endosperm ploidy (with the exception of the Onagraceae) appear to be exceedingly rare. From this perspective, the many homoplastic origins of bisporic and tetrasporic female gametophytes and higher-ploidy central cells and, importantly, the unidirectional nature of these evolutionary transitions, should be taken as strong evidence that the predictions of the heterozygosity, ploidy, and conflict hypotheses have been borne out over the course of flowering plant evolutionary history.

The "Origin of the Fittest" and the "Survival of the Fittest": The Relation between Female Gametophyte Development and Endosperm Genetics

Edward Cope, the late nineteenth-century neo-Lamarckian, first coined the phrase "origin of the fittest" to describe his focus on the then mysterious source of variation on which natural selection was hypothesized to act (Cope 1887). It is, however, an apt phrase to use in the context of our analysis. The relation between female gametophyte development (and its variation) and endosperm (and its genetic constitution) should be seen as one between the origin of structural novelty (variation in mature female gametophyte construction) and its downstream consequences on the relative fitness (survival) of these novelties, as expressed in the biology of endosperm.

For more than a century, biologists have viewed the structural diversity of angiosperm female gametophytes as trivial variants of the reproductive process. We hope that we have demonstrated that this diversity is the direct result of developmental evolution that is tightly linked to the genetic constitution of endosperm. Variation in female gametophyte developmental patterns represents the raw material that ultimately generates variation in endosperm genetics. Variation in endosperm genetic patterns presents a diversity of functional phenotypes that are subject to natural selection. Thus, culling among the possible variants in female gametophyte development will be a function not of selection directly acting on the phenotypes of female gametophytes but rather of selection on the resultant phenotypes of endosperms.

After 130 million years of angiosperm evolution and diversification, we can discern seven basic genetic types of endosperm: monosporic diploid, monosporic triploid, bisporic triploid, tetrasporic triploid, tetrasporic pentaploid, tetrasporic nonaploid, and tetrasporic decapentaploid. Each of these distinctive types of endosperm has arisen through developmental modifications of female gametophyte ontogeny. In many cases, there is compelling evidence for the homoplastic origins of a particular mature female gametophyte type and its linked endosperm genetic type (e.g., Fritillaria-type and tetrasporic pentaploid endosperm). In several cases, the evolution of different apomorphic female gametophyte ontogenies (e.g., Fritillaria-type and Penaea-type) has led to identical endosperm genetic constructs (e.g., tetrasporic pentaploid). In any case, we hope that we have shown that embedded in figures such as those of Maheshwari (1950; fig. 1) is a wealth of information waiting to be connected to a world of developmental biology, genetic theory, and the amazing diversity of plant mating systems.

Acknowledgments

We thank Pamela Diggle, Paula Rudall, and an anonymous reviewer for excellent suggestions for the improvement of this manuscript. This work was supported by grants from the National Science Foundation to W. E. Friedman (IOB-0446191) and J. H. Williams (DEB-0640792).

Appendix

Expected Endosperm Heterozygosity

Table A1 shows the gamete (diploid central cell and haploid sperm) frequencies in a population and the probabilities of all six possible genotypic classes that can arise when the central cell is polyploid and derived from more than one megaspore. The diploid central cell frequencies are written as the product of haploid egg and sperm gamete frequencies in the previous generation (here, generation $n - 1$) and/or the second-division segregation rate, q, in the present generation, n. This assumes Hardy-Weinberg (HW) fertilization probabilities in generation $n - 1$. In the case of inbreeding, one should substitute observed sporophyte genotypic frequencies for the HW theoretical frequencies given in the central cell gamete frequency column. Diploid central cell gamete frequencies are equivalent to female sporophytic genotypic frequencies in the case of five-ploid or higher tetrasporic endosperms because all four megaspores are represented in the central cell. However, in bisporic 3N and tetrasporic 3N endosperms, only two megaspores are represented; hence, average central cell heterozygosity will always be reduced from the maximum possible because of recombination.

If a population is randomly mating, then one can assume that the haploid sperm and egg allele frequencies in generation $n - 1$ are the same as haploid sperm allele frequencies in the current generation, n, and thus, $x_{n-1} = x_n$ and $y_{n-1} = y_n$. For bisporic 3N and tetrasporic 3N endosperms, table A1 shows that only two of the six classes are homozygous, so the probability of heterozygosity can be written as one minus the probability that the two alleles in a dyad (bisporic 3N),

or in separate dyads (tetrasporic 3N), are the same. Then from table A1, for bisporic triploid endosperm,

$$H = 1 - [x^3 + x^2y(1 - q)] - [y^3 + y^2x(1 - q)], \quad (A1)$$

$$H = 1 - x^3 - y^3 - xy(1 - q), \quad (A2)$$

$$H = 2xy + qxy, \quad (A3)$$

which is equivalent to

$$H = H_e + \frac{qH_e}{2}, \quad (A4)$$

where H_e is the expected HW heterozygosity of embryos and $0 \leq q \leq 2/3$.

In some tetrasporic female gametophytes (Adoxa-type), only two nuclei are contributed to the central cell and to endosperm (tetrasporic 3N). These two maternal nuclei apparently derive from different dyads. From table A1, for tetrasporic triploid endosperm,

$$H = 1 - \left[x^3 + x^2y\left(\frac{q}{2}\right)\right] - \left[y^3 + y^2x\left(\frac{q}{2}\right)\right], \quad (A5)$$

$$H = 1 - x^3 - y^3 - xy\left(\frac{q}{2}\right), \quad (A6)$$

$$H = 3xy - xy\left(\frac{q}{2}\right), \quad (A7)$$

which is equivalent to

$$H = \frac{3H_e}{2} - \frac{qH_e}{4}, \tag{A8}$$

where H_e and q are as in equation (A4).

For the cases when all four megaspores are represented in the central cell, recombination does not affect central cell heterozygosity, and thus,

$$H = 1 - x^3 - y^3, \tag{A9}$$

$$H = 3xy, \tag{A10}$$

$$H = \frac{3H_e}{2}. \tag{A11}$$

Expected endosperm heterozygosity in endosperms derived from monosporic female gametophytes is equivalent to that of the embryo. Thus, heterozygosity of monosporic (diploid and triploid) endosperm is equivalent to

$$H = 2xy = H_e. \tag{A12}$$

Table A1

Expected Heterozygosity of Endosperm Derived from Polysporic Female Gametophytes

Central cell genotypes	Sperm genotypes				
	Genotype A_1 (frequency $= x_n$)		Genotype A_2 (frequency $= y_n$)		
	Endosperm class	Frequency	Endosperm class	Frequency	
A_1A_1[a]	$A_1A_1A_1$	$(x_{n-1}^2 + x_{n-1}y_{n-1}S_{\text{hom}})x_n$	$A_1A_1A_2$	$(x_{n-1}^2 + x_{n-1}y_{n-1}S_{\text{hom}})y_n$	
A_1A_2[b]	$\underline{A_1A_2A_1}$	$2x_{n-1}y_{n-1}S_{\text{het}}(x_n)$	$\underline{A_1A_2A_2}$	$2x_{n-1}y_{n-1}S_{\text{het}}(y_n)$	
A_2A_2[c]	$\underline{A_2A_2A_1}$	$(y_{n-1}^2 + x_{n-1}y_{n-1}S_{\text{hom}})x_n$	$A_2A_2A_2$	$(y_{n-1}^2 + x_{n-1}y_{n-1}S_{\text{hom}})y_n$	

Note. The genotypic frequencies of six endosperm classes are the products of sperm and central cell genotypic frequencies. Here, x and y are the frequencies of alleles A_1 and A_2, respectively, from generation n or $n - 1$. Central cells (endosperm precursor cells) are diploid, and their genotypic frequencies are determined by egg/sperm allele frequencies in generation $n - 1$ and the second-division segregation rate, q. For convenience, the generic term "S" is used to designate a fraction of central cell genotypes that arise from second-division segregation of sporophytic alleles. For bisporic 3N endosperm, $S_{\text{het}} = q$ and $S_{\text{hom}} = (1 - q)$; for tetrasporic 3N endosperm, $S_{\text{het}} = 1 - q/2$ and $S_{\text{hom}} = q/2$; and for all other tetrasporic endosperms, $S_{\text{het}} = 1$ and $S_{\text{hom}} = 0$. Heterozygous endosperm classes are indicated by underlining.

[a] Frequency $= x_{n-1}^2 + x_{n-1}y_{n-1}S_{\text{hom}}$.

[b] Frequency $= 2x_{n-1}y_{n-1}S_{\text{het}}$.

[c] Frequency $= y_{n-1}^2 + x_{n-1}y_{n-1}S_{\text{hom}}$.

Literature Cited

Arias T, JH Williams Forthcoming Embryology of *Manekia naranjoana* (Piperaceae) and the origin of tetrasporic, sixteen-nucleate female gametophytes in Piperales. Am J Bot.

Armbruster WS 1996 Exaptation, adaptation, and homplasy: evolution of ecological traits in *Dalechampia* vines. Pages 227–243 *in* MJ Sanderson, L Hufford, eds. Homoplasy: the recurrence of similarity in evolution. Academic Press, New York.

Battaglia E 1989 The evolution of the female gametophyte of angiosperms: an interpretive key. Ann Bot (Rome) 47:7–144.

Brink RA, DC Cooper 1940 Double fertilization and development of the seed in angiosperms. Bot Gaz 102:1–25.

——— 1947 The endosperm in seed development. Bot Rev 13:423–541.

Bulmer MG 1986 Genetic models of endosperm evolution in higher plants. Pages 743–763 *in* S Karlin, E Nevo, eds. Evolutionary process and theory. Academic Press, New York.

Charnov EL 1979 Simultaneous hermaphroditism and sexual selection. Proc Natl Acad Sci USA 76:2480–2484.

Chiarugi A 1927 Il gametofito femmineo delle angiosperme nei suoi vari tipi di costruzione e di sviluppo. Nuovo G Bot Ital (NS) 34:1–133.

Cook RE 1981 Plant parenthood. Nat Hist 90:30–35.

Cope ED 1887 The origin of the fittest: essays on evolution. Appleton, New York.

D'Amato FD 1984 Role of polyploidy in reproductive organs and tissues. Pages 519–543 *in* BM Johri, ed. Embryology of angiosperms. Springer, New York.

Donoghue MJ, SM Scheiner 1992 The evolution of endosperm: a phylogenetic account. Pages 356–389 *in* R Wyatt, ed. Ecology and evolution of plant reproduction. Chapman & Hall, New York.

Favre-Duchartre M 1977 Eight interpretations of the embryo sac. Phytomorphology 27:407–418.

Fincham JRS 1994 Genetic analysis. Blackwell, London.

Friedman WE 1995 Organismal duplication, inclusive fitness theory, and altruism: understanding the evolution of endosperm and the angiosperm reproductive syndrome. Proc Natl Acad Sci USA 92:3913–3917.

——— 2006*a* Embryological evidence for developmental lability during early angiosperm evolution. Nature 441:337–340.

——— 2006*b* Sex among the flowers. Natl Hist 115:48–53.

Friedman WE, WN Gallup, JH Williams 2003 Female gametophyte development in *Kadsura*: implications for Schisandraceae, Austrobaileyales, and the early evolution of flowering plants. Int J Plant Sci 164(suppl):S293–S305.

Friedman WE, JH Williams 2003 Modularity of the angiosperm female gametophyte and its bearing on the early evolution of endosperm in flowering plants. Evolution 57:216–230.

———— 2004 Developmental evolution of the sexual process in ancient flowering plant lineages. Plant Cell 16(suppl):S119–S132.

Galati BG 1985 Estudios embriologicos en *Cabomba australis* (Nymphaeaceae). I. La esporogenesis y las generaciones sexuadas. Bol Soc Argent Bot 24:29–47.

Galbraith DW, KR Harkins, S Knapp 1991 Systemic endopolyploidy in *Arabidopsis thaliana*. Plant Physiol 96:985–989.

Gifford EM, AS Foster 1987 Morphology and evolution of vascular plants. WH Freeman, New York.

Haig D 1986 Conflict among megaspores. J Theor Biol 123:471–480.

———— 1987 Kin conflict in seed plants. Trends Ecol Evol 2:337–340.

———— 1990 New perspectives on the angiosperm female gametophyte. Bot Rev 56:236–274.

Haig D, M Westoby 1988 Inclusive fitness, seed resources, and maternal care. Pages 60–79 *in* J Lovett Doust, L Lovett Doust, eds. Plant reproductive ecology. Oxford University Press, Oxford.

———— 1989*a* Parent-specific gene expression and the triploid endosperm. Am Nat 134:147–155.

———— 1989*b* Selective forces in the emergence of the seed habit. Biol J Linn Soc 38:215–238.

Härdling R, P Nilsson 1999 Parent-offspring conflict and the evolution of seed provisioning in angiosperms. Oikos 84:27–34.

———— 2001 A model of triploid endosperm evolution driven by parent-offspring conflict. Oikos 92:417–423.

Ishikawa M 1918 Studies on the embryo sac and fertilization in *Oenothera*. Ann Bot 32:277–317.

Johnson DS 1900 The embryo sac of *Peperomia pellucida*. Proc Am Assoc Adv Sci 49:279–280.

———— 1914 Studies of the development of the Piperaceae. II. The structure and seed-development of *Peperomia hispidula*. Am J Bot 1:357–397.

Johri BM, KB Ambegaokar, PS Srivastava 1992 Comparative embryology of angiosperms. McGraw-Hill, New York.

Law R, C Cannings 1984 Genetic analysis of conflicts arising during development of seeds in the angiospermophyta. Proc R Soc B 221:53–70.

Maheshwari P 1950 An introduction to the embryology of angiosperms. McGraw-Hill, New York.

Palser BG 1975 The bases of angiosperm phylogeny: embryology. Ann Mo Bot Gard 62:621–646.

Queller DC 1983 Kin selection and conflict in seed maturation. J Theor Biol 100:153–172.

———— 1989 Inclusive fitness in a nutshell. Oxf Surv Evol Biol 6:73–109.

———— 1994 Male-female conflict and parent-offspring conflict. Am Nat 144(suppl):S84–S99.

Samuels J 1912 Études sur le développement du sac embryonnaire et sur la fécondation du *Gunnera macrophylla* Bl. Arch Zellforsch 8:52–120.

Schnarf K 1929 Embryologie der angiospermen. Borntraeger, Berlin.

Stebbins GL 1974 Flowering plants: evolution above the species level. Harvard University Press, Cambridge, MA.

———— 1976 Seeds, seedlings, and the origin of angiosperms. Pages 300–311 *in* CB Beck, ed. Origin and early evolution of angiosperms. Columbia University Press, New York.

Stephens EL 1909 The embryo sac and embryo of certain Penaeaceae. Ann Bot 23:363–378.

Takhtajan A 1991 Evolutionary trends in flowering plants. Columbia University Press, New York.

Tiffney BH 1981 Diversity and major events in the evolution of land plants. Pages 193–230 *in* KJ Niklas, ed. Paleobotany, paleoecology, and evolution. Praegar, New York.

Tobe H, PH Raven 1986 A comparative study of the embryology of *Ludwigia* (Onagraceae): characteristics, variation, and relationships. Ann Mo Bot Gard 73:768–787.

Wake DB 1991 Homoplasy: the result of natural selection, or evidence of design limitations? Am Nat 138:543–567.

Westoby M, B Rice 1982 Evolution of the seed plants and inclusive fitness of plant tissues. Evolution 36:713–724.

Williams JH, WE Friedman 2002 Identification of diploid endosperm in an early angiosperm lineage. Nature 415:522–526.

———— 2004 The four-celled female gametophyte of *Illicium* (Illiciaceae; Austrobaileyales): implications for understanding the origin and early evolution of monocots, eumagnoliids, and eudicots. Am J Bot 91:332–351.

Willson MF, N Burley 1983 Mate choice in plants: tactics, mechanisms, and consequences. Princeton University Press, Princeton, NJ.

Yoshida O 1962 Embryologische Studien über *Schisandra chinensis* Bailey. J Coll Arts Sci Chiba Univ 3:459–462.

Int. J. Plant Sci. 169(1):93–104. 2008.
1058-5893/2008/16901-0008$15.00   DOI: 10.1086/523362

LOSS OF SELF-INCOMPATIBILITY AND ITS EVOLUTIONARY CONSEQUENCES

Boris Igic,* Russell Lande,† and Joshua R. Kohn†

*Department of Biological Sciences (M/C 067), University of Illinois at Chicago, 840 West Taylor Street, Chicago, Illinois 60607, U.S.A.; and †Section of Ecology, Behavior, and Evolution, Division of Biological Sciences, University of California, San Diego, 9500 Gilman Drive, La Jolla, California 92093, U.S.A.

We review and analyze the available literature on the frequency and distribution of self-incompatibility (SI) among angiosperms and find that SI is reported in more than 100 families and occurs in an estimated 39% of species. SI frequently has been lost but rarely has been gained during angiosperm diversification, and there is no evidence that any particular system of SI, once lost, has been regained. Irreversible loss of SI systems is thought to occur because transitions to self-compatibility (SC) are accompanied by collapse of variation at the S-locus and by accumulation of loss-of-function mutations at multiple loci involved in the incompatibility response. The asymmetry in transitions implies either that SI is declining in frequency or that it provides a macroevolutionary advantage. We present a model in which the loss of SI is irreversible and species can be SI, SC but outcrossing, or predominantly selfing. Increased diversification rates of SI relative to SC taxa are required to maintain SI at equilibrium, while transition rates between states, together with state-specific diversification rates, govern the frequency distribution of breeding-system states. We review empirical studies about the causes and consequences of the loss of SI, paying particular attention to the model systems *Arabidopsis* and *Solanum* sect. *Lycopersicon*. In both groups, losses of SI have been recent and were accompanied by loss of most or all of the functional variation at the S-locus. Multiple loss-of-function mutations are commonly found. Some evidence indicates that mutations causing SC strongly increase the selfing rate and that SC species have lower genetic diversity than their SI relatives, perhaps causing an increase in the extinction rate.

Keywords: self-incompatibility, ancestral state reconstruction, irreversibility, mating system, outcrossing, selfing, S-locus.

> The evolutionary pathway from obligate outcrossing based upon self-incompatibility to predominant self-fertilization has probably been followed by more different lines of evolution in flowering plants than has any other. (Stebbins 1974, p. 51)

Introduction

The vast majority of flowering plants are simultaneous hermaphrodites (Yampolsky and Yampolsky 1922). Despite many potential advantages of self-fertilization (Darwin 1876; Fisher 1941; Baker 1955, 1967; Schoen et al. 1996; Schoen and Busch 2007), most angiosperms possess some mechanism that greatly reduces or prevents it. Current estimates indicate that outcrossing is enforced by self-incompatibility (SI) or dioecy in approximately half of all angiosperm species (Igic and Kohn 2006). Here we are primarily concerned with SI, broadly defined as any postpollination prezygotic mechanism that prevents self-fertilization. We focus on the well-described homomorphic SI mechanisms in which the molecular bases for self-pollen recognition and rejection are at least partially understood (Takayama and Isogai 2005; McClure 2006), the history of a particular mechanism can be traced (Igic and Kohn 2001; Steinbachs and Holsinger 2002; Castric and Vekemans 2004; Igic et al. 2006), and the causes and consequences of its loss can be documented.

We first review the phylogenetic distribution of SI systems among angiosperms. SI systems have arisen many times, though one form of homomorphic SI, the RNase-based gametophytic system found in the Solanaceae, Plantaginaceae, and Rosaceae, is the ancestral state for the majority of dicots (Igic and Kohn 2001; Steinbachs and Holsinger 2002). We show that, while SI clearly has multiple origins, losses of SI vastly outnumber gains. Reconstructing ancestral states and estimating rates of transition between SI and self-compatibility (SC) states has been a goal of many studies (e.g., Weller et al. 1995; Barrett et al. 1996; Igic et al. 2004; Mast et al. 2006; Ferrer and Good-Avila 2007). Here we caution against such practices using the currently available tools. Phylogenetic methods for estimating rates of gain and loss of characters based on the states of extant taxa (e.g., Pagel 1994, 1997) rely on an assumption that differences in character state have no effect on lineage diversification rates (Igic et al. 2006; Maddison 2006). If this assumption is violated, as we show is likely for SI, estimates of transition rates and ancestral states will often be misleading, as has been demonstrated for the Solanaceae (Igic et al. 2004, 2006).

We also review the current state of knowledge about the frequency of SI in angiosperms. Because losses of SI appear far more common than gains, the high frequency of SI among

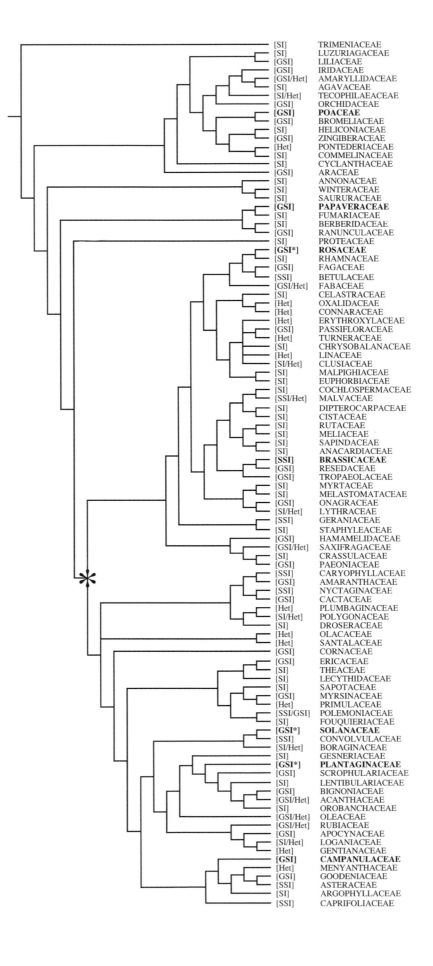

[SI]	TRIMENIACEAE
[SI]	LUZURIAGACEAE
[GSI]	LILIACEAE
[GSI]	IRIDACEAE
[GSI/Het]	AMARYLLIDACEAE
[SI]	AGAVACEAE
[SI/Het]	TECOPHILAEACEAE
[GSI]	ORCHIDACEAE
[GSI]	**POACEAE**
[GSI]	BROMELIACEAE
[SI]	HELICONIACEAE
[GSI]	ZINGIBERACEAE
[Het]	PONTEDERIACEAE
[SI]	COMMELINACEAE
[SI]	CYCLANTHACEAE
[GSI]	ARACEAE
[SI]	ANNONACEAE
[SI]	WINTERACEAE
[SI]	SAURURACEAE
[GSI]	**PAPAVERACEAE**
[SI]	FUMARIACEAE
[SI]	BERBERIDACEAE
[GSI]	RANUNCULACEAE
[SI]	PROTEACEAE
[GSI*]	**ROSACEAE**
[SI]	RHAMNACEAE
[GSI]	FAGACEAE
[SSI]	BETULACEAE
[GSI/Het]	FABACEAE
[SI]	CELASTRACEAE
[Het]	OXALIDACEAE
[Het]	CONNARACEAE
[Het]	ERYTHROXYLACEAE
[GSI]	PASSIFLORACEAE
[Het]	TURNERACEAE
[SI]	CHRYSOBALANACEAE
[Het]	LINACEAE
[SI/Het]	CLUSIACEAE
[SI]	MALPIGHIACEAE
[SI]	EUPHORBIACEAE
[SI]	COCHLOSPERMACEAE
[SSI/Het]	MALVACEAE
[SI]	DIPTEROCARPACEAE
[SI]	CISTACEAE
[SI]	RUTACEAE
[SI]	MELIACEAE
[SI]	SAPINDACEAE
[SI]	ANACARDIACEAE
[SSI]	**BRASSICACEAE**
[GSI]	RESEDACEAE
[GSI]	TROPAEOLACEAE
[SI]	MYRTACEAE
[SI]	MELASTOMATACEAE
[GSI]	ONAGRACEAE
[SI/Het]	LYTHRACEAE
[SSI]	GERANIACEAE
[SI]	STAPHYLEACEAE
[GSI]	HAMAMELIDACEAE
[GSI/Het]	SAXIFRAGACEAE
[SI]	CRASSULACEAE
[GSI]	PAEONIACEAE
[SSI]	CARYOPHYLLACEAE
[GSI]	AMARANTHACEAE
[SSI]	NYCTAGINACEAE
[GSI]	CACTACEAE
[Het]	PLUMBAGINACEAE
[SI/Het]	POLYGONACEAE
[SI]	DROSERACEAE
[Het]	OLACACEAE
[Het]	SANTALACEAE
[GSI]	CORNACEAE
[GSI]	ERICACEAE
[SI]	THEACEAE
[SI]	LECYTHIDACEAE
[SI]	SAPOTACEAE
[GSI]	MYRSINACEAE
[Het]	PRIMULACEAE
[SSI/GSI]	POLEMONIACEAE
[SI]	FOUQUIERIACEAE
[GSI*]	**SOLANACEAE**
[SSI]	CONVOLVULACEAE
[SI/Het]	BORAGINACEAE
[SI]	GESNERIACEAE
[GSI*]	**PLANTAGINACEAE**
[GSI]	SCROPHULARIACEAE
[SI]	LENTIBULARIACEAE
[GSI]	BIGNONIACEAE
[GSI/Het]	ACANTHACEAE
[SI]	OROBANCHACEAE
[GSI/Het]	OLEACEAE
[GSI/Het]	RUBIACEAE
[GSI]	APOCYNACEAE
[SI/Het]	LOGANIACEAE
[Het]	GENTIANACEAE
[GSI]	**CAMPANULACEAE**
[Het]	MENYANTHACEAE
[GSI]	GOODENIACEAE
[SSI]	ASTERACEAE
[SI]	ARGOPHYLLACEAE
[SSI]	CAPRIFOLIACEAE

angiosperm species suggests that SI lineages may often be associated with higher net diversification rates (Igic et al. 2004). We expand an earlier model (Igic et al. 2004) to involve transitions between three states: SI, SC with predominant outcrossing, or predominant selfing. This model can lead to stable maintenance of SI only if it is associated with increased diversification. The balance of transition rates and state-specific diversification rates determines the relative frequency distribution of states.

Once lost, any system of homomorphic SI is difficult to regain, for at least two reasons. First, polymorphism at the S-locus is expected to collapse over time when rendered selectively neutral by the transition to SC. Preexisting polymorphism is needed for the evolution of these systems (Charlesworth and Charlesworth 1979). Second, all well-characterized homomorphic SI systems rely on the coordinated functions of several genes, both linked and unlinked to the S-locus (Takayama and Isogai 2005; McClure 2006). Additional loss-of-function mutations are expected to accumulate after the transition to SC, increasing the difficulty of reversal. We assess the validity of these predictions by reviewing what is known about S-locus polymorphism and the mutations responsible for SC in model systems where SC has recently evolved, the genus *Arabidopsis* and *Solanum* sect. *Lycopersicon* (the wild and cultivated tomatoes). Motivated by the observation that SI is likely to be associated with higher net diversification rates, we also examine the consequences of the transition to SC on the mating system, genomewide genetic variation, and the evolutionary potential of lineages.

The Distribution of SI

At least 100 flowering plant families reportedly contain SI species (fig. 1). This is probably a conservative estimate because the appropriate studies (see Charlesworth 1985) to demonstrate SI are rarely performed or are infrequently reported. The expression of SI cannot be described monolithically because the underlying mechanisms differ widely among groups. Textbook classifications are most often based on the presence or absence of variation in morphology between mutually compatible flowers (heteromorphic or homomorphic) and the genetic mode of action (gametophytic or sporophytic). Less frequently, the site of expression (stigma, style, or ovary) and the number of loci involved in determining the self-recognition phenotype are also reported (de Nettancourt 1977). Presently, species in at least 25 plant families are known to express heteromorphic SI (fig. 1; Ganders 1979; Gibbs 1990; Barrett 1992; Steinbachs and Holsinger 2002), and at least 94 families express some manner of homomorphic SI (Gibbs 1990; Weller

et al. 1995; Steinbachs and Holsinger 2002; B. Igic, unpublished data). Within groups with homomorphic SI, the sporophytic genetic mode of control (SSI), in which the female evaluates the diploid paternal genome of each pollen grain, is found in 10 families. The gametophytic mode of SI (GSI), in which the haploid genome of a pollen grain is evaluated by the female, occurs in 36 families. In 47 families, homomorphic SI is reported, but without sufficient information to ascertain the genetic mode of action. Although heteromorphic and homomorphic SI coexist within 12 families, the classification of homomorphic SI within families is apparently invariant, with the exception of Polemoniaceae (Levin 1993; Goodwillie 1997). Invariance at the family level is often the basis for the assumption that SI is homologous within relatively old monophyletic groups (i.e., families), an assumption that holds true for the few families in which the molecular basis of incompatibility is known and multiple genera have been studied (Rosaceae, Solanaceae, Plantaginaceae, and Brassicaceae; reviewed in Castric and Vekemans 2004). Although classifications such as SSI and GSI can be didactically helpful, families sharing the same form of SI cannot be assumed to do so by homology (Gibbs 1986).

However, in one case, that of homomorphic, RNase/F-box-controlled (McClure 2006) GSI found in three distantly related eudicot families (Rosaceae, Solanaceae, and Plantaginaceae), the shared molecular basis of incompatibility, coupled with the structural and phylogenetic relationships of some of the underlying genes, strongly implies homology (Igic and Kohn 2001; Steinbachs and Holsinger 2002; Ushijima et al. 2003; Qiao et al. 2004; Sijacic et al. 2004). This finding rejects the view (e.g., Weller et al. 1995) that the ancestor of most eudicots (ca. 90–100 million years ago) was SC. Instead, it implies that the common ancestor of ca. 75% of dicots was GSI and that SSI arose independently at least 10 times within higher eudicots (fig. 1; Igic and Kohn 2001; Steinbachs and Holsinger 2002). Furthermore, the phylogenetically scattered occurrence of the families that possess species with heteromorphic SI together with the shared ancestry of GSI in the Rosid and Asterid lineages both strongly suggest that heteromorphic SI evolved at least 22 times in flowering plants (fig. 1; see also Barrett 1992). In addition, homomorphic GSI evolved independently at least four times. Nonhomologous GSI systems, employing different molecular mechanisms, are known to operate outside the Rosid/Asterid clade in the Papaveraceae (Foote et al. 1994) and Poaceae (Baumann et al. 2000). Within the Asterids, the Campanulaceae are now also suspected to possess a form of GSI that does not rely on the RNase/F-box mechanism (S. Good-Avila and A. Stephenson, personal communication). Overall, some form of SI evolved independently at least 35 times in angiosperms.

Fig. 1 Distribution of self-incompatibility (SI) in 105 families of angiosperms. The hypothesis of relationships among families derives from Davies et al. (2004), as implemented in *Phylomatic* (Webb and Donoghue 2005). The status of families was collected from literature searches by Gibbs (1990), Weller et al. (1995), Steinbachs and Holsinger (2002), and B. Igic (unpublished data). SI system designations are encoded as follows: [GSI] = homomorphic gametophytic SI, [SSI] = homomorphic sporophytic SI, [SI] = homomorphic SI, uncharacterized mode of action, and [Het] = heteromorphic SI. Any combination indicates the presence of multiple systems within the designated family. The extent of evidence for SI is variable. Note that Rosaceae, Solanaceae, and Plantaginaceae share the homologous RNase-based GSI mechanism (each is marked by an asterisk, as is the ancestral node). Families with SI systems whose genetic basis is known and molecular basis is at least partially characterized are in bold. Although the molecular basis for Campanulaceae is unknown, it appears not to be controlled by the RNase-based system.

SI to SC Transitions

Although new forms of SI have evolved many times, on simple mechanistic grounds, mutations are far more likely to cause the loss of SI than its gain (reviewed in Stone 2002). The characterized homomorphic SI systems use distinct genes for pollen and pistil recognition phenotypes, and many accessory genes are also necessary for the proper function of SI. Thus loss-of-function mutations at many loci can lead to the breakdown of SI. The loss of SI releases other participating loci from selective pressure, making further degradation of the system likely. By comparison to the loss of SI, the buildup of the requisite pathways and variability for self-recognition and rejection appears considerably more difficult. This makes the loss of SI a likely example of an evolutionary transition that is extremely difficult to reverse (Marshall et al. 1994; Igic et al. 2006). Without exception, large-scale studies performed in SI species find individuals or populations that recently became SC, and many studies show that the breakdown occurred more than once within species (e.g., Rick and Chetelat 1991; Tsukamoto et al. 1999, 2003a, 2003b; Sherman-Broyles et al. 2007; Shimizu et al., forthcoming). By comparison, a convincing case for the recent development of one or more forms of SI within any species would be far more unusual.

In addition to these verbal arguments for why loss of SI should be common and reverse transitions extremely difficult, there is strong genetic evidence that, at the level of at least one large family, the loss of SI has been frequent and irreversible. In the Solanaceae, a family of some 2600 species, of which ca. 40% are SI (Whalen and Anderson 1981; Igic et al. 2006), widespread shared ancestral polymorphism at the S-locus implies an uninterrupted ancestry of SI for all species from which S-alleles have been sampled (for a more complete discussion, see Igic et al. 2006). When inference from shared ancestral polymorphism is used to inform reconstructions of the rates of gain and loss of SI, the loss of SI is found to be effectively irreversible (rate of gain = 0 cannot be rejected; Igic et al. 2006). However, in the absence of information from shared ancestral polymorphism, the opposite and incorrect conclusion is reached, that transitions in both directions have occurred.

Why do reconstructions of transition rates fail when applied to the question of SI when only the character states of extant taxa are used? Recent work suggests that current reconstruction methods (Sanderson 1993; Maddison 1994; Pagel 1994; Schluter et al. 1997) can fail if net diversification rates vary with character state (Igic et al. 2006; Maddison 2006). These methods implicitly assume that the character states being reconstructed do not influence rates of diversification (Igic et al. 2006; Maddison 2006). Simulations show that violation of the assumption of equal diversification rates leads to reconstructions that falsely reject the character transition rates used to simulate the data (Maddison 2006). Reconstructed transition rates overestimate the rate of transition to the character state that confers higher diversification. We show below that there are strong reasons to believe that net diversification rates are higher in SI than in SC taxa. If this is true, reconstructions that ignore differences in diversification rates (e.g., Ferrer and Good-Avila 2007) will falsely inflate the rate of transition from SC to SI. Similar caution applies to reconstructions involving other character state transitions, when alternative states may cause

differential net diversification, including the evolution of selfing versus outcrossing, specialists versus generalists, and sexuality versus asexuality, among many others.

Although the number of gains of SI in angiosperms may appear surprisingly high, losses vastly outnumber gains. For instance, SI has been lost a minimum of 60 times in the Solanaceae (Igic et al. 2006) and eight to nine times in the tomatoes (*Solanum* sect. *Lycopersicon*) alone (fig. 2). Like the Solanaceae, most families appear to have a single homologous mechanism of SI (but see Goodwillie 1997). Within-family homology has also been verified in the Brassicaceae, Plantaginaceae, and Rosaceae, where the molecular basis of incompatibility is known (reviewed in Castric and Vekemans 2004). Each of these families possesses a large and phylogenetically widespread number of SC species (Heilbuth 2000). Consequently, we expect that the pattern of frequent losses of SI is common.

The Frequency of SI

For several reasons, our knowledge of the frequency of SI among angiosperm species is inadequate. The estimate of the frequency of SI (ca. 40%; table 1) comes from 27 published surveys of breeding systems in New World plant communities. This is somewhat lower than an earlier estimate (ca. 50%) by Darlington and Mather (1949), though it is far higher than the representation of SI species in the studies of the distribution of outcrossing rates (Igic and Kohn 2006). In community surveys, sampling schemes and experimental methods differ, with some authors exclusively choosing woody or herbaceous species and often employing different criteria for the classification of SI and SC species. We recoded the data using a consistent cutoff for the value of the index of SI (ISI, the relative success of selfed vs. outcrossed seed or fruit set; Bawa 1974) of 0.2. Changing this value to 0.1 does not substantially change the trends reported below, nor does the application of

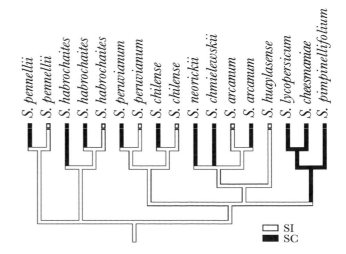

Fig. 2 Evolution of breeding systems in *Solanum* sect. *Lycopersicon*. The phylogenetic hypothesis is modified from Spooner et al. (2005). Data on breeding-system status are derived from Kondo et al. (2002a), Peralta and Spooner (2001), Rick and Chetelat (1991), and B. Igic (unpublished data). SI = self-incompatible; SC = self-compatible.

Table 1

Frequency of Self-Incompatibility (SI) in Different Plant Communities

Flora	n	SI (%)	Habit	Study
Temperate:				
Canadian forest herbs	12	33.3	Herbaceous	Barrett and Helenurm 1987
Canadian salt marsh	17	23.5	Herbaceous	Pojar 1974
Canadian bog	28	25.0	Herbaceous	Pojar 1974
Canadian subalpine meadow	37	40.5	Herbaceous	Pojar 1974
New England shrubs	12	8.3	Woody	Rathcke 1988
North Carolina forest wildflowers	11	27.3	Herbaceous	Motten 1986
Arizona desert	14	92.9	Woody	Neff et al. 1977
Argentine deserts	12	58.3	Woody	Neff et al. 1977
Patagonian alpine flora	124	28.0	Herbaceous[a]	Arroyo and Squeo 1990
Chilean temperate dry forest	37	34.4	Mixed	Arroyo and Uslar 1993
Chilean valvidian forest	39	30.0	Mixed	Riveros et al. 1996
Argentine chaco forest	15	60.0	Mixed	Aizen and Feinsinger 1994
Argentine chaco forest	32	43.8	Mixed	Morales and Galetto 2003
Argentine chaco understory	7	85.7	Woody	Bianchi et al. 2000
Tropical:				
Mexican deciduous forest	33	65.4	Mixed	Bullock 1985
Costa Rican dry forest	34	61.7	Woody	Bawa 1974
Costa Rican lowland forest	64	43.9	Mixed	Kress and Beach 1994
Brazilian Caatinga, semiarid	36	56.7	Mixed	Machado et al. 2006
Brazilian savanna, near Brasilia	30	70.6	Woody	Oliveira and Gibbs 2000
Venezuelan tropical dry forest	49	50.7	Mixed	Jaimes and Ramírez 1999
Venezuelan palm swamp	25	19.4	Mixed	Ramírez and Brito 1990
Venezuelan cloud forest	25	37.0	Mixed	Sobrevila and Arroyo 1982
Island:				
Chiloé Island, Chile	20	45.0	Woody	Smith-Ramírez et al. 2005
Galapagos Islands, Ecuador	51	1.9	Mixed	McMullen 1990
Juan Fernandez Islands, Chile	22	13.6	Mixed	Anderson et al. 2001
Jamaican montane forest	8	9.9	Woody	Tanner 1982
New Zealand	47	18.2	Mixed	Newstrom and Robertson 2005

Note. Studies are considered to be tropical if they were performed between the Tropics of Cancer and Capricorn, and if not, they were considered temperate. Of the island studies, only that of McMullen (1990) is tropical by this criterion.

[a] Two woody species were examined in the study but excluded in our analyses.

sample size weighting, which lowers the frequency estimate to 36.7%.

At least three trends can be gleaned from the published surveys of the frequency of SI in natural plant communities (table 1; fig. 3). First, species on oceanic islands are far less likely to be SI (Wilcoxon test $W = 94$, $P < 0.02$), providing yet another line of evidence for "Baker's Rule," the predicted association of isolated and peripheral habitats with SC (Baker 1955, 1967). This holds true even though the sample size for the islands (five surveys) is small and despite the fact that one island group, the Chiloé Islands, was connected with the mainland for much of the Pleistocene (Villagrán 1988). Second, studies of woody taxa generally find a higher frequency of SI than those focused on herbaceous taxa ($W = 6$, $P = 0.065$; island studies excluded). Because of the possibility of increased rates of somatic mutation, Scofield and Schultz (2006) suggest that woody species may be unable to purge genetic load, perhaps selecting for the maintenance of SI. The trend for lower frequency of SI in herbaceous taxa, however, is highly confounded with the third trend; SI is potentially more common in tropical than in temperate species (fig. 3; $F = 3.253$, $P = 0.086$). This possibility was recognized by Dobzhansky (1950), who posited that repeated glacial advances and successive continental changes between arid and pluvial climates ensured more recent

colonization of temperate areas. Consequently, the higher latitudes could harbor a greater proportion of species with derived opportunistic traits (such as selfing or SC, e.g.), which may provide a temporary advantage at the price of future adaptive potential. Dobzhansky (1950) clearly viewed the sacrifice of outcrossing for reproductive assurance by selfing as a doomed evolutionary gambit that is more prevalent in the temperate zones: "Although some plant species native in the tropics have also become stranded in these evolutionary blind alleys, the incidence of such species is higher in and near the regions which were glaciated" (p. 219). He proposed that there should be a latitudinal gradient in the proportion of outcrossing plants. While his 1950 paper contains few specifics, he is rarely credited with the ideas that appear to have anticipated or coexisted with those of Baker (1955) and Stebbins (1957, 1974), positing that recently colonized areas would contain more selfers, that transitions from SI to SC are common, and that the transition from outcrossing to selfing is an evolutionary dead end.

Several potentially confounded variables, such as the proportions of woody/herbaceous, tropical/temperate, and island/mainland species, are generally unknown. Consequently, our estimate of the frequency of SI, while clearly showing that it is common, will certainly be revised as more information becomes available. Continental- or global-scale databases on species

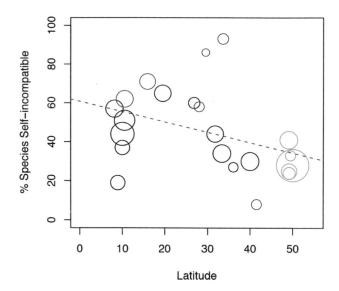

Fig. 3 Relationship between the frequency of self-incompatibility in community surveys and latitude. Circled areas are scaled indicators of sample sizes (see table 1), color coded to reflect the primary makeup of the surveyed floras: green = herbaceous, red = woody, black = mixed.

growth form and range will be helpful in determining these relative proportions. The present data on the frequency of SI in nature are far from definitive, and additional community surveys are badly needed.

The Model

We have shown above that SI is phylogenetically widespread, quite frequent among species, and its loss is irreversible except in the relatively rare instances when novel SI systems arise. Common and essentially irreversible loss of SI implies that SI is either declining in frequency or that it confers a macroevolutionary advantage. We previously presented a deterministic equilibrium model for the stable coexistence of SI and SC taxa (Igic et al. 2004). Here, we expand that model to include two distinct SC groups, those that are predominant outcrossers and those that are selfers. Our model is similar to Nunney's (1989) model for the evolutionary maintenance of sex, and it has similarities with the early portions of the model of Schoen and Busch (2007) for the evolution of self-fertilization in a metapopulation. It assumes exponential growth in the number of species and only allows for irreversible transitions from SI to SC outcrossing and from SC outcrossing to SC selfing, a useful heuristic approach. We do not mean to imply that SC outcrossing taxa inevitably become highly selfing, but as they cannot return to SI, they and their descendants can either remain SC but outcrossing or evolve to become highly selfing. We assume the highly selfing state to be absorbing because, if SI confers an increase in diversification rate relative to both SC groups, it is reasonable that it does so because selfing is often an evolutionary dead end. The advantage of this three-state model over our previous model (Igic et al. 2004) is that it allows for SC taxa that vary in outcrossing rate and that it can also predict the frequency of various mating system states.

Let N_I, N_O, and N_S be the numbers of species with the three character states SI, SC outcrossing, and SC selfing, respectively. Given that each state has an associated exponential rate parameter (r_I, r_O, and r_S) defined for each state as the speciation rate minus the extinction rate, and the only appreciable transitions occur from SI to SC outcrossing (l_I) and from SC outcrossing to SC selfing (l_O), the change in expected numbers in each group over time can be expressed as

$$\frac{dN_I}{dt} = (r_I - l_I)N_I,$$
$$\frac{dN_O}{dt} = l_I N_I + (r_O - l_O)N_O,$$
$$\frac{dN_S}{dt} = l_O N_O + r_S N_S.$$

If the starting condition is a single SI lineage, the results depend on which of the following net diversification rates is largest: r_S, $r_O - l_O$, or $r_I - l_I$. Assuming that at least one of these quantities is positive, if r_S is largest, then both N_I and N_O approach zero frequency, while only N_S increases at the rate r_S. If $r_O - l_O$ is the largest, then N_I asymptotically approaches zero frequency, while N_O and N_S increase at the rate $r_O - l_O$, attaining the ratios

$$(N_O, N_S) \propto \left(1, \frac{l_O}{r_O - l_O - r_S}\right).$$

Finally, if $r_I - l_I$ is the largest, then all three types asymptotically increase in numbers at the rate $r_I - l_I$, attaining the ratios

$$(N_I, N_O, N_S) \propto$$
$$\left[1, \frac{l_I}{r_I - l_I - r_O + l_O}, \frac{l_O l_I}{(r_I - l_I - r_S)(r_I - l_I - r_O + l_O)}\right]. \quad (1)$$

SI systems have persisted for many millions of years (Igic and Kohn 2001; Igic et al. 2006), and transition rates to SI are extremely small in comparison to transitions from SI to SC. Consequently, the most likely reason for the abundance of SI species in nature is that the conditions for equation (1) are met. Only under this condition are appreciable frequencies of SI species maintained at equilibrium. This condition requires that the net diversification rate (speciation − extinction − loss) of SI lineages exceed that of SC outcrossing or selfing lineages.

Despite Stebbins's (1957) conjecture that selfing is an evolutionary dead end, lineage-specific net diversification rates have rarely been advanced as factors governing the current distribution of plant mating systems (Igic and Kohn 2006). While recent microevolutionary models of plant mating systems examine the stability of mixed mating (Goodwillie et al. 2005 and references therein; Porcher and Lande 2005a), even those that predict only predominant outcrossing and selfing as stable states (Lande and Schemske 1985) do not predict the relative frequencies of selfers and outcrossers. This simple model makes explicit the relation between state-specific diversification and transition rates, and the equilibrium proportions of mating system states. Our model can be elaborated to capture additional complexities of mating system evolution and to more closely approximate the true distribution of mating systems when known.

The Loss of SI

While phylogenetic analyses and macroevolutionary models are important tools for understanding broad evolutionary patterns, microevolutionary studies can focus on the causes and consequences of particular mating system shifts. For single-locus multiple-allele SI systems, several predictions can be made about the near-term effects of the loss of SI. First, if mutations to SC substantially reduce the outcrossing rate, they are unlikely to be selectively neutral. Within a population, the fate of SC mutations will be determined principally by the magnitudes of inbreeding depression, pollen discounting, and the power of selection favoring reproductive assurance (reviewed in Barrett 2002; Schoen and Busch 2007). Drift is probably unimportant except for cases of long-distance dispersal, and, even there, selection will play a role by favoring individuals capable of uniparental reproduction when plant or pollinator abundance is low. Second, many SI species display multiple flowers simultaneously, both within and among inflorescences. The opportunity for geitonogamy suggests that a mutation causing loss of SI will often substantially reduce the primary outcrossing rate, even in the absence of subsequent changes in floral features such as size that might further lead to the evolution of a selfing syndrome. Third, transition to SC renders the extraordinary number of S-alleles in a population selectively neutral. After transition to SC, alleles at the S-locus will no longer be protected by negative frequency-dependent selection. Fixation on a single S-allele is expected to occur in $4N_e$ generations (N_e = effective population size; Hudson 1990), on average, if the mutation causing SC is unlinked to the S-locus and more rapidly if loss of SI is caused by a selective sweep of a nonfunctional S-allele. Following loss of SI, particularly where the mating system proceeds toward predominant selfing, genomic changes in the levels and distribution of genetic variation associated with the outcrossing to selfing transition may ensue (Glemin et al. 2006; Wright et al. 2007).

The best-studied system for investigating a particular transition from SI to SC is the genus *Arabidopsis*, where the SI sister species *A. lyrata* and *A. halleri* are the closest relatives of the highly selfing *A. thaliana* (Koch et al. 2000, 2001). Divergence of the lineage leading to SC *A. thaliana* from its SI sister group is thought to have occurred ca. 5 million years ago (Koch et al. 2000, 2001; Ramos-Onsins et al. 2004). However, when the mating system changed relative to this divergence has been an open question. SI *Arabidopsis* species have the SSI system found in other Brassicaceae (reviewed in Takayama and Isogai 2005). The stylar recognition component in this system is specified by the allele encoding the S-locus receptor kinase (*SRK*), while the pollen component is specified by the allele for the S-locus cystein-rich protein (*SCR*). The relationships among these taxa and the wealth of genomic information available in *Arabidopsis* make this system particularly advantageous for addressing questions about the mutation(s) that caused the transition to SC in *A. thaliana*, estimating when this transition took place, and dissecting events both at the S-locus and elsewhere in the genome that took place after the shift in compatibility system.

Many accessions of *A. thaliana* have a nonfunctional allele at the *SCR* locus, termed *SCR1*. Shimizu et al. (2004) found that sequence polymorphism within this allele was far lower than at many reference loci in *A. thaliana*. They estimated that the conversion to SC was very recent (>300,000 yr BP) and involved a selective sweep by the *SCR1* allele. It is interesting that polymorphism persists at the *SRK* locus, though it is much reduced relative to SI congeners. Three distinct *SRK* alleles were found, with levels of divergence among them similar to functional S-alleles in related SI species.

Bechsgaard et al. (2006) estimated how long the three alleles at *SRK* had been under neutral rather than diversifying selection. A switch to neutrality would occur on loss of SI. Again, conversion to SC was estimated as very recent (<420,000 yr BP). Neither molecular analysis (Shimizu et al. 2004; Bechsgaard et al. 2006) can reject 0 yr BP for the loss of SI. Analysis of *SRK* alleles in the SI species of *Arabidopsis* confirmed that they contain close orthologues of each of the three *SRK* alleles found in *A. thaliana* (Bechsgaard et al. 2006).

Nasrallah et al. (2004), however, suggested that additional polymorphism might also exist at the *SCR* locus in *A. thaliana* because the *SCR1* allele could not be detected by Southern blotting in 17 of 27 accessions tested. Polymorphism at *SCR* casts doubt on the hypothesis that a selective sweep by the nonfunctional *SCR1* allele caused the transition to SC, undermining the findings of Shimizu et al. (2004). Nasrallah et al. (2002, 2004) also transformed *A. thaliana*, inserting functional pollen-specificity (*SCR*) and style-specificity (*SRK*) alleles from *A. lyrata*. In one accession, full SI was restored, indicating that SC was due solely to mutations in the *SRK* and/or the *SCR* alleles. In another six accessions, such transformations failed to restore SI in mature flowers, though these transformed accessions showed varying levels of SI when pollinations were performed on developing flower buds (Nasrallah et al. 2004). The gene, which is S linked, that underlies variation in the timing and duration of the SI response following transformation has recently been cloned (Liu et al. 2007). Variants in this locus either represent additional mutations affecting SI that have accumulated in most populations subsequent to the loss of SI or may represent an ancestry in which most populations of *A. thaliana* exhibited transient (also called pseudo or partial) SI before conversion to full SC. Transient SI is the phenomenon in which flowers lose their SI response over time, allowing self-fertilization after opportunities for outcrossing have diminished (Levin 1996; Good-Avila and Stephenson 2002; Vallejo-Marin and Uyenoyama 2004).

Sherman-Broyles et al. (2007) analyzed the S-haplotype of the accession of *A. thaliana* that was restorable to SI following transformation. Surprisingly, this S-haplotype was derived from recombination between two different S-locus haplotypes. It is unclear whether the recombination event caused or followed the transition to SC; however, the date of the recombination event appears to have been recent (<200,000 yr BP; Sherman-Broyles et al. 2007).

The remaining issue of potential additional polymorphism at the *SCR* locus of *A. thaliana* has recently been addressed (Sherman-Broyles et al. 2007; Shimizu et al., forthcoming). It appears that several deletions are responsible for the failure of some European accessions of *A. thaliana* to provide a signal on Southern blots probed with the *SCR1* allele. Sequencing of the flanking regions shows that these deletions occurred in the *SCR1* allele and that 96% of 286 European accessions examined contain either the nonfunctional *SCR1* allele or its deletion mutants (Shimizu et al., forthcoming). The remaining accessions fail to amplify when primers designed for the *SCR1* allele or its flanking regions are used, perhaps because of additional deletions, as in the recombinant haplotype that lacks *SCR* altogether (Sherman-Broyles et al. 2007; Shimizu et al., forthcoming). A different *A. thaliana SRK* allele, known from the Cape Verde Islands, is linked to a second *SCR* allele, indicating a possible independent origin of selfing. No *SCR* allele has yet been identified that is linked to the third *SRK* allele, which is more commonly found in Asian accessions. It is possible that variation in *SRK* reflects independent transitions to selfing in different glacial refugia (Mediterranean, Asia), with an additional origin of selfing in the Cape Verde Islands (Sherman-Broyles et al. 2007; Shimizu et al., forthcoming).

Taken together, these studies agree on several points. First, while some doubt remains about the causal mutations, conversion to SC in the lineage leading to *A. thaliana* appears to have occurred quite recently, within the past 0.5 million years, and perhaps as recently as the last glacial cycle (Shimizu et al. 2004). Therefore, the lineage leading to *A. thaliana* was probably SI for >90% of the time since the most recent ancestor shared with extant SI *Arabidopsis* species. Second, while polymorphism at the S-locus has not yet completely disappeared, it is certainly far reduced relative to related SI species, with only three alleles so far found at the style locus. Third, all the floral evolution considered part of the selfing syndrome displayed by *A. thaliana* (much-reduced flower size, pollen and stigma presentation positions and schedules that increase autogamy) would almost certainly have been detrimental before the loss of SI, implying rapid evolution of floral characters following transition to SC. Fourth, in addition to the nonfunctional *SCR* and *SRK* alleles, most populations have other mutations affecting SI (Nasrallah et al. 2004).

The primary effect of the loss of SI on the mating system of *A. thaliana* is difficult to ascertain because of extensive subsequent floral evolution resulting in the selfing syndrome. Even though many cases are known where homomorphic SI and SC plants coexist in populations (e.g., Tsukamoto et al. 1999; Mable et al. 2005; Stone et al. 2006), no studies compare the outcrossing rates of SI and SC forms in the same population. Studies of this sort would be useful because they would allow estimation of the direct effect of loss-of-function mutations on the mating system without the confounding effects of subsequent changes in floral form.

In contrast to its obligately outcrossing relatives, *A. thaliana* currently outcrosses ca. 1% of the time (Abbott and Gomes 1989). Individuals of *A. thaliana* are highly homozygous, and within-population nucleotide variation is lower in *A. thaliana* than in its SI sister group (Wright et al. 2003, 2007; Ramos-Onsins et al. 2004). However, species-wide variation in *A. thaliana* remains quite high (Nordborg et al. 2005), reflecting its broad geographic range and large species-wide population size. Single populations contain a substantial fraction (33%) of the total genetic variation found in the species, and populations within a region contain an additional 35% of the total variation. Whether persistence of large amounts of genetic variation within and among populations of *A. thaliana* is due to the recent shift to selfing, large local and species-wide population sizes, the complex history of postglacial colonization and admixture of populations, or other factors remains a topic of intense research (see Wright et al. 2007).

The genus *Solanum*, sect. *Lycopersicon* (the wild and the cultivated tomatoes), provides another excellent opportunity for studies of recent transitions to SC. The SI species in this group use the well-characterized S-RNase/F-box GSI mechanism (McClure 2006) that operates elsewhere in the Solanaceae. Repeated recent transitions to SC have occurred (fig. 2), with SI/SC polymorphism found among populations of all well-studied SI species. Among SC tomato species, a number of different defects in the SI mechanism are known (Kondo et al. 2002a, 2002b). All SC species examined have much-reduced levels of S-RNase activity in their styles, but some (*S. pimpinellifolium*, *S. cheesmaniae*, and *S. lycopersicum*) appear to lack the S-RNase gene altogether, while others (*S. neorickii*, *S. habrochaites*, and *S. chmielewskii*) have S-RNase alleles, though different ones in different species. All SC species display reduced transcription of the HT-B gene (Kondo et al. 2002a), which is unlinked to the S-locus but encodes a protein that is essential for SI in Solanaceae (McClure et al. 1999). Because the phylogenetic relationships among SI and SC species are difficult to resolve with certainty (Spooner et al. 2005) and the species in question may hybridize, it may be too early to propose a scenario to explain the order of events leading to SC in all species and populations. Nevertheless, the multiple defects seen in many taxa and the limited implied age of SC in any of these lineages indicate that additional mutations accumulate once SI is lost.

Several transitions from obligate or predominant outcrossing to selfing are thought to have occurred in the tomato group (fig. 2), but few quantitative studies of outcrossing rates have been performed. Rick et al. (1978) experimentally confirmed low rates of outcrossing in *S. pimpinellifolium* (mean $t = 0.135$), which is almost entirely selfing in the southern part of its range, with some large-flowered individuals from the northern part of the range exhibiting outcrossing rates as high as $t = 0.4$. Based on its diminutive flower size, very low stigma exertion, and autogamy in glasshouse-grown individuals, it is commonly thought that *S. neorickii* is also primarily selfing. The cultivated tomato (*S. lycopersicum*) and its close relatives (*S. cheesmaniae* and *S. galapagensis*) each share similar selfing attributes.

Convergence on the selfing syndrome following loss of SI appears to be rapid. This small group shows correlated variation in mating systems and flower sizes not only among closely related species but also within species, where interpopulation variation in SI status is found. For example, two populations in the principally SI *S. habrochaites*, which occur at the northern and southern geographical limits of the species, independently acquired SC (Rick et al. 1979; Rick and Chetelat 1991). These SC populations also show marked reductions in flower size. Similarly, a SC accession of *S. arcanum* (LA2157) has smaller flowers than other accessions from this species (Rick 1982). It seems that the breakdown of SI can be rapidly followed by the evolutionary transition to selfing.

Recent studies of the population genetic consequences of transitions to SC in the tomato group (Baudry et al. 2001; Roselius et al. 2005; Städler et al. 2005) find far more striking reductions in genetic variation between SI and SC taxa than has been found in *Arabidopsis*. For instance, silent nucleotide diversity in the SC and partially selfing *S. pimpinellifolium* and *S. chmielewskii* was less than half that of the three SI species examined (*S.*

peruvianum, *S. habrochaites*, and *S. chilense*; Roselius et al. 2005). Because complete selfing theoretically reduces N_e by a factor of only one-half relative to obligate outcrossing, demographic and ecological shifts in addition to the transition in mating system must account for some of these reductions in genetic variability. Roselius et al. (2005) examined only single-population samples of each species, so species-wide effects remain unknown. Opportunities abound to document both population- and species-wide shifts in patterns of genetic variation associated with transitions from SI to SC in *Solanum* sect. *Lycopersicon*.

Several studies find interpopulation or interindividual variation in the presence or strength of SI (Tsukamoto et al. 1999; Good-Avila and Stephenson 2002; Mable et al. 2005; Stone et al. 2006). In some cases, SI may be best characterized as a quantitative trait whose strength is affected by various genetic factors, both linked and unlinked to the *S*-locus. Under some conditions, partial SI may be evolutionarily stable (Vallejo-Marin and Uyenoyama 2004). However, the effect of variation in the presence or strength of SI on the outcrossing rate of populations has rarely been measured. In perhaps the most complete study, Mable et al. (2005) found variation within and among *A. lyrata* populations from the southern Great Lakes region of North America in the ability of plants to set fruit and seed on selfing. They also estimated the outcrossing rates of populations that displayed different frequencies of self fertile individuals, a unique practice among studies of this type. Populations with higher frequencies of SC in glasshouse studies showed markedly lower outcrossing rates and reduced genetic diversity than more SI populations. Three populations with low frequencies of SC individuals had outcrossing rates not statistically different from 1. Two populations, in which the frequency of SC was much higher, had outcrossing rates well below 0.5. Because no differences in flower size were noted, this would appear to reflect differences in the mating system resulting from differences in the presence or strength of SI, though differences among populations in pollination services may also play a role. Because even the more selfing populations contained substantial fractions of individuals that were strongly SI, this is a minimum estimate of the difference in individual outcrossing rate caused by a switch to SC in this species.

The existence of populations containing individuals that vary in the presence or strength of SI brings the inevitable question of whether such populations represent stable polymorphisms (and stable mixed mating) or whether they represent transitory situations in which genes causing SC will either disappear following their temporary invasion or will spread to fixation. Porcher and Lande (2005b) showed theoretically that mutations causing SC readily invade GSI populations under many parameter combinations. However, the conditions for stable polymorphisms are much more restrictive than conditions for fixation. Given the number of parameters involved in their model, empirical demonstration of the stability of mixed populations appears unlikely, though experiments on the fitness of SC individuals when rare and frequent might be a reasonable avenue of future research.

Several authors studying mixed populations of SI and SC individuals (e.g., Brennan et al. 2006; Stone et al. 2006) have pointed out that the balance between colonization ability of SC forms and the increased fitness of outcrossed offspring following population establishment could lead to the frequent observation of mixtures of SI and SC individuals within populations. Metapopulation models (Pannell and Barrett 1998; Schoen and Busch 2007) indicate that selection for reproductive assurance intensifies whenever local population persistence, the number of migrants to unoccupied sites, or the proportion of occupied sites is low. Therefore, SC forms would be more likely to occur and persist in otherwise SI species wherever local population turnover is high.

Conclusions

While mixtures of SI and SC individuals may be observable in many species, the phylogenetic record of well-characterized SI mechanisms indicates the frequent, complete, and irreversible loss of SI. The evidence for irreversibility is broadly shared ancestral polymorphism at the *S*-locus among SI species and genera in various families (Castric and Vekemans 2004; Igic et al. 2004, 2006). This implies a continuous ancestry of SI for all SI species sharing ancestral *S*-locus polymorphism. We show that SC species, even those recently derived from an SI ancestor, often harbor severely reduced variation at the *S*-locus. Most SC populations and some species are fixed on a single *S*-allele despite these recent transitions to SC. They also tend to harbor multiple loss-of-function mutations, making reversion to SI difficult. Because of the loss of *S*-locus diversity, recovery of the same SI mechanism would leave a genealogical imprint on the *S*-locus for tens of millions of years. To date, only one example of a severe historical bottleneck at the *S*-locus is known, the bottleneck shared by the Solanaceae genera *Witheringia* and *Physalis* (Richman et al. 1996; Richman and Kohn 1999, 2000; Lu 2001; Stone and Pierce 2005). In that case, three or four allelic lineages predate the bottleneck, providing no evidence that SI was ever completely lost.

A large fraction of angiosperms are SI, even if the current estimate of the frequency of SI species suffers from bias. Coupled with the evidence for frequent transitions to SC, the high frequency of SI among angiosperm species suggests that SI provides a macroevolutionary advantage. It is probably not a coincidence that in both *Arabidopsis* and the tomatoes, evidence of multiple recent transitions to SC from SI ancestors is found. If frequent transitions have been the rule since the origin of each SI mechanism, nearly all species would be expected to be SC in the absence of differences in net diversification rates between SI and SC lineages.

Clearly, not all SC lineages are doomed, particularly when they do not shift to predominant self-fertilization. This is best supported by the evidence for the evolution of nonhomologous SI systems and the existence of relatively large and old monophyletic groups that seem to lack SI altogether (e.g., the Cucurbitaceae). Where sufficient rates of outcrossing can be maintained by alternative mechanisms, SC lineages may proliferate. In addition, shifts from increased to decreased rates of self-fertilization are known, if not particularly frequent, in the literature (Barrett and Shore 1987; Olmstead 1990; Takebayashi and Morrell 2001). However, we know of no large and old angiosperm families where self-fertilization is the predominant mode of reproduction.

Our simple model posits unidirectional mating system shifts from obligate outcrossing (SI) to predominant outcrossing and then to predominant selfing. It is capable of producing a variety of distributions of breeding-system states among species even if the direction of mating system change, once SI is lost, usually leads to selfing. For too long the U-shaped distribution (a dearth of mixed-mating species with many outcrossers and selfers) has been the main issue discussed in studies of the distribution of outcrossing rates. Given variation in diversification rates among lineages with different mating systems, a variety of distributions is possible, and symmetric bimodality is, in fact, rather unlikely.

Acknowledgments

We thank S. Good-Avila, K. Shimizu, and A. Stephenson for sharing unpublished data. A. Angert, V. Zeldovich, S. C. H. Barrett, and two anonymous reviewers provided comments that improved the manuscript. Support for this work was provided by National Science Foundation grants DEB-0108173 and DEB-0639984 to J. R. Kohn, DEB-0309184 to B. Igic and J. R. Kohn, and DEB-0313653 to R. Lande.

Literature Cited

Abbott RJ, M Gomes 1989 Population genetic structure and outcrossing rate of *Arabidopsis thaliana* (L.) Heynth. Heredity 62: 411–418.

Aizen MA, P Feinsinger 1994 Forest fragmentation, pollination, and plant reproduction in a Chaco dry forest, Argentina. Ecology 75: 330–351.

Anderson GJ, G Bernardello, TF Stuessy, DJ Crawford 2001 Breeding system and pollination of selected plants endemic to Juan Fernandez Islands. Am J Bot 88:220–233.

Arroyo MTK, F Squeo 1990 Relationship between plant breeding systems and pollination. Pages 205–227 *in* S Kawano, ed. Biological approaches and evolutionary trends in plants. Academic Press, London.

Arroyo MTK, P Uslar 1993 Breeding systems in a temperate Mediterranean-type climate montane sclerophyllous forest in central Chile. Bot J Linn Soc 111:83–102.

Baker HG 1955 Self-compatibility and establishment after "long-distance" dispersal. Evolution 9:347–348.

——— 1967 Support for Baker's Law as a rule. Evolution 21:853–856.

Barrett SCH 1992 Heterostylous genetic polymorphisms: model systems for evolutionary analysis. Pages 1–29 *in* SCH Barrett, ed. Evolution and function of heterostyly. Monographs on theoretical and applied genetics. Springer, Berlin.

——— 2002 The evolution of plant sexual diversity. Nat Rev Genet 3:274–284.

Barrett SCH, LD Harder, AC Worley 1996 The comparative biology of pollination and mating in flowering plants. Philos Trans R Soc B 351:1271–1280.

Barrett SCH, K Helenurm 1987 The reproductive ecology of boreal forest herbs. I. Breeding systems and pollination. Can J Bot 65: 2036–2046.

Barrett SCH, JS Shore 1987 Variation and evolution of breeding systems in the *Turnera ulmifolia* L. complex (Turneraceae). Evolution 41:340–354.

Baudry E, C Kerdelhué, H Innan, W Stephan 2001 Species and recombination effects on DNA variability in the tomato genus. Genetics 158:1725–1735.

Baumann U, J Juttner, X Bian, P Langridge 2000 Self-incompatibility in the grasses. Ann Bot 85:203–209.

Bawa KS 1974 Breeding systems of tree species of a lowland tropical community. Evolution 28:85–92.

Bechsgaard JS, V Castric, D Charlesworth, X Vekemans, MH Schierup 2006 The transition to self-compatibility in *Arabidopsis thaliana* and evolution within S-haplotypes over 10 Myr. Mol Biol Evol 23:1741–1750.

Bianchi MB, PE Gibbs, DE Prado, JL Vespini 2000 Studies on the breeding systems of understory species of a Chaco woodland in ME Argentina. Flora 195:339–348.

Brennan AC, SA Harris, SJ Hiscock 2006 Modes and rates of selfing and associated inbreeding depression in the self-incompatible plant *Senecio squalidus* (Asteraceae): a successful colonizing species in the British Isles. New Phytol 168:475–486.

Bullock SH 1985 Breeding systems in the flora of a tropical deciduous forest in Mexico. Biotropica 17:287–301.

Castric V, X Vekemans 2004 Plant self-incompatibility in natural populations: a critical assessment of recent theoretical and empirical advances. Mol Ecol 13:2873–2889.

Charlesworth D 1985 Distribution of dioecy and self-incompatibility in angiosperms. Pages 237–268 *in* JJ Greenwood, PH Harvey, M Slatkin, eds. Evolution: essays in honour of John Maynard Smith. Cambridge University Press, Cambridge.

Charlesworth D, B Charlesworth 1979 The evolution and breakdown of S-allele systems. Heredity 43:41–55.

Darlington CD, K Mather 1949 The elements of genetics. Macmillan, New York.

Darwin C 1876 The effects of cross and self fertilisation in the vegetable kingdom. J Murray, London.

Davies TJ, TG Barraclough, MW Chase, PS Soltis, DE Soltis, V Savolainen 2004 Darwin's abominable mystery: insights from a supertree of the angiosperms. Proc Natl Acad Sci USA 101:1904–1909.

de Nettancourt D 1977 Incompatibility in angiosperms. Springer, Berlin.

Dobzhansky T 1950 Evolution in the tropics. Am Sci 38:209–221.

Ferrer MM, SV Good-Avila 2007 Macrophylogenetic analysis of the gain and loss of self-incompatibility in the Asteraceae. New Phytol 173:401–414.

Fisher RA 1941 Average excess and average effect of an allelic substitution. Ann Eugen 11:53–63.

Foote HCC, JP Ride, VE Franklin-Tong, EA Walker, MJ Lawrence, FCH Franklin 1994 Cloning and expression of a distinctive class of self-incompatibility (S) gene from *Papaver rhoeas* L. Proc Natl Acad Sci USA 91:2265–2269.

Ganders FR 1979 The biology of heterostyly. N Z J Bot 17:607–635.

Gibbs PE 1986 Do homomorphic and heteromorphic self-incompatibility systems have the same sporophytic mechanism? Plant Syst Evol 154: 285–323.

——— 1990 Self-incompatibility in flowering plants: a Neotropical perspective. Rev Bras Bot 13:125–136.

Glemin S, E Bazin, D Charlesworth 2006 Impact of mating systems on patterns of sequence polymorphism in flowering plants. Proc R Soc B 273:3011–3019.

Good-Avila SV, AG Stephenson 2002 The inheritance of modifiers conferring self-fertility in the partially self-incompatible perennial, *Campanula rapunculoides* L. (Campanulaceae). Evolution 56:263–272.

Goodwillie C 1997 The genetic control of self-incompatibility in *Linanthus parviflorus* (Polemoniaceae). Heredity 79:424–432.

Goodwillie C, S Kalisz, CG Eckert 2005 The evolutionary enigma of mixed mating systems in plants: occurrence, theoretical explanations, and empirical evidence. Annu Rev Ecol Syst 11:15–39.

Heilbuth JC 2000 Lower species richness in dioecious clades. Am Nat 156:221–241.

Hudson RR 1990 Gene genealogies and the coalescent process. Pages 1–44 *in* D Futuyma, J Antonovics, eds. Oxford surveys in evolutionary biology. Vol 7. Oxford University Press, New York.

Igic B, L Bohs, JR Kohn 2004 Historical inferences from the self-incompatibility locus. New Phytol 161:97–105.

——— 2006 Ancient polymorphism reveals unidirectional breeding system shifts. Proc Natl Acad Sci USA 103:1359–1363.

Igic B, JR Kohn 2001 Evolutionary relationships among self-incompatibility RNases. Proc Natl Acad Sci USA 98:13167–13171.

——— 2006 The distribution of plant mating systems: study bias against obligately outcrossing species. Evolution 60:1098–1103.

Jaimes I, N Ramírez 1999 Breeding systems in a secondary deciduous forest in Venezuela: the importance of life form, habitat, and pollination specificity. Plant Syst Evol 215:23–36.

Koch MA, B Haubold, T Mitchell-Olds 2000 Comparative evolutionary analysis of chalcone synthase and alcohol dehydrogenase loci in Arabidopsis, Arabis, and related genera (Brassicaceae). Mol Biol Evol 17:1483–1498.

——— 2001 Molecular systematics of the Brassicaceae: evidence from coding plastidic matK and nuclear Chs sequences. Am J Bot 88: 534–544.

Kondo K, M Yamamoto, R Itahashi, T Sato, H Egashira, T Hattori, Y Kowyama 2002a Insights into the evolution of self-compatibility in Lycopersicon from a study of stylar factors. Plant J 30:143–153.

Kondo K, M Yamamoto, DP Matton, T Sato, M Hirai, S Norioka, T Hattori, Y Kowyama 2002b Cultivated tomato has defects in both S-RNase and HT genes required for stylar function of self-incompatibility. Plant J 29:627–636.

Kress WJ, JH Beach 1994 Flowering plant reproductive systems. Pages 161–185 in LA McDade, KS Bawa, HA Hespenheide, GS Hartshorn, eds. La Selva: ecology and natural history of a Neotropical rainforest. University of Chicago Press, Chicago.

Lande R, DW Schemske 1985 The evolution of self-fertilization and inbreeding depression in plants. I. Genetic models. Evolution 39:24–40.

Levin DA 1993 S-gene polymorphism in Phlox drummondii. Heredity 71:193–198.

——— 1996 The evolutionary significance of pseudo-self-incompatibility. Am Nat 148:321–332.

Liu P, S Sherman-Broyles, ME Nasrallah, JB Nasrallah 2007 A cryptic modifier causing transient self-incompatibility in Arabidopsis thaliana. Curr Biol 17:734–740.

Lu Y 2001 Roles of lineage sorting and phylogenetic relationship in the genetic diversity at the self-incompatibility locus of Solanaceae. Heredity 86:195–205.

Mable BK, AV Robertson, S Dart, C Di Berardo, L Witham 2005 Breakdown of self-incompatibility in the perennial Arabidopsis lyrata. Evolution 59:1437–1448.

Machado IC, AV Lopes, M Sazima 2006 Plant sexual systems and a review of the breeding system studies in the Caatinga, a Brazilian tropical dry forest. Ann Bot 97:277–287.

Maddison DR 1994 Phylogenetic methods for inferring the evolutionary history and processes of change in discretely valued characters. Annu Rev Entomol 39:267–292.

Maddison WP 2006 Confounding asymmetries in evolutionary diversification and character change. Evolution 60:1743–1746.

Marshall CR, EC Raff, RA Raff 1994 Dollo's Law and the death and resurrection of genes. Proc Natl Acad Sci USA 91:12283–12287.

Mast AR, S Kelso, E Conti 2006 Are any primroses (Primula) primitively monomorphic? New Phytol 171:605–616.

McClure BA 2006 New views of S-RNase-based self-incompatibility. Curr Opin Plant Biol 9:639–646.

McClure BA, B Mou, S Canevascini, R Bernatzky 1999 A small asparagine-rich protein required for S-allele-specific pollen rejection in Nicotiana. Proc Natl Acad Sci USA 96:13548–13553.

McMullen CK 1990 Reproductive biology of Galápagos Islands angiosperms. Monogr Syst Bot 32:35–45.

Morales CL, L Galetto 2003 Influence of compatibility system and life form on plant reproductive success. Plant Biol 5:567–573.

Motten AF 1986 Pollination ecology of the spring wildflower community of a temperate deciduous forest. Ecol Monogr 56:21–42.

Nasrallah ME, P Liu, JB Nasrallah 2002 Generation of self-incompatible Arabidopsis thaliana by transfer of two S locus genes from A. lyrata. Science 297:247–249.

Nasrallah ME, P Liu, S Sherman-Broyles, NA Boggs, JB Nasrallah 2004 Natural variation in expression of self-incompatibility in Arabidopsis thaliana: implications for the evolution of selfing. Proc Natl Acad Sci USA 101:16070–16074.

Neff JL, BB Simpson, AR Moldenke 1977 Flowers, flower visitor system. Pages 204–224 in GH Orians, OT Solbrig, eds. Convergent evolution in warm deserts. Dowden, Hutchinson & Ross, Stroudsburg, PA.

Newstrom L, A Robertson 2005 Progress in understanding pollination systems in New Zealand. N Z J Bot 43:1–59.

Nordborg M, TT Hu, Y Ishino, J Jhaveri, C Toomajian, H Zheng, E Bakker, et al 2005 The pattern of polymorphism in Arabidopsis thaliana. PLoS Biol 3:1289–1299.

Nunney L 1989 The maintenance of sex by group selection. Evolution 43:245–257.

Oliveira PE, PE Gibbs 2000 Reproductive biology of woody plants in a cerrado community of central Brazil. Flora 195:311–329.

Olmstead RG 1990 Biological and historical factors influencing genetic diversity in the Scuttelaria angustifolia complex (Labiatae). Evolution 44:54–70.

Pagel M 1994 Detecting correlated evolution on phylogenies: a general method for the comparative analysis of discrete characters. Proc R Soc B 255:37–45.

——— 1997 Inferring evolutionary processes from phylogenies. Zool Scr 26:331–348.

Pannell JR, SCH Barrett 1998 Baker's Law revisited: reproductive assurance in a metapopulation. Evolution 53:657–668.

Peralta IE, DM Spooner 2001 Granule-bound starch synthase (GBSSI) gene phylogeny of wild tomatoes (Solanum L. section Lycopersicon [Mill.] Wettst. subsection Lycopersicon). Am J Bot 88: 1888–1902.

Pojar J 1974 Reproductive dynamics of four plant communities of southwestern British Columbia. Can J Bot 52:1819–1834.

Porcher E, R Lande 2005a The evolution of self-fertilization and inbreeding depression under pollen discounting and pollen limitation. J Evol Biol 18:497–508.

——— 2005b Loss of gametophytic self-incompatibility with the evolution of inbreeding depression. Evolution 59:46–60.

Qiao H, F Wang, L Zhao, J Zhou, Z Lai, Y Zhang, TP Robbins, Y Xue 2004 The F-box protein AhSLF-S_2 controls the pollen function of S-RNase-based self-incompatibility. Plant Cell 16:2307–2322.

Ramírez N, Y Brito 1990 Reproductive biology of a tropical palm swamp community in the Venezuelan Llanos. Am J Bot 77:1260–1271.

Ramos-Onsins SE, BE Stranger, T Mitchell-Olds, M Aguadé 2004 Multilocus analysis of variation and speciation in the closely related species Arabidopsis halleri and A. lyrata. Genetics 166:373–388.

Rathcke B 1988 Flowering phenologies in a shrub community: competition and constraints. J Ecol 76:975–994.

Richman AD, JR Kohn 1999 Self-incompatibility alleles from Physalis: implications for historical inference from balanced polymorphisms. Proc Natl Acad Sci USA 96:168–172.

——— 2000 Evolutionary genetics of self-incompatibility in the Solanaceae. Plant Mol Biol 42:169–179.

Richman AD, MK Uyenoyama, JR Kohn 1996 Contrasting patterns of allelic diversity and gene genealogy at the self-incompatibility locus in two species of Solanaceae. Science 273:1212–1216.

Rick CM 1982 Genetic relationships between self-incompatibility and floral traits in the tomato species. Biol Zbl 101:185–198.

Rick CM, R Chetelat 1991 The breakdown of self-incompatibility in Lycopersicon hirsutum. Pages 253–256 in JG Hawkes, RN Lester,

M Nee, N Estrada, eds. Solanaceae III: taxonomy, chemistry, evolution. Kew, Richmond, Surrey.

Rick CM, JF Fobes, SD Tanksley 1979 Evolution of mating systems in *Lycopersicon hirsutum* as deduced from genetic variation in electrophoretic and morphological characters. Plant Syst Evol 132:279–298.

Rick CM, M Holle, RW Thorp 1978 Rates of cross-pollination in *Lycopersicon pimpinellifolium*: impact of genetic variation in floral characters. Plant Syst Evol 129:31–44.

Riveros M, AM Humaña, MT Kalin Arroyo 1996 Sistemas de reproducción en especies del bosque valdiviano (40° latitud sur). Phyton 58:167–176.

Roselius K, W Stephan, T Städler 2005 The relationship of nucleotide polymorphism, recombination rate and selection in wild tomato species. Genetics 171:753–763.

Sanderson MJ 1993 Reversibility in evolution: a maximum likelihood approach to character gain and loss bias in phylogenies. Evolution 47:236–252.

Schluter D, T Price, AO Mooers, D Ludwig 1997 Likelihood of ancestor states in adaptive radiation. Evolution 51:1699–1711.

Schoen DJ, JW Busch 2008 On the evolution of self-fertilization in a metapopulation. Int J Plant Sci 169:119–127.

Schoen DJ, MT Morgan, T Bataillon 1996 How does self-pollination evolve? inferences from floral ecology and molecular genetic variation. Philos Trans R Soc B 351:1281–1290.

Scofield DG, ST Schultz 2006 Mitosis, stature and evolution of plant mating systems: low-Φ and high-Φ plants. Proc R Soc B 273:275–282.

Sherman-Broyles S, N Boggs, A Farkas, P Liu, J Vrebalov, ME Nasrallah, JB Nasrallah 2007 S locus genes and the evolution of self-fertility in *Arabidopsis thaliana*. Plant Cell 19:94–106.

Shimizu KK, JM Cork, AL Caicedo, CA Mays, RC Moore, KM Olsen, S Ruzsa, et al 2004 Darwinian selection on a selfing locus. Science 306:2081–2084.

Shimizu KK, R Shimizu-Inatsugi, T Tsuchimatsu, MD Purugganan Forthcoming Independent origins of self-compatibility in *Arabidopsis thaliana*. Mol Ecol.

Sijacic P, X Wang, AL Skirpan, Y Wang, PE Dowd, AG McCubbin, S Huang, T-h Kao 2004 Identification of the pollen determinant of S-RNase-mediated self-incompatibility. Nature 429:302–305.

Smith-Ramírez C, P Martinez, M Nuñez, C González, JJ Armesto 2005 Diversity, flower visitation frequency and generalism of pollinators in temperate rain forests of Chiloé Island, Chile. Bot J Linn Soc 147:399–416.

Sobrevila C, MTK Arroyo 1982 Breeding systems in a montane tropical cloud forest in Venezuela. Plant Syst Evol 140:19–37.

Spooner DM, IE Peralta, S Knapp 2005 Comparison of AFLPs with other markers for phylogenetic inference in wild tomatoes [*Solanum* L. section *Lycopersicon* (Mill.) Wettst.]. Taxon 54:43–61.

Städler T, K Roselius, W Stephan 2005 Genealogical footprints of speciation processes in wild tomatoes: demography and evidence for historical gene flow. Evolution 59:1268–1279.

Stebbins GL 1957 Self-fertilization and population variability in the higher plants. Am Nat 91:337–354.

——— 1974 Flowering plants: evolution above the species level. Belknap, Cambridge, MA.

Steinbachs JE, KE Holsinger 2002 S-RNase-mediated gametophytic self-incompatibility is ancestral in eudicots. Mol Biol Evol 19:825–829.

Stone JL 2002 Molecular mechanisms underlying the breakdown of gametophytic self-incompatibility. Q Rev Biol 77:17–32.

Stone JL, SE Pierce 2005 Rapid recent radiation of S-RNase lineages in *Witheringia solanacea* (Solanaceae). Heredity 94:547–555.

Stone JL, MA Sasuclark, CP Blomberg 2006 Variation in the self-incompatibility response within and among populations of the tropical shrub *Witheringia solanacea* (Solanaceae). Am J Bot 93:592–598.

Takayama S, A Isogai 2005 Self-incompatibility in plants. Annu Rev Plant Biol 58:467–489.

Takebayashi N, PL Morrell 2001 Is self-fertilization an evolutionary dead end? revisiting an old hypothesis with genetic theories and a macroevolutionary approach. Am J Bot 88:1143–1150.

Tanner EVJ 1982 Species diversity and reproductive mechanism in Jamaican trees. Biol J Linn Soc 18:263–278.

Tsukamoto T, T Ando, H Kokobun, H Watanabe, M Masada, X Zhu, E Marchesi, T-h Kao 1999 Breakdown of self-incompatibility in a natural population of *Petunia axillaris* (Solanaceae) in Uruguay containing both self-incompatible and self-compatible plants. Sex Plant Reprod 12:6–13.

Tsukamoto T, T Ando, H Kokobun, H Watanabe, T Sato, M Masada, E Marchesi, T-h Kao 2003*a* Breakdown of self-incompatibility in a natural population of *Petunia axillaris* caused by a modifier locus that suppresses the expression of an S-RNase gene. Sex Plant Reprod 15:255–263.

Tsukamoto T, T Ando, K Takahashi, T Omori, H Watanabe, H Kokobun, E Marchesi, T-h Kao 2003*b* Breakdown of self-incompatibility in a natural population of *Petunia axillaris* caused by loss of pollen function. Plant Physiol 131:1903–1912.

Ushijima K, H Sassa, AM Dandekar, TM Gradziel, R Tao, H Hirano 2003 Structural and transcriptional analysis of the self-incompatibility locus of almond: identification of a pollen-expressed F-box gene with haplotype-specific polymorphism. Plant Cell 15:771–781.

Vallejo-Marin M, MK Uyenoyama 2004 On the evolutionary costs of self-incompatibility: incomplete reproductive compensation due to pollen limitation. Evolution 58:1924–1935.

Villagrán C 1988 Late Quaternary vegetation of southern Isla Grande de Chiloé, Chile. Quat Res 29:294–306.

Webb CO, MJ Donoghue 2005 Phylomatic: tree assembly for applied phylogenetics. Mol Ecol Notes 5:181–183.

Weller SG, MJ Donoghue, D Charlesworth 1995 The evolution of self-incompatibility in flowering plants: a phylogenetic approach. Pages 355–382 *in* PC Hoch, AG Stephenson, eds. Experimental and molecular approaches to plant biosystematics. Vol 53. Missouri Botanical Garden, St. Louis.

Whalen MD, GJ Anderson 1981 Distribution of gametophytic self-incompatibility and infrageneric classification in *Solanum*. Taxon 30:761–767.

Wright SI, B Lauga, D Charlesworth 2003 Subdivision and haplotype structure in natural populations of *Arabidopsis lyrata*. Mol Ecol 12:1247–1263.

Wright SI, RW Ness, JP Foxe, SCH Barrett 2008 Genomic consequences of outcrossing and selfing in plants. Int J Plant Sci 169:105–118.

Yampolsky C, H Yampolsky 1922 Distribution of sex forms in the phanerogamic flora. Bibl Genet 3:1–62.

Note Added in Proof

A recent article by Tang et al. argues that the transition from SI to SC in *A. thaliana* is considerably older than previous reports indicate. (Tang C, C Toomajian, S Sherman-Broyles, V Plagnol, Y-L Guo, TT Hu, RM Clark, et al 2007 The evolution of selfing in *Arabidopsis thaliana*. Science 317:1070–1072.)

Int. J. Plant Sci. 169(1):105–118. 2008.
1058-5893/2008/16901-0009$15.00   DOI: 10.1086/523366

GENOMIC CONSEQUENCES OF OUTCROSSING AND SELFING IN PLANTS

Stephen I. Wright,* Rob W. Ness,† John Paul Foxe,* and Spencer C. H. Barrett†

*Department of Biology, York University, 4700 Keele Street, Toronto, Ontario M3J 1P3, Canada; and
†Department of Ecology and Evolutionary Biology, University of Toronto, 25 Willcocks Street,
Toronto, Ontario M5S 3B2, Canada

Evolutionary transitions from outcrossing to selfing are expected to cause a reduction in the effective population size and a corresponding increase in fixation rates of slightly deleterious mutations and decrease in fixation of advantageous mutations. Despite these predictions, evidence from genomic data does not suggest a significant reduction in the efficacy of selection associated with high levels of self-fertilization. Here, we discuss opportunities for selfing populations to avoid an irreversible decline in fitness toward extinction and the implications for genome evolution. Most directly, large population sizes and the purging of deleterious recessive mutations can reduce genetic loads and slow the effects of genetic drift. Theory suggests that recombination rates may also evolve in response to the evolution of mating system, which can offset the harmful effects of inbreeding. Cytological data supporting the evolution of higher recombination rates in selfing species should be supplemented with genetic and molecular methods for estimating this parameter. Mutation rates may also evolve to be higher in selfing plants as a result of hitchhiking with advantageous mutations, although this is unlikely to lead to increased fitness. Finally, the abundance and activity of selfish genetic elements may also be reduced in selfing lineages, reducing the accumulation of transposable elements, B chromosomes, biased gene conversion, and the spread of cytoplasmic male sterility mutations. This reduction in genomic conflict can increase mean fitness, reduce deleterious mutation rates, and reduce genome size. We show, using comparative data, that highly selfing plants have genomes significantly smaller than those of outcrossing relatives, consistent with reduced activity and spread of repetitive elements in inbred plants. We discuss opportunities for tests of theory as plant genomic data accumulate and argue that a genomic perspective on reproductive transitions in a phylogenetic context should provide important insights into the diversity of reproductive systems in flowering plants.

Keywords: inbreeding, genome evolution, recombination, transposable elements, genome size, deleterious mutation.

Introduction

Flowering plants exhibit a spectacular diversity in reproductive systems, and this can have important effects on the amount and structuring of genetic variation within and among populations (Hamrick and Godt 1996; Glemin et al. 2006). Reproductive transitions, such as the shift in mating system from outcrossing to selfing, tend to increase the extent of linkage disequilibrium, the degree of association among polymorphic sites. When we consider the probability of fixation of mutations subject to natural selection, the strength, efficacy, and sign of selection acting on mutations can be influenced by the extent of linkage disequilibrium with other sites in the genome. Therefore, evolutionary transitions in reproductive systems should play a central role in genome evolution. However, little is known about the genomic consequences of plant reproductive diversity and how transitions in sexual systems and patterns of mating may influence genome evolution.

At last count, there are 41 large-scale plant genome sequencing projects under way (http://www.ncbi.nlm.nih.gov/genomes/leuks.cgi), with more to follow in the next few years. Although many of these projects focus on crop plants, increasing atten-

tion is being focused on ecological and evolutionary model systems, including *Mimulus guttatus*, *Aquilegia formosa*, *Thellungiella halophila*, *Capsella rubella*, and *Arabidopsis lyrata* (http://www.jgi.doe.gov/sequencing/allinoneseqplans.php). As our ability to compare patterns of genome structure and evolution accelerates, a theoretical and empirical framework for understanding plant genome evolution in wild plant populations becomes increasingly important. Moreover, because many of these nondomesticated species display extensive variation in reproductive traits and patterns of mating, there are likely to be rich opportunities for investigating the genomic consequences of reproductive diversity in flowering plants.

Although extensive research has focused on genome evolution in polyploids (Chen 2007), relatively little attention has focused on the potential for mating system transitions to restructure genomes, perhaps because the effects of polyploidy are immediate and more conducive to experimental manipulation (Husband et al. 2008). In contrast, mating system transitions are expected to lead to shifts in the selective dynamics of genomic elements over evolutionary timescales. Nevertheless, such changes could be equally important for understanding the evolutionary dynamics of plant genomes.

Mating system variation has several important effects on the genetic properties of populations (fig. 1), which we treat here only briefly because they have been covered in several recent

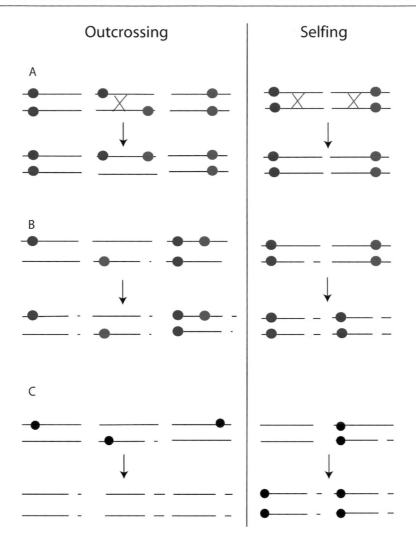

Fig. 1 Expected effects of outcrossing and selfing on patterns of genetic variation and molecular evolution. *A*, Effects on linkage disequilibrium. The blue and red circles represent neutral polymorphic mutations, and each pair of lines represents a diploid individual. In outcrossing species, polymorphic sites will often be found in heterozygotes, and crossing over between polymorphic sites can generate novel haplotypes, breaking up linkage disequilibrium. With selfing, although physical crossing over occurs, the low level of heterozygosity leads to an effective reduction in recombination rate, maintaining linkage disequilibrium. *B*, Effects on diversity. The yellow circle represents an advantageous mutation arising in one chromosome. In outcrossers, the effectively high rate of recombination can uncouple the fate of the advantageous mutation from linked variation, maintaining neutral variation as the advantageous mutation gets fixed. With selfing, the fixation of the advantageous mutation is accompanied by fixation of linked neutral variants, due to hitchhiking. *C*, Effects on the fixation of deleterious mutations. Deleterious mutations are shown in black. With outcrossing, the deleterious mutations can be eliminated by selection independently of the fixation of an advantageous mutation. With selfing, fixation of the advantageous mutation can be accompanied by fixation of a linked deleterious mutation, provided that the net selection coefficient is highest on the linkage group with both the advantageous and the deleterious mutations.

reviews (Charlesworth and Wright 2001; Glemin et al. 2006). Most directly, homozygosity increases as a function of selfing rate, and this reduces the effective size of a population (N_e) as a result of a reduction in the number of distinct alleles. The effective size of a completely selfing population is reduced twofold as a result of homozygosity (Charlesworth et al. 1993; Nordborg 2000). Further, because of homozygosity, crossing over rarely occurs between heterozygous sites, increasing linkage disequilibrium among loci (figs. 1*A*, 2; see Nordborg 2000). This results in stronger effects of genetic hitchhiking, in the form of selective sweeps of positively selected mutations (fig. 1*B*) and back-

ground selection acting against deleterious mutations (reviewed in Charlesworth and Wright 2001), which further reduce N_e in the affected regions. Linkage among weakly selected sites with opposing selective forces can also interfere with the ability of selection to act efficiently (McVean and Charlesworth 2000). All of these forces reduce N_e and may be further exaggerated by life-history characteristics associated with selfing that promote population subdivision, isolation, and genetic bottlenecks. Finally, highly selfing species may experience reduced levels of between-species introgression (Sweigart and Willis 2003), leading to further reductions in genetic diversity.

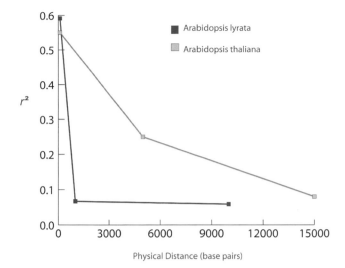

Fig. 2 Decay of linkage disequilibrium with physical distance in the highly selfing *Arabidopsis thaliana* and the self-incompatible *Arabidopsis lyrata*. The Y-axis depicts the average squared correlation coefficient between pairs of polymorphic sites, and the X-axis represents their physical distance. These data, as well as data from species-wide samples, are reported by Wright et al. (2006) and Nordborg et al. (2005) for *A. lyrata* and *A. thaliana*, respectively.

Together, these processes should lead to a decrease in the efficacy of natural selection and an increase in the fixation rate of slightly deleterious mutations (fig. 1C), with important consequences for evolution at the genome level (Lynch and Conery 2003). Over evolutionary time, increased deleterious mutation accumulation can be important in causing species extinction (Lynch et al. 1995), and this may, in part, explain the lack of persistence of selfing lineages, as revealed from comparative and phylogenetic studies (reviewed in Takebayashi and Morrell 2001; Igic et al. 2008).

There is clear and consistent evidence for a reduction in levels of within-population neutral diversity in highly selfing species (Schoen and Brown 1991; Glemin et al. 2006). In contrast, to date, studies of molecular evolution have found only limited evidence for elevated fixation of deleterious mutations in selfing species (Wright et al. 2002; Glemin et al. 2006). One possible explanation for this is the inadequacy of sampling to date. There is simply a need to collect data from very large numbers of consistently sampled loci, given the inherent stochasticity of deleterious mutation accumulation. This will soon become feasible with the rapid expansion of whole-genome data. Alternatively, the timescale for the evolution of selfing may be too recent in many lineages for substantial genomic changes to have occurred. However, it is also possible that many selfing lineages, particularly those successful model systems that have been the focus of research, can avoid long-term fitness decline through compensatory mechanisms.

In this review, we examine the potential for avoidance of fitness decline associated with the evolution of self-fertilization. First, elevated homozygosity, particularly in large populations, can lead to the purging of recessive, strongly deleterious mutations and can enhance the fixation of recessive advantageous mutations. Second, recombination rates are predicted to

evolve after the evolution of selfing, which can directly reduce the harmful effects of suppressed recombination. Third, mutation rates may evolve to be higher; although this could increase rates of adaptation, it may also contribute to mutational meltdown and thus is unlikely to lead to increased fitness in selfing lineages. Finally, selfish genetic elements, which represent a major class of deleterious genomic mutation, may be more effectively selected against in highly selfing populations. Theory predicts that the spread and effects of transposable elements (TEs), B chromosomes, biased gene conversion (BGC), and cytoplasmic male sterility (CMS) mutations should all be reduced in highly selfing species. Because these elements can spread in outcrossing species despite fitness costs, their elimination in selfing lineages can increase mean fitness and reduce genome size and deleterious mutation rates. We now consider these issues in turn.

Inbreeding and the Efficacy of Natural Selection

Reductions in effective population size in inbreeders due to the factors discussed previously are expected to elevate fixation rates of deleterious mutations and decrease fixation rates of advantageous mutations. However, the degree to which polymorphism and/or substitution rates are affected by selfing will depend strongly on the distribution of mutational effects on fitness, which remains very poorly understood (Wang et al. 1999; Weinreich and Rand 2000; Eyre-Walker and Keightley 2007). In particular, high homozygosity due to inbreeding can lead to greater expression of deleterious recessive mutations, leading to their elimination, a phenomenon referred to as purging (Crnokrak and Barrett 2002; see Schoen and Busch 2008). Because of this, selfing leads to a rapid elimination of highly recessive, strongly deleterious mutations from the population. Models show that deleterious mutation accumulation in inbreeding taxa is dominated by slightly deleterious mutations that are closer to additive in their fitness effects (Wang et al. 1999; S. Glemin, unpublished manuscript). If a large fraction of deleterious mutations are recessive and strongly selected, the effects of purging may thus dominate over slightly deleterious mutation accumulation, at least in the context of studies of molecular evolution. Furthermore, recessive advantageous mutations can be fixed more effectively in highly selfing populations (Charlesworth 1992; S. Glemin, unpublished manuscript), increasing rates of adaptive evolution. In general, the predictions for molecular evolution rely on the assumption of the presence of a large class of mutations that are weakly selected and that are nearly additive in their fitness effects, mitigating any influence of homozygosity in purging deleterious recessive mutations and enhancing the fixation of recessive advantageous mutations (S. Glemin, unpublished manuscript).

In addition to the parameters associated with deleterious and advantageous mutations, population sizes can also be important in predicting the rate of fitness decline. Because the extent of linkage disequilibrium is determined by $4N_e r(1 - s)$, where r is the rate of recombination and s is the selfing rate (Nordborg 2000), high selfing rates can be partially compensated for by large population sizes, and this will reduce any expected effects on diversity and molecular evolution. Even in highly selfing populations with large population sizes, linkage disequilibrium may not extend over large genomic regions, and the efficacy of natural selection may not be low.

Reduced effectiveness of selection on amino-acid-altering mutations can be inferred by increases in the proportion of nonsynonymous relative to synonymous substitutions (Wright et al. 2002) or polymorphisms (Bustamante et al. 2002). In addition, a reduced efficacy of selection on biased codon usage can be detected via elevation of unpreferred over preferred synonymous substitutions and polymorphisms (Marais et al. 2004). A comparison of substitution rates and codon bias between the highly selfing *Arabidopsis thaliana* and the self-incompatible *Arabidopsis lyrata* failed to find evidence for an effect of mating system for both amino acid substitutions and patterns of codon bias (Wright et al. 2002). This suggests that there is not a major decline in effectiveness of selection, at least in these taxa. Similarly, in a comparative analysis of nuclear and chloroplast DNA sequence data from multiple species, Glemin et al. (2006) found no effect of inbreeding on codon usage, with the exception of comparisons in the Poaceae. In this family, differences in base composition were observed for both synonymous sites and noncoding sites, suggesting that this result is unlikely to be caused by differences in selection on codon usage bias.

Although the above comparisons indicate little effect of inbreeding on the effectiveness of selection, these tests focused on effects on substitution rates between species. Such tests would not detect elevation of slightly deleterious mutations if selfing evolved recently, nor could they reveal any reduction in positive selection on amino acids in selfers because they assume that nonsynonymous substitutions are predominantly slightly deleterious in their effects. A high proportion of rare nonsynonymous polymorphisms has been reported in the inbreeding wild barley *Hordeum vulgare* ssp. *spontaneum* (Cummings and Clegg 1998) and *A. thaliana* (Bustamante et al. 2002; Nordborg et al. 2005), although explicit comparisons of polymorphism and divergence with related outcrossing species have not, as yet, been undertaken. Using a Bayesian analysis of the ratios of polymorphism to divergence for amino acid replacement and synonymous changes, Bustamante et al. (2002) found evidence for excess amino acid polymorphism and a reduction in positive selection in selfers. However, that study compared *A. thaliana* with *Drosophila melanogaster*, and clearly, more detailed analyses of polymorphism and divergence for close relatives with contrasting mating systems are needed.

In their analyses of sequence variation in plant species with contrasting mating systems, Glemin et al. (2006) detected a weak but significant elevation in the ratio of amino acid to synonymous polymorphism in species with high selfing rates. This suggests that there may be an elevated frequency of mildly deleterious polymorphisms segregating in selfing species. However, much of the data they reviewed are restricted to a small number of taxa and/or loci, and direct comparisons of polymorphism and divergence for the same loci are sparse. Large-scale sampling of polymorphism and divergence among close relatives will be important in further assessing whether inbreeding plays a significant role in coding sequence evolution. Currently, comparisons of patterns of amino acid polymorphism and divergence in *Arabidopsis* do not suggest a significant difference between selfing and outcrossing species (J. P. Foxe, H. Zheng, M. Nordborg, B. S. Gaut, and S. I. Wright, unpublished data).

Limited evidence suggests that there may be an elevation of nonsynonymous polymorphism with increased selfing. However, there is little to indicate that inbreeding leads to a sufficient decline in the effectiveness of selection to influence substitution rates and genome evolution. This suggests that there may be relatively few sites experiencing sufficiently weak additive selection to be affected by changes in selfing rate. As a result, the power to detect subtle changes in the patterns of molecular evolution may be too low, given the sample of genes studied to date.

In addition, large population sizes may reduce linkage disequilibrium in the selfing taxa studied. *Arabidopsis thaliana* is a predominantly selfing annual with a near-worldwide distribution. This successful colonizing species exhibits high nucleotide diversity and retains a fairly rapid decay of linkage disequilibrium with physical distance in rangewide population samples, despite very high selfing rates (Nordborg et al. 2005). Although within-population diversity is much lower in *A. thaliana* than in populations of the self-incompatible *A. lyrata* (Savolainen et al. 2000; Wright et al. 2003*b*; Ramos-Onsins et al. 2004), this does not necessarily imply a general reduction in the efficacy of selection. If selection against slightly deleterious mutations is effective in a metapopulation-type structure, the effective size across populations in *A. thaliana* could counteract reduced effective recombination rates. Furthermore, outcrossing populations can also experience population bottlenecks, reducing variation and increasing linkage disequilibrium to levels comparable to those of inbreeders (Wright et al. 2003*b*). A very broad taxonomic comparison is illustrative; humans have an order of magnitude lower level of variability and much more extensive linkage disequilibrium than do worldwide samples of *A. thaliana* (Nordborg et al. 2005), and they show a comparable, or more severe, genome-wide excess of slightly deleterious amino acid polymorphism (Bustamante et al. 2005). Similarly, low linkage disequilibrium has recently been demonstrated in highly selfing wild barley populations (Morrell et al. 2005). In contrast, recent comparisons of outcrossing and selfing species of *Caenorhabditis* identified very extensive levels of linkage disequilibrium in the selfing *C. elegans*, compared with outcrossing congeners (Cutter et al. 2006). Thus, variation in the population history of a species may contribute to variation in fixation rates at weakly selected sites as much as, or more than, mating system, swamping out any effect of the latter.

The role of purging and large population sizes in eliminating deleterious mutations in selfing populations may explain the contrast between the results to date for selfing species and those found for nonrecombining Y chromosomes in plants and animals, where recurrent evidence for deleterious mutation accumulation has been found (Charlesworth and Charlesworth 2000). In inbreeders, deleterious mutations are fully expressed as homozygotes, so fixation is directly dependent on the selection coefficient. In contrast, because of the persistence of functional gene copies on the X chromosome, Y chromosome fixation probabilities will be determined by hS, the product of the selection coefficient (S) and the dominance coefficient (h). In the extreme case, mutations that are completely recessive in their fitness effects can fix neutrally on Y chromosomes, despite strong selection coefficients in homozygotes. Thus, if deleterious mutations are strongly recessive, the permanent heterozygosity of the Y chromosome brings a much larger proportion of deleterious mutations into a parameter space that allows their fixation. Perhaps even more important, the non-recombining portion of Y chromosomes exhibits near-complete recombination suppression, whereas highly selfing species can more easily escape the linkage effects, for example, by having larger population sizes. The complete suppression of recombination in Y chromosomes leads to much stronger hitchhiking than is experienced even with very high selfing in large populations.

Evolution of Recombination Rate

The increase in homozygosity as a result of selfing reduces opportunities for recombination to break down associations among alleles by crossing over between heterozygous loci (Nordborg 2000). This increases linkage disequilibrium and decreases the effective rate of recombination. However, few species that engage in high levels of selfing are exclusively selfing (Barrett and Eckert 1990; Igic and Kohn 2006), and therefore, some opportunity exists for recombination during rare outcrossing events. It is therefore possible that natural selection may favor modifiers that increase rates of physical recombination or crossing over in selfers to offset the effects of inbreeding. Although theory predicts that mating systems may play an important role in the evolution of recombination rates, this area has received relatively little empirical attention.

Recombination is generally thought to be advantageous because it breaks down associations between alleles (linkage disequilibrium) and is favored under several non–mutually exclusive conditions, for example, with weak negative epistatic associations among mutations (Feldman et al. 1980; Kondrashov 1982, 1988; Barton 1995), when directional selection negatively covaries between habitats (Lenormand and Otto 2000), and when interference among selected sites is strong, such as in populations with small effective population size (Fisher 1930; Muller 1932; Otto and Barton 1997, 2001; Barton and Otto 2005; Keightley and Otto 2006). Under these conditions, recombination increases the variance in fitness and can introduce higher-fitness alleles on the same genetic background.

Models exploring the fate of recombination modifiers have mostly assumed random mating, and little attention has been given to how the mating system may alter the outcome of these models. Simulations indicate that hitchhiking between a recombination modifier and a pair of selectively important loci is stronger with selfing and that this difference can, under some conditions, favor the evolution of higher recombination rates. In contrast, the recombination rate is generally driven downward with random mating (Charlesworth et al. 1977, 1979; Holsinger and Feldman 1983a). Supporting these results, Roze and Lenormand (2005) generated an analytical model and found that even small amounts of selfing can greatly increase the range of parameters under which selection favors increased recombination.

In small populations, genetic drift is primarily responsible for generating negative linkage disequilibrium that favors recombination modifiers, increasing rates of crossing over (Otto and Barton 2001). The strength of indirect selection is stronger when linkage among loci is tight and population size is small; with tight linkage, a broader parameter space of population size favors recombination modifiers (Barton and Otto 2005). Although not modeled explicitly, these results suggest that with

self-fertilization, there may be stronger selection on recombination modifiers. Although details of the models vary, there is a general agreement that a fairly broad set of conditions favor an increase in the recombination rate with increased selfing. Support for these predictions is provided by comparisons of chiasma frequency in plant species with contrasting mating systems (Ross-Ibarra 2004; Roze and Lenormand 2005).

An alternative method for estimating the rate of crossing over per base pair is to integrate genetic linkage maps with physical maps. This allows for estimates of the average frequency of crossing over per physical length and a more detailed view of rate heterogeneity in different genomic regions, although the accuracy of cytological versus genetic estimates of recombination rate has been subject to debate (Nilsson et al. 1993). Such integrated maps are available for several model systems and agricultural taxa. However, the only detailed comparison of map-based recombination rates in closely related species with contrasting mating systems is between *Arabidopsis thaliana* and *Arabidopsis lyrata* (Kuittinen et al. 2004; Hansson et al. 2006; Kawabe et al. 2006). As predicted, overall rates of recombination per unit physical length are higher in *A. thaliana*. The degree of difference in recombination rate between the two *Arabidopsis* species varies across different genomic regions.

In the *Arabidopsis* comparison, the contrast in overall rate of recombination is complicated by a shift in genome size. *Arabidopsis thaliana* has a reduced genome size, and there is a general trend that rates of recombination per base pair decrease with increasing genome size (Ross-Ibarra 2006). Therefore, it is difficult to rule out an effect of genome size difference rather than direct selection on recombination modifiers. *Arabidopsis lyrata* has a larger genome and potentially more heterochromatic, nonrecombining DNA than *A. thaliana*. Therefore, the difference in average recombination rates across large regions could result from other aspects of genome evolution in addition to selection on recombination modifiers, such as differences in the accumulation of TEs. Nevertheless, using phylogenetically independent contrasts of 142 species with genome size as a covariate, Ross-Ibarra (2006) detected elevated chiasma frequencies in selfers, suggesting higher rates of recombination than in related outcrossers. In addition to overall recombination frequencies, outcrossing rates have also been shown to correlate with the ratio of recombination rates in female function relative to male function. Lenormand and Dutheil (2005) reported that selfing species tend to have a higher male : female ratio of recombination rates than outcrossing species. Their interpretation is that because pollen competition is reduced in selfing species, this relaxes selection against recombination breaking up favorable epistatic allelic combinations in pollen. Given that most estimates of chiasma frequencies have been determined in pollen, it will be important to test whether sex-averaged recombination rates are generally elevated in selfing species.

Relations between Mating System and Genomic Mutation Rate

Beneficial mutations are ultimately necessary for novel adaptations to evolve. However, the relative input of beneficial and deleterious mutations determines the evolution of genome-wide mutation rate, which may be strongly influenced by mating system. In a random-mating population, a modifier that increases the mutation rate should be selected against because it introduces more deleterious than beneficial mutations. Therefore, the fact that estimates of mutation rate are all above zero presumably reflects physical constraints on further reductions or a trade-off between fitness effects of new deleterious mutations and the amount of energy required for higher-fidelity DNA replication and repair (see Sniegowski et al. 2000). However, in selfing and asexual populations, strong linkage between a modifier that increases the mutation rate and a resulting beneficial mutation may result in indirect selection for the modifier. It is therefore possible to achieve an equilibrium rate of mutation that is above zero without invoking any physical constraints. The equilibrium value depends both on the extent of linkage disequilibrium and on the relative frequency and effect of deleterious and beneficial mutations. However, while many models have investigated the effects of indirect selection in asexual populations (Kimura 1967; Leigh 1970; Eshel 1973; Painter 1975; Woodcock and Higgs 1996; Orr 2000), only one has explicitly modeled the effect of self-fertilization (Holsinger and Feldman 1983b). Furthermore, there is a paucity of empirical tests of these predictions. We summarize advances made in understanding the effect of mating system, specifically selfing, on the evolution of genomic mutation rate.

The effective rate of recombination in highly selfing populations is low. Therefore, a modifier that increases the mutation rate will have a higher probability of linkage to a beneficial mutation and will be indirectly selected. In a highly selfing population, a modifier locus that alters the mutation rate at a second locus is predicted to reach a nonzero equilibrium mutation rate when overdominance favors the novel heterozygote genotype (Holsinger and Feldman 1983b). This result is robust for a nontrivial amount of outcrossing (<10%), making it more relevant to natural systems, which, as discussed earlier, are rarely, if ever, completely selfing. Johnson (1999) considered a mutation modifier that generates mutations of varying effect across a genetic map instead of one that acts on a single selectively important locus. With smaller map sizes, beneficial mutations will have a stronger influence on the fate of a modifier than previously predicted. Interestingly, the strength of linkage disequilibrium, which is of primary importance, is much higher for selfers than for random-mating populations. This suggests that if a high proportion of mutations are beneficial, they may be more important in predominantly or partially selfing populations. However, if beneficial mutations are very rare and weakly selected, then deleterious mutations will cause stronger negative selection against increased mutation rates in selfing (or asexual) populations than in outcrossing populations (Kondrashov 1995; McVean and Hurst 1997; Dawson 1998). To distinguish between these possibilities, the distribution of mutational effects is required, yet empirical estimates of the relative proportion of beneficial and deleterious mutations are largely unavailable.

Rates of deleterious mutations between predominately outcrossing *Amsinckia douglasiana* and selfing *Amsinckia gloriosa* have been compared in a mutation accumulation (MA) experiment (Schoen 2005). Although both lineages showed declines in fitness because of an accumulation of mutations, there was no significant difference associated with the mating system. Only three other MA experiments have been conducted in plants,

two with *Arabidopsis thaliana* (Schultz et al. 1999; Shaw et al. 2000) and another with *Triticum durum* (hard wheat; Bataillon et al. 2000). A problem inherent to MA studies is the difficulty of conducting experiments with organisms with long generation times. Because most mutations are of small effect, many generations are required for the power necessary to detect differences among groups. MA studies are primarily concerned with the rate of deleterious mutations and therefore underestimate the total mutation rate across the genome. Further, the exact measure of U (deleterious mutations per genome per generation) is dependent on the underlying distribution of mutational effects, and estimates of U can vary by orders of magnitude, depending on the distribution that is assumed (Shaw et al. 2000, 2002).

An alternative method of estimating mutation rates is by comparing nucleotide sequences of divergent lineages. The level of neutral divergence between two lineages is equal to $2\mu t$, where μ is the mutation rate (per nucleotide per generation) and t is the number of generations since divergence (Kimura 1968). However, this assumes a constant mutation rate and is therefore not useful for testing whether there are rate differences between the lineages. The relative rates test (Sarich and Wilson 1967; Wu and Li 1985; Tajima 1993) can be used to test for heterogeneity of rates among lineages. While there are a number of studies that apply this technique to noncoding and silent sites of nuclear loci in plants (e.g., Gaut et al. 1996, 1999; Wright et al. 2002; Senchina et al. 2003), there are none that do so for enough loci to estimate a mean mutation rate across the genome. In fact, substantial heterogeneity of mutation rates among loci in the above studies demonstrates the necessity of using many loci to estimate a genomic mutation parameter. A parsimony-based study of substitution rates at 23 genes failed to detect a significant difference in neutral mutation rate between *A. thaliana* and *Arabidopsis lyrata* (Wright et al. 2002). A follow-up comparison did suggest a slight but significant elevation of synonymous substitutions in *A. thaliana* using a sample of 83 genes (Wright 2003). This contrast is complicated by differences in generation time because *A. thaliana* is annual and *A. lyrata* is biennial or perennial. However, Whittle and Johnston (2003) found no association of generation time with mutation rate in a comparison of 24 phylogenetically independent pairs of annual and perennial plants. Complete uncoupling of generation time from mating system will require broader-scale comparisons in a phylogenetic context.

Could higher mutation rates enhance the ability of selfers to adapt, counteracting the harmful effects of inbreeding? Both theory and data suggest that this is unlikely. In particular, because most mutations are deleterious, elevated mutation rates are likely to be transient outcomes of hitchhiking with beneficial mutations rather than long-term shifts that increase adaptation (Sniegowski et al. 2000). Overall, higher mutation rates should enhance the mutational meltdown predicted in inbreeding populations rather than reduce its deleterious consequences.

Mating Systems and Genetic Conflicts

"Selfish" genetic elements that enhance their own transmission, despite null or negative fitness consequences for the genome, represent a dominant component driving evolutionary divergence in eukaryotic genomes (Burt and Trivers 2006).

The persistence and spread of selfish genetic elements can be generally understood as a balance between their transmission advantage and any deleterious effects related to their activity. If the rate of spread of selfish genetic elements predominates over reduced fitness, such elements can persist and even become fixed in natural populations. Their activity can also represent a significant contribution to spontaneous mutation rates and genetic load (Lai et al. 1994; Houle and Nuzhdin 2004).

As we will discuss, transitions in mating system, including increased rates of selfing, are generally expected to have an important effect on the outcome of genomic conflicts. In general, highly inbred or asexual taxa are expected to experience reduced intragenomic conflict because the lack of outcrossing prevents the spread of selfish genetic elements into other genetic backgrounds, thereby increasing the variance in fitness and leading to stronger purifying selection (Cavalier-Smith 1980; Hickey 1982). Furthermore, because selfish genetic elements will remain in greater linkage disequilibrium with replicate elements causing harmful mutations, there may be selection on the elements themselves for reduced activity, as models have shown for the evolution of transposition rates (Charlesworth and Langley 1996). Finally, because many selfish genetic elements increase transmission rates by gaining overrepresentation in gametes when competing with alternative alleles, the presence of high homozygosity in selfing populations acts as a further deterrent to alleles that make use of meiotic drive to increase in frequency. With homozygosity, the competition among alleles that leads to drive is absent because individuals have two copies of either the driving allele or the wild-type allele but not both. Therefore, selfing ensures a second level of homogeneity not experienced by asexuals and should act further to inhibit the spread of selfish elements. Given the potential importance of selfish genetic element activity for mean population fitness and deleterious mutation rates, selective elimination of selfish elements by inbreeders could offset the predicted decline in fitness associated with selfing.

Transposable Elements

Equilibrium models of TE evolution indicate that stable copy numbers can be maintained in populations by a balance between transposition increasing copy number and the action of negative selection removing insertions from the population (Charlesworth and Langley 1989). In inbreeders, computer simulations (Wright and Schoen 1999) and analytical models (Morgan 2001) show that the spread and accumulation of TEs can be inhibited by the lack of outcrossing as a result of the reduced spread of elements between individuals, as well as the purging of insertions with deleterious recessive effects on fitness. As mentioned previously, self-regulated transposition is also more likely to evolve in inbreeders (Charlesworth and Langley 1996), bringing down the copy number compared with that of outcrossing relatives.

By contrast, if natural selection acts predominantly against TE insertions through ectopic (i.e., between-element) recombination events, causing chromosomal rearrangements in heterozygotes, there may be a strong relaxation of natural selection against TEs in selfers (Charlesworth and Charlesworth 1995). This could lead to rapid accumulation and pop-

ulation fixation (Wright and Schoen 1999; Morgan 2001). The net outcome will depend on the underlying nature of selection on TEs and the history of selfing. For example, recent transitions to inbreeding may relax selection against ectopic exchange and may reduce effective population size, leading to increased frequencies and fixation of insertions. However, over the long term, the inability of new insertions with deleterious effects to spread through selfing populations may limit a rampant accumulation process. Simulations show that even under the ectopic exchange model, stochastic loss of elements from selfing populations is more frequent, potentially resulting in a net loss of TEs from selfing genomes (Wright and Schoen 1999). With a recent transition to selfing, this could lead to an effect in one direction on polymorphism patterns (i.e., increased TE frequencies at individual sites inherited from the ancestor), without a large copy number increase or with, perhaps, a decrease in copy number in selfers. Note that in either case, TE activity reduces fitness less in inbreeding than in outbreeding populations.

Early results suggest a decline in TE abundance in selfing species, consistent with the basic predictions of deleterious insertion models. Morgan (2001) reviewed evidence from a number of species pairs, showing a general reduction in copy number in inbreeding species. Furthermore, given the strong correlation between TE abundance and genome size, preliminary evidence for reduced genome size in selfers is at least consistent with inbreeding playing a role in the elimination of selfish genetic elements. In a population genetic study, Ac-like TE insertions were at higher frequencies in populations of the selfing *Arabidopsis thaliana* compared with *Arabidopsis lyrata* (Wright et al. 2001). This suggests a relaxation of natural selection, but this was not accompanied by an accumulation of new element insertions, consistent with inhibition of new element activity. Similar results were obtained in a study of retrotransposable elements in tomato (Tam et al. 2007), although a lack of evidence for purifying selection for most transposons, even in outcrossing taxa, suggests that the patterns may reflect primarily neutral insertion polymorphism. Analysis of TE distributions in *A. thaliana* also suggested that recombination rate heterogeneity does not appear to influence TE accumulation in this species, as has been observed in outcrossing genomes (Wright et al. 2003a), and this pattern is also apparent in the selfing nematode *C. elegans* (Duret et al. 2000). Selection against ectopic recombination may be a weak force in most selfing genomes.

As with inbreeding, asexual species are expected to show reduced TE activity and abundance, particularly as a result of the inability of active elements to spread among individuals (Hickey 1982). In contrast to this prediction, analysis of sequence evolution of three retrotransposon families in four asexual plant species identified many copies and evidence for selective constraint on TEs in these taxa, suggesting that elements remain active (Docking et al. 2006). However, given the likely recent evolution of asexuality in these plant species (Docking et al. 2006), simulations indicated that the evidence for residual selective constraint in TEs was not unexpected, even if long-term reductions in activity are occurring. More direct comparisons of abundance and polymorphism patterns between related sexual and obligate asexual taxa would be useful to provide a direct test of the effects of transitions to asexuality on TE evolution.

B Chromosomes

B chromosomes are nonessential chromosomes found in addition to the basic set of chromosomes. They have now been identified in more than 2000 species (Burt and Trivers 2006), including more than a thousand plant species (Jones 1995), are often morphologically distinct, are usually smaller than essential chromosomes, and show numerical variation within and between individuals. B chromosomes may be neutral, positive, or, more often, harmful in their effects and are maintained in the genome via a meiotic drive mechanism, ensuring a greater representation in gametes than expected by chance (Camacho 2006).

Models have shown a strong effect of outcrossing rate on the equilibrium frequencies of B chromosomes as a result of reduced transmission and greater variance in fitness. For example, Burt and Trivers (1998) found that outcrossing rates below 50% lead to complete elimination of B chromosomes from populations, reducing significant fractions of "junk DNA" and any fitness costs associated with B chromosome maintenance. A comparative survey of the distribution of B chromosomes in 353 plant species from the United Kingdom (12.5% of which contained B chromosomes), using phylogenetically independent contrasts, confirmed a strong and consistent effect of mating system on the presence of B chromosomes (Burt and Trivers 1998). Three independent analyses performed to test for an association between the presence of B chromosomes and mating system revealed a positive correlation between outcrossing and the presence of B chromosomes in 16 out of 19 taxonomic contrasts. Three predominantly inbreeding species, namely, *Desmazaria rigidum*, *Poa annua*, and *Luzula campestris*, were found to contain B chromosomes. In these cases, it is possible that B chromosomes are beneficial. Alternatively, the timescale of mating system evolution may have been too recent for selection to have effectively purged these genetic elements. Theoretical work incorporating the transition to selfing will be important to better understand the timescales required for the selective elimination of B chromosomes after a shift in mating system.

Biased Gene Conversion

Gene conversion is the nonreciprocal copying of one stretch of DNA into another during recombination (Marais 2003). It has been argued that the genetic systems involved in this repair are biased and can lead to transmission distortion in favor of GC bases. For example, if an individual is heterozygous at a site for a G/T polymorphism, BGC will lead to an overrepresentation of G over T gametes, thus leading to biased transmission. The net effect is a selective advantage of GC over AT bases, with the selection coefficient being determined by the rate of BGC (Marais 2003). Several studies to date, using patterns of genome structure, population genetic data, and evidence from DNA mismatch repair processes, have found evidence for BGC in yeast (Birdsell 2002), *Drosophila* (Galtier et al. 2006), and mammals (Duret et al. 2006). However, to date, there is little evidence for the presence of BGC in plant genomes.

Because of the transmission advantage inherent in the process, GC bases under this model are effectively selfish genetic elements, where there is a fixation bias toward GC bases as a result of biased transmission. Because the process leads to an increased fixation probability of GC bases, it can have important consequences for base composition evolution. GC-BGC could lead to reduced fitness at sites under weak selection favoring A-T nucleotides, such as plant introns (Ko et al. 1998).

In an outcrossing species, the efficacy of BGC is given by the product $N_e \gamma c$, where γ is the probability per generation that a given site is affected by a gene conversion tract L and c is the bias in favor of the GC allele (Nagylaki 1983a, 1983b; Galtier et al. 2001). However, because BGC will occur only in heterozygotes, the effective rate of this process will be reduced dramatically in highly selfing species. Analytical results have shown that the strength of BGC, and thus GC content, will be reduced as a direct product of the selfing rate; because BGC will occur only in heterozygotes, an organism with an outcrossing rate of 1% will experience only 1% the level of BGC of an equivalent highly outcrossing species (Marais et al. 2004). As a result, the mating system could play a role in eliminating any effect of this process on base composition.

If BGC plays a major role in structuring genomic base composition, inbreeding species should generally evolve to become more AT rich, particularly at sites less constrained by other forms of selection, e.g., synonymous sites. In their analyses of genomic diversity in angiosperms, Glemin et al. (2006) examined the GC content of 10 species with contrasting mating systems. Although no effects were identified in most comparisons, outcrossing species in the Poaceae were found to have GC content significantly higher than that of selfers, as measured by total GC, GC at third-codon positions, and GC in introns. Given the consistent elevation of GC across noncoding and coding sites, this pattern is unlikely to be explained by a difference between species in the efficacy of natural selection on codon usage bias. Instead, it is more likely to be the result of contrasting rates of BGC.

Similarly, an analysis of base composition in the Brassicaceae revealed a consistent reduction in GC content at third-codon positions in *A. thaliana* compared with outcrossing *Brassica oleracea* and *A. lyrata* (Wright et al. 2007). Analysis of base composition evolution controlling for gene expression and codon preferences indicated that the contrasting patterns were unlikely to result from differences in the effectiveness of selection on codon usage. Instead, the pattern appeared to be the direct result of a consistent decline in GC richness in *A. thaliana*, potentially as a result of a reduction in BGC following the recent evolution of selfing in this lineage (Nasrallah et al. 2004). Although alternative explanations, including shifts in the patterns of mutation bias, are possible, the difference is consistent with expectations under BGC. Comparisons of polymorphism and divergence for GC→AT versus AT→GC changes should help untangle the role of BGC versus mutation bias. In particular, if BGC is strong, AT→GC changes are expected to segregate at higher frequencies than GC→AT changes. This should result in higher frequencies and fixation rates of AT→GC changes in outcrossing species compared with selfing species, under the model of BGC. By contrast, no such effect is expected if the difference is driven by changes in mutation bias.

In addition to an overall reduction in GC content, the reduced opportunity for BGC could act to reduce heterogeneity in base composition across the genome in selfers. In particular, variation in GC content may be driven, in part, by variation in rates of recombination, assuming a tight correlation between estimated rates of recombination and rates of BGC. However, in a highly selfing species, the overall reduction in BGC should result in a weaker relation. Consistent with this, recombination rate and GC content are not positively correlated in *A. thaliana*, as they are in other genomes (Marais et al. 2004). In the future, it will be important to test for such a correlation using the complete genomes of related selfing and outcrossing taxa.

Cytoplasmic Male Sterility

Because of the predominant maternal inheritance of mitochondria and chloroplasts, cytoplasmic mutations reducing male fertility will be selectively favored in these genomes if they cause even a slight increase in female fitness, even if total reproductive output is reduced. However, increased female frequencies will often favor the spread of nuclear restorers suppressing male sterility. CMS is a maternally inherited condition leading to male infertility as a result of an inability to produce viable pollen. CMS has been documented in more than 150 plant species and can arise spontaneously in natural populations or after interspecific hybridization (Schnable et al. 1998). Although theoretical work on the population dynamics of plant cytonuclear conflict has focused on gynodioecious species (reviewed in Saur Jacobs and Wade 2003), any outcrossing hermaphroditic species is potentially susceptible to mitochondrial mutations that increase female fertility at a cost to pollen fertility. Indeed, the discovery of CMS in many interspecific crosses is suggestive of a hidden history of cytonuclear coevolution. With high selfing, the selective advantage for pollen sterility mutations in the cytoplasm disappears because CMS mutants reduce female fertility as well. Because of this, we expect much less selection on cytonuclear interactions in highly selfing species, and thus the spread of fitness-reducing pollen sterility mutations should be reduced or eliminated.

Theoretical models of gynodioecy (reviewed in Saur Jacobs and Wade 2003) suggest that fixation of both CMS and restorer alleles is a common evolutionary outcome, with fixation of the restorer allele bringing the population back to hermaphroditism. Evidence for the exposure of "cryptic" CMS through interspecific crosses of hermaphroditic plants is consistent with this hypothesis. If distinct restorer and CMS types have been fixed in different species or populations, wide crosses reexpose this evolutionary history. There are several conditions under which CMS and non-CMS (or restored and unrestored) individuals may be maintained in a population, leading to stable gynodioecy, including pollen limitation experienced by females as they increase in frequency and a fitness cost of the restorer allele (Budar et al. 2003). However, historical "epidemics" of CMS and restoration may be the common outcome, and outcrossing hermaphrodites may have a hidden history of CMS-restorer evolution.

In interspecific crosses, cryptic CMS should be exposed when the maternal parent is outcrossing but not when it is highly selfing. Recently, Fishman and Willis (2006) report cryptic CMS revealed in *Mimulus guttatus* (an outcrosser) × *Mimulus nasutus* (highly selfing) hybrids. They found that hybrid sterility was differentially expressed in the *M. guttatus* cytoplasmic background and that pollenless anther phenotypes

were recovered in F$_2$ hybrids with *M. guttatus* cytoplasm but not in the reciprocal hybrids. The lack of CMS phenotypes in the reciprocal F$_2$ hybrids is consistent with predictions from theoretical models because we would not expect to find any evidence for a CMS mutant in a highly inbreeding species. Does cryptic CMS exposed in interspecific crosses generally derive from highly or partially outcrossing species, as opposed to highly selfing taxa? This question is difficult to address because of the limited numbers of species for which quantitative estimates of outcrossing rate are available. It would be interesting to systematically examine pairs of closely related taxa with contrasting selfing rates to test for the degree to which CMS is exposed in interspecific crosses.

Because CMS most often results from novel chimeric mitochondrial proteins (Delph et al. 2007), we might expect the rate of mitochondrial genome structure evolution to be elevated in outcrossers compared to selfers. In addition, CMS-restorer dynamics may lead to a proliferation of members of the pentatricopeptide repeat (PPR) gene family acting as restorers in the nuclear genome. Evidence from genetic mapping of fertility restorers indicates that novel restorers often represent new or mutated forms of a multigene family of PPR genes (Brown et al. 2003; Wang et al. 2006).

In addition to CMS dynamics, antagonistic coevolution between the sex functions should generally be reduced in highly selfing populations as a result of the absence of conflict when the maternal and paternal parents are the same. Brandvain and Haig (2005) hypothesize that this effect will tend to cause asymmetric hybridization success because sexual conflict will tend to lead to growth suppression of offspring by the maternal parent and promote offspring growth by the paternal parent. The net effect in interspecific hybrids is that crosses will be more successful when the selfing species is the maternal parent than in the reciprocal case because sexual conflict is reduced in the inbreeder. A review of the literature generally supports this model and suggests that genomic imprinting driven by sexual conflict has declined in selfing species (Brandvain and Haig 2005).

Evolution of Genome Size

There is more than 1200-fold variation in nuclear DNA content (C value) in the angiosperms alone. A variety of phenotypic traits have been shown to be associated with C value, including cellular characteristics such as nucleus and cell size (Mirsky and Ris 1951; Price et al. 1973), duration of cell division (Van't Hof and Sparrow 1963; Van't Hof 1965; Bennett 1972), seed size (Beaulieu et al. 2007), and annual or biennial life form (Bennett 1972; Vinogradov 2001). It is possible that genome size has direct effects on these traits and is selected in association with these phenotypes. On the other hand, an alternative class of explanation invokes genetic drift as the primary determinant of genome size. Species with small effective population sizes will experience reduced efficacy of natural selection, leading to the accumulation of slightly deleterious insertions, resulting in increased genome size (Lynch and Conery 2003). Finally, genome size evolution may reflect the selective dynamics associated with TE evolution and other genomic conflicts and/or differences among species in the amount and pattern of DNA repair.

In highly selfing species, genome loss may predominate. Proposed evolutionary forces predicted to reduce genome size in selfers include mechanistic explanations, such as a lower accumulation of TEs and other selfish genetic elements or increased fixation of large underdominant deletions as a result of higher homozygosity in selfers (Charlesworth 1992). In addition, faster cell replication and generation time, both of which are negatively associated with C value, may be selected for in selfing species, leading to an association between selfing and smaller genomes. Annuals tend to have reduced genome size (Bennett 1972), so this would confound effects of selfing and life history because annuals often display higher rates of selfing than perennials (Barrett et al. 1996). Alternatively, if increased genome size is driven by reduced effective population size (Lynch and Conery 2003) or if relaxation of selection against ectopic recombination between TEs drives accumulation in selfers, we would expect increased genome size in selfing species.

There is preliminary evidence that mating system may have an effect on genome size evolution. In *Veronica*, selfers show a significant reduction in genome size, with mating system showing a stronger effect than between annual and perennial sister taxa (Albach and Greilhuber 2004). In their analysis of B chromosomes and genome size, Trivers et al. (2004) also found a strong positive correlation between outcrossing and genome size among plants in the United Kingdom (correcting for B chromosome presence/absence), although this effect was lost when they used phylogenetically independent contrasts.

A limitation of many comparative studies in relation to mating system is the lack of quantitative estimates of selfing rate. Many comparisons rely on descriptions from floras, usually based on floral morphology. Because most models predict a strong effect of inbreeding only with very high selfing rates, analyses based on morphological inferences could mask effects of mating system transitions. Here, we reexplore the effect of inbreeding on genome size by presenting a comparison of C value and genome size in pairs of outcrossing and highly selfing congeners. We refer to DNA amount in a postreplication nucleus; this is equivalent to the 4C DNA amount (C value). However, it is important to distinguish this measure from genome size, defined as the amount of DNA in a monoploid chromosome set. We therefore report genome size as the 4C DNA amount divided by the ploidy of the sample, consistent with the rationale provided by Bennett et al. (1998). A species was considered highly selfing if the outcrossing rate was less than 10% ($t_m < 0.10$). We combined two databases of outcrossing rate from Barrett and Eckert (1990) and modified by Igic and Kohn (2006) with the Kew C value database (http://www.rbgkew.org.uk/cval) and selected those species for which we had both a C value and an outcrossing estimate or evidence of self-incompatibility. We then identified pairs of congeners with contrasting mating systems. From this, there were 14 pairs from nine genera representing eight families, including both eudicots and monocots. A sign test was used to test the prediction that predominantly selfing species have smaller genomes.

While there was no significant pattern of C value reduction in selfers uncorrected for ploidy, there was a significant reduction in genome size in selfing species (fig. 3). This pattern is due to four pairs in which the selfer is polyploid and therefore

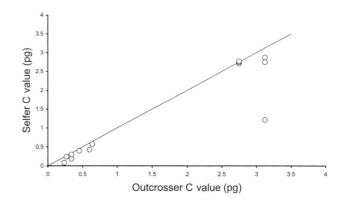

Fig. 3 C value comparisons for pairs of outcrossing (*X*-axis) and selfing (*Y*-axis) congeners. Each point represents the genome sizes for a single pair of congeners estimated as the 4C value divided by the ploidy. The line is the 1 : 1 line of equality. Note that *Bromus arvenis* and *Hordeum bulbosum* are used as the outcrossing species in comparisons with all selfing congeners in their respective genera. Species pairs are listed with the outcrosser, followed by the corresponding selfer and their respective genome size (in pg): *Amaranthus cruentus* (1.06), *Amaranthus hypochondriacus* (0.96); *Arabidopsis lyrata* (0.94), *Arabidopsis thaliana* (0.32); *Bromus arvensis* (12.5), *Bromus japonicus* (11); *Bromus rubens* (4.85); *Bromus squarrosus* (11.5); *Cerastium arvense* (1.36), *C. arvense fontanum* (0.73); *Glycine argyrea* (2.56), *Glycine soja* (2.3); *Hordeum bulbosum* (11), *Hordeum jubatum* (10.8); *Hordeum spontaneum* (11), *Hordeum vulgare* (11.1); *Phaseolus coccineus* (1.36), *Phaseolus vulgaris* (1.2); *Plantago lanceolata* (2.4), *Plantago major* (1.7); *Senecio squalidus* (1.8), *Senecio vulgaris* (1.58). Genome size data from the Kew C value database (Leitch and Bennett 2004; Angiosperm DNA C value database, http://www.rbgkew.org.uk/cval, ver. 5.0, Dec. 2004). Information on mating systems of species from Barrett and Eckert (1990), modified by Igic and Kohn (2006).

has a much higher DNA amount. In all four of these pairs, the actual genome size is reduced in the polyploid selfer, and there is a significant reduction in genome size across all pairs after controlling for ploidy (sign test, $P = 0.006$). This is consistent with a reduction in genome size in response to either polyploidization or selfing or both. Previous work has shown a general reduction in genome size in polyploids (Leitch and Bennett 2004), likely reflecting DNA loss following whole-genome duplication (Ku et al. 2000). If we exclude pairs with polyploid selfers from the analysis, both C value and genome size are significantly reduced in selfers, suggesting a reduction in predominantly selfing species, independent of ploidal level ($P = 0.016$). Further, after excluding those pairs with annual and perennial congeners, the pattern of reduced genome size in selfers remains significant ($P = 0.016$).

A weakness of our analysis is that in the contrasts involving monocots, two outcrossing species were used repeatedly in comparisons with different selfing species. If selfing evolved independently in each lineage, these should represent independent contrasts. If the selfing species have a shared evolutionary history, the analysis will introduce nonindependence. However, if we take the average contrasts for these groups as single data points, there is still a significant reduction in genome size in the selfing species in our data ($P < 0.01$).

These preliminary results are consistent with a reduction in genome size in self-fertilizing species, but they should be interpreted with caution because of the small sample size. Nevertheless, this is the first confirmation of a reduced genome size in selfers using taxonomically paired contrasts. It suggests that changes in selection associated with genomic conflict and/or life-history evolution may be more significant for driving genome evolution in selfers than reduced effective population size. The fact that there are four polyploid selfers out of the 14 contrasts may indicate that selection against increased DNA amount is not driving the pattern. Instead, changes to the monoploid genome size, such as a reduction in TEs, are possibly driving this trend. Also, the general association between polyploidy and selfing (see Husband et al. 2008) may have obscured the correlation between selfing rate and genome size in previous work, particularly if some plants in the C value database were incorrectly assumed to be diploid. The role of inbreeding in genome size evolution deserves more thorough investigation as more data become available, but to date, the results are consistent with the prediction of increased fitness and reduced deleterious mutation rates via genomic conflict in inbreeding taxa.

Conclusions and Future Directions

The preliminary evidence suggests a role for mating system differences in genome evolution. However, the relation between mating system transitions to selfing and reduced fitness associated with the harmful effects of linkage is not clear. Highly inbred taxa may, in some cases, avoid deleterious mutation accumulation through purging of deleterious mutations, large population sizes, the evolution of higher recombination rates, and the elimination of selfish genetic elements. Although there have been few rigorous comparative or experimental tests of the predictions and patterns outlined in this article, future investigations should become increasingly tractable. Completion of the genome of *Arabidposis lyrata*, along with that of the outgroup *Capsella rubella*, will present the first opportunity for large-scale comparison of substitution patterns and genome evolution between closely related species (*Arabidopsis thaliana* vs. *A. lyrata*) with contrasting mating systems. If molecular evolutionary rates and genome structure in these species are examined on a whole-genome scale, there will be extraordinary power to detect differences among species.

Although whole-genome analysis in *Arabidopsis* will be informative, taxonomic replication and a broader sampling of species, particularly using ecological model systems, will be crucial for robust generalizations to be made. This is because studies involving a single comparison are confounded with the history of the lineages involved. Despite independent evolutionary and coalescent history of genes across the *Arabidopsis* genome, providing a form of evolutionary replication, the entire genome has been influenced by the evolutionary and demographic history of the species sampled. Recent evidence suggests historical bottlenecks in some populations of *A. lyrata* (Wright et al. 2003b; Ramos-Onsins et al. 2004), large species-wide effective population size in *A. thaliana* (Nordborg et al. 2005), and the recent and potentially independent evolution of selfing from standing variation in *A. thaliana* (Nasrallah et al. 2004). Collectively, these all point to a complicated demographic history that may reduce the equivalence of the taxa. More widespread taxonomic replication would allow for a detailed picture of the role of mating system evolution and its interaction with ecology and demography.

If many mating system transitions have been recent, especially in short-lived herbaceous taxa, we will need a better theoretical understanding of the timescales and dynamics of genomic changes associated with transition to selfing. Nearly all of the theory discussed in this article assumes selfing populations at long-term equilibrium, and models have not considered the transition to selfing. Such theoretical work will be important in understanding the factors determining the success and/or extinction of selfing lineages. Species in which there is evidence for multiple independent shifts from outcrossing to selfing provide opportunities to examine the genomic consequences of mating system transitions and enable investigation of the interaction of mating patterns and demographic factors (e.g., *Amsinckia spectabilis* [Schoen et al. 1997] and *Eichhornia paniculata* [Husband and Barrett 1993]). Polymorphism and divergence data should also be integrated with genetic information on population structure, gene flow, effective population size, and phylogeographic history. It is critical to consider these effects because while mating system alone can influence effective population size through the processes outlined in our review, diverse ecological factors can also play a role in shaping population genetic structure. A more comprehensive understanding of genome evolution in plants will require information on the interactions among demography, life history, and mating system and how these govern genetic parameters.

Acknowledgments

We thank S. Glemin for allowing us to cite unpublished work, two anonymous reviewers for helpful comments on the manuscript, and the Natural Sciences and Engineering Council of Canada for Discovery Grants to S. I. Wright and S. C. H. Barrett that funded this work. S. I. Wright also thanks Deborah Charlesworth and Magnus Nordborg for numerous discussions related to these issues.

Literature Cited

Albach DC, J Greilhuber 2004 Genome size variation and evolution in *Veronica*. Ann Bot 94:897–911.

Barrett SCH, CG Eckert 1990 Variation and evolution of mating systems in seed plants. Pages 229–254 *in* S Kawano, ed. Biological approaches and evolutionary trends in plants. Academic Press, Tokyo.

Barrett SCH, LD Harder, AC Worley 1996 The comparative biology of pollination and mating in flowering plants. Philos Trans R Soc B 351:1271–1280.

Barton NH 1995 A general model for the evolution of recombination. Genet Res 65:123–144.

Barton NH, SP Otto 2005 Evolution of recombination due to random drift. Genetics 169:2353–2370.

Bataillon T, P Roumet, S Poirier, JL David 2000 Measuring genome-wide spontaneous mutation in *Triticum durum*: implications for long-term selection response and management of genetic resources. Pages 251–257 *in* A Gallais, C Dillmann, I Goldringer, eds. Eucarpia: quantitative genetics and breeding methods—the way ahead. INRA Editions, Versailles.

Beaulieu JM, AT Moles, IJ Leitch, MD Bennett, JB Dickie, CA Knight 2007 Correlated evolution of genome size and seed mass. New Phytol 173:422–437.

Bennett MD 1972 Nuclear DNA content and minimum generation time in herbaceous plants. Proc R Soc B 181:109–135.

Bennett MD, IJ Leitch, L Hanson 1998 DNA amounts in two samples of angiosperm weeds. Ann Bot 82(suppl):121–134.

Birdsell JA 2002 Integrating genomics, bioinformatics, and classical genetics to study the effects of recombination on genome evolution. Mol Biol Evol 19:1181–1197.

Brandvain Y, D Haig 2005 Divergent mating systems and parental conflict as a barrier to hybridization in flowering plants. Am Nat 166:330–338.

Brown GG, N Formanová, H Jin, R Wargachuk, C Dendy, P Patil, M Laforest, J Zhang, WY Cheung, B Landry 2003 The radish *rfo* restorer gene of *Ogura* cytoplasmic male sterility encodes a protein with multiple pentatricopeptide repeats. Plant J 35:262–272.

Budar F, P Touzet, R De Paepe 2003 The nucleo-mitochondrial conflict in cytoplasmic male sterilities revisited. Genetica 117:3–16.

Burt A, R Trivers 1998 Selfish DNA and breeding system in flowering plants. Proc R Soc B 265:141–146.

——— 2006 Genes in conflict: the biology of selfish genetic elements. Belknap Press of Harvard University Press, Cambridge, MA.

Bustamante CD, A Fledel-Alon, S Williamson, R Nielsen, M Todd Hubisz, S Glanowski, DM Tanenbaum, et al 2005 Natural selection on protein-coding genes in the human genome. Nature 437: 1153–1157.

Bustamante CD, R Nielsen, SA Sawyer, KM Olsen, MD Purugganan, DL Hartl 2002 The cost of inbreeding in *Arabidopsis*. Nature 416: 531–534.

Camacho JPM 2006 B chromosomes. Pages 224–273 *in* TR Gregory, ed. The evolution of the genome. Elsevier Academic, Burlington, MA.

Cavalier-Smith T 1980 How selfish is DNA? Nature 285:617–618.

Charlesworth B 1992 Evolutionary rates in partially self-fertilizing species. Am Nat 140:126–148.

Charlesworth B, D Charlesworth 2000 The degeneration of Y chromosomes. Philos Trans R Soc B 355:1563–1572.

Charlesworth B, CH Langley 1989 The population genetics of *Drosophila* transposable elements. Annu Rev Genet 23:251–287.

——— 1996 The evolution of self-regulated transposition of transposable elements. Genetics 112:359–383.

Charlesworth B, MT Morgan, D Charlesworth 1993 The effects of deleterious mutations on neutral molecular variation. Genetics 134: 1289–1303.

Charlesworth D, B Charlesworth 1995 Transposable elements in inbreeding and outbreeding populations. Genetics 140:415–417.

Charlesworth D, B Charlesworth, C Strobeck 1977 Effects of selfing on selection for recombination. Genetics 86:213–226.

——— 1979 Selection for recombination in partially self-fertilizing populations. Genetics 93:237–244.

Charlesworth D, S Wright 2001 Breeding systems and genome evolution. Curr Opin Genet Dev 11:685–690.

Chen M 2007 Genetic and epigenetic mechanisms for gene expression and phenotypic variation in plant polyploids. Annu Rev Plant Biol 58:377–406.

Crnokrak P, SCH Barrett 2002 Purging the genetic load: a review of the experimental evidence. Evolution 56:2347–2358.

Cummings MP, MT Clegg 1998 Nucleotide sequence diversity at the alcohol dehydrogenase 1 locus in wild barely (*Hordeum vulgare* ssp.

spontaneum): an evaluation of the background selection hypothesis. Proc Natl Acad Sci USA 95:5637–5642.

Cutter AD, JD Wasmuth, ML Blaxter 2006 The evolution of biased codon and amino acid usage in nematode genomes. Mol Biol Evol 23:2303–2315.

Dawson KJ 1998 Evolutionarily stable mutation rates. J Theor Biol 194:143–157.

Delph LF, P Touzet, MF Bailey 2007 Merging theory and mechanism in studies of gynodioecy. Trends Ecol Evol 22:17–24.

Docking TR, FE Saade, MC Elliott, DJ Schoen 2006 Retrotransposon sequence variation in four asexual plant species. J Mol Evol 62: 375–387.

Duret L, A Eyre-Walker, N Galtier 2006 A new perspective on isochore evolution. Gene 385:71–74.

Duret L, G Marais, C Biemont 2000 Transposons but not retrotransposons are located preferentially in regions of high recombination rate in *Caenorhabditis elegans*. Genetics 156:1661–1669.

Eshel I 1973 Clone-selection and optimal rates of mutation. J Appl Prob 10:728–738.

Eyre-Walker A, PD Keightley 2007 The distribution of fitness effects of new mutations. Nat Rev Genet 8:610–618.

Feldman MW, FB Christiansen, LD Brooks 1980 Evolution of recombination in a constant environment. Proc Natl Acad Sci USA 77: 4838–4841.

Fisher RA 1930 The genetical theory of natural selection. Oxford University Press, Oxford.

Fishman L, JH Willis 2006 A cytonuclear incompatibility causes anther sterility in *Mimulus* hybrids. Evolution 60:1372–1381.

Galtier N, E Bazin, N Bierne 2006 GC-biased segregation of noncoding polymorphisms in *Drosophila*. Genetics 172:221–228.

Galtier N, G Piganeau, D Mouchiroud, L Duret 2001 GC-content evolution in mammalian genomes: the biased gene conversion hypothesis. Genetics 159:907–911.

Gaut BS, BR Morton, BC McCaig, MT Clegg 1996 Substitution rate comparisons between grasses and palms: synonymous rate differences at the nuclear gene *adh* parallel rate differences at the plastid gene *rbc*L. Proc Natl Acad Sci USA 93:10274–10279.

Gaut BS, AS Peek, BR Morton, MR Duvall, MT Clegg 1999 Patterns of genetic diversification within the *adh* gene family in the grasses (Poaceae). Mol Biol Evol 16:1086–1097.

Glemin S, E Bazin, D Charlesworth 2006 Impact of mating systems on patterns of sequence polymorphism in flowering plants. Proc R Soc B 273:3011–3019.

Hamrick JL, MJ Godt 1996 Effects of life history traits on genetic diversity in plant species. Philos Trans R Soc B 351:1291–1298.

Hansson B, A Kawabe, S Preuss, H Kuittinen, D Charlesworth 2006 Comparative gene mapping in *Arabidopsis lyrata* chromosomes 1 and 2 and the corresponding *A. thaliana* chromosome 1: recombination rates, rearrangements and centromere location. Genet Res 87:75–85.

Hickey DA 1982 Selfish DNA: a sexually transmitted nuclear parasite. Genetics 101:519–531.

Holsinger KE, MW Feldman 1983a Linkage modification with mixed random mating and selfing: a numerical study. Genetics 103:323–333.

——— 1983b Modifiers of mutation rate: evolutionary optimum with complete selfing. Proc Natl Acad Sci USA 80:6732–6734.

Houle D, S Nuzhdin 2004 Mutation accumulation and the effect of copia insertions in *Drosophila melanogaster*. Genet Res 83:7–18.

Husband BC, SCH Barrett 1993 Multiple origins of self-fertilization in tristylous *Eichhornia paniculata* (Pontederiaceae): inferences from style morph and isozyme variation. J Evol Biol 6:591–608.

Husband BC, B Ozimec, SL Martin, L Pollock 2008 Mating consequences of polyploid evolution in flowering plants: current trends and insights from synthetic polyploids. Int J Plant Sci 169:195–206.

Igic B, JR Kohn 2006 The distribution of plant mating systems: study bias against obligately outcrossing species. Evolution 60:1098–1103.

Igic B, R Lande, JR Kohn 2008 Loss of self-incompatibility and its evolutionary consequences. Int J Plant Sci 169:93–104.

Johnson T 1999 Beneficial mutations, hitchhiking and the evolution of mutation rates in sexual populations. Genetics 151:1621–1631.

Jones RN 1995 B chromosomes in plants. New Phytol 131:411–434.

Kawabe A, B Hansson, A Forrest, J Hagenblad, D Charlesworth 2006 Comparative gene mapping in *Arabidopsis lyrata* chromosomes 6 and 7 and *A. thaliana* chromosome IV: evolutionary history, rearrangements and local recombination rates. Genet Res 88:45–56.

Keightley PD, SP Otto 2006 Interference among deleterious mutations favours sex and recombination in finite populations. Nature 443:89–92.

Kimura M 1967 On evolutionary adjustment of spontaneous mutation rates. Genet Res 9:23–34.

——— 1968 Evolutionary rate at the molecular level. Nature 217:624–626.

Ko CH, V Brendel, RD Taylor, V Walbot 1998 U-richness is a defining feature of plant introns and may function as an intron recognition signal in maize. Plant Mol Biol 36:573–583.

Kondrashov AS 1982 Selection against harmful mutations in large sexual and asexual populations. Genet Res 40:325–332.

——— 1988 Deleterious mutations and the evolution of sexual reproduction. Nature 336:435–440.

——— 1995 Modifiers of mutation-selection balance: general approach and the evolution of mutation rates. Genet Res 66:53–69.

Ku H-M, T Vision, J Liu, SD Tanksley 2000 Comparing sequenced segments of the tomato and *Arabidopsis* genomes: large-scale duplication followed by selective gene loss creates a network of synteny. Proc Natl Acad Sci USA 97:9121–9126.

Kuittinen H, AA de Haan, C Vogl, S Oikarinen, J Leppala, M Koch, T Mitchell-Olds, CH Langley, O Savolainen 2004 Comparing the linkage maps of the close relatives *Arabidopsis lyrata* and *A. thaliana*. Genetics 168:1575–1584.

Lai CG, RF Lyman, AD Long, CH Langley, TFC Mackay 1994 Naturally occurring variation in bristle number and DNA polymorphisms at the scabrous locus of *Drosophila melanogaster*. Science 266:1697–1702.

Leigh EG 1970 Natural selection and mutability. Am Nat 104:301–305.

Leitch I, M Bennett 2004 Genome downsizing in polyploid plants. Biol J Linn Soc 82:651–663.

Lenormand T, J Dutheil 2005 Recombination difference between sexes: a role for haploid selection. PLoS Biol 3:396–403.

Lenormand T, SP Otto 2000 The evolution of recombination in a heterogeneous environment. Genetics 156:423–438.

Lynch M, JS Conery 2003 The origins of genome complexity. Science 302:1401–1404.

Lynch M, J Conery, R Bürger 1995 Mutation accumulation and the extinction of small populations. Am Nat 146:489–518.

Marais G 2003 Biased gene conversion: implications for genome and sex evolution. Trends Genet 19:330–338.

Marais G, B Charlesworth, SI Wright 2004 Recombination and base composition: the case of the highly self-fertilizing plant *Arabidopsis thaliana*. Genome Biol 5:R45.

McVean GT, B Charlesworth 2000 The effects of Hill-Robertson interference between weakly selected mutations on patterns of molecular evolution and variation. Genetics 155:929–944.

McVean GT, LD Hurst 1997 Evidence for a selectively favourable reduction in the mutation rate of the X chromosome. Nature 386:388–392.

Mirsky AE, H Ris 1951 The DNA content of animal cells and its evolutionary significance. J Gen Physiol 34:451–462.

Morgan MT 2001 Transposable element number in mixed mating populations. Genet Res 77:261–275.

Morrell PL, DM Toleno, KE Lundy, MT Clegg 2005 Low levels of linkage disequilibrium in wild barley (*Hordeum vulgare* ssp. *sponta-

neum*) despite high rates of self-fertilization. Proc Natl Acad Sci USA 27:3289–3294.

Muller HJ 1932 Some genetic aspects of sex. Am Nat 66:118–138.

Nagylaki T 1983a Evolution of a finite population under gene conversion. Proc Natl Acad Sci USA 80:6278–6281.

——— 1983b Evolution of a large population under gene conversion. Proc Natl Acad Sci USA 80:5941–5945.

Nasrallah ME, P Liu, S Sherman-Broyles, NA Boggs, JB Nasrallah 2004 Natural variation in expression of self-incompatibility in *Arabidopsis thaliana*: implications for the evolution of selfing. Proc Natl Acad Sci USA 101:16070–16074.

Nilsson NO, T Sall, BO Bengtsson 1993 Chiasma and recombination data in plants: are they compatible? Trends Genet 9:344–348.

Nordborg M 2000 Linkage disequilibrium, gene trees and selfing: ancestral recombination graph with partial self-fertilization. Genetics 154:923–939.

Nordborg M, TT Hu, Y Ishino, J Jhaveri, C Toomajian, H Zheng, E Bakker, et al 2005 The pattern of polymorphism in *Arabidopsis thaliana*. PLoS Biology 3:1289–1299.

Orr HA 2000 The rate of adaptation in asexuals. Genetics 155:961–968.

Otto SP, NH Barton 1997 The evolution of recombination: removing the limits to natural selection. Genetics 147:879–906.

——— 2001 Selection for recombination in small populations. Evolution 55:1921–1931.

Painter PR 1975 Mutator genes and selection for mutation rate in bacteria. Genetics 79:649–660.

Price HJ, AH Sparrow, AF Nauman 1973 Correlations between nuclear volume, cell volume and DNA content in meristematic cells of herbaceous angiosperms. Experientia 29:1028–1029.

Ramos-Onsins SE, BE Stranger, T Mitchell-Olds, M Aguade 2004 Multilocus analysis of variation and speciation in the closely related species *Arabidopsis halleri* and *A. lyrata*. Genetics 166:373–388.

Ross-Ibarra J 2004 The evolution of recombination under domestication: a test of two hypotheses. Am Nat 163:105–112.

——— 2006 Genome size and recombination in angiosperms: a second look. J Evol Biol 20:800–803.

Roze D, T Lenormand 2005 Self-fertilization and the evolution of recombination. Genetics 170:841–857.

Sarich VM, AC Wilson 1967 Immunological time-scale for hominid evolution. Science 150:1200–1203.

Saur Jacobs M, MJ Wade 2003 A synthetic review of the theory of gynodioecy. Am Nat 161:837–851.

Savolainen O, CH Langley, BP Lazzaro, H Freville 2000 Contrasting patterns of nucleotide polymorphism at the alcohol dehydrogenase locus in the outcrossing *Arabidopsis lyrata* and the selfing *Arabidopsis thaliana*. Mol Biol Evol 17:645–655.

Schnable P, S Wise, P Roger 1998 The molecular basis of cytoplasmic male sterility and fertility restoration. Trends Plant Sci 3:175–180.

Schoen DJ 2005 Deleterious mutation in related species of the plant genus *Amsinckia* with contrasting mating systems. Evolution 59:2370–2377.

Schoen DJ, AHD Brown 1991 Intraspecific variation in population gene diversity and effective population size correlates with the mating system in plants. Proc Natl Acad Sci USA 88:4494–4497.

Schoen DJ, JW Busch 2008 On the evolution of self-fertilization in a metapopulation. Int J Plant Sci 169:119–127.

Schoen DJ, MO Johnston, A L'Heureux, JV Marsolais 1997 Evolutionary history of the mating system in *Amsinckia* (Boraginaceae). Evolution 51:1090–1099.

Schultz ST, M Lynch, JH Willis 1999 Spontaneous deleterious mutation in *Arabidopsis thaliana*. Proc Natl Acad Sci USA 96:11393–11398.

Senchina DS, I Alvarez, RC Cronn, B Liu, J Rong, RD Noyes, AH Paterson, RA Wing, TA Wilkins, JF Wendel 2003 Rate variation among nuclear genes and the age of polyploidy in *Gossypium*. Mol Biol Evol 20:633–643.

Shaw FH, CJ Geyer, RG Shaw 2002 A comprehensive model of mutations affecting fitness and inferences for *Arabidopsis thaliana*. Evolution 56:453–463.

Shaw RG, DL Byers, E Darmo 2000 Spontaneous mutational effects on reproductive traits of *Arabidopsis thaliana*. Genetics 155:369–378.

Sniegowski PD, PJ Gerrish, T Johnson, A Shaver 2000 The evolution of mutation rates: separating causes from consequences. Bioessays 22:1057–1066.

Sweigart AL, JH Willis 2003 Patterns of nucleotide diversity in two species of *Mimulus* are affected by mating system and asymmetric introgression. Evolution 57:2490–2506.

Tajima F 1993 Simple methods for testing the molecular evolutionary clock hypothesis. Genetics 135:599–607.

Takebayashi N, PL Morrell 2001 Is self-fertilization an evolutionary dead end? revisiting an old hypothesis with genetic theories and a macroevolutionary approach. Am J Bot 88:1143–1150.

Tam SM, M Causse, C Garchery, H Burck, C Mhiri, MA Grandbastien 2007 The distribution of copia-type retrotransposons and the evolutionary history of tomato and related wild species. J Evol Biol 20:1056–1072.

Trivers R, A Burt, BG Palestis 2004 B-chromosomes and genome size in flowering plants. Genome 47:1–8.

Van't Hof J 1965 Relationships between mitotic cycle duration, S period duration and average rate of DNA synthesis in root meristem cells of several plants. Exp Cell Res 39:48–58.

Van't Hof J, AH Sparrow 1963 A relationship between DNA content, nuclear volume, and minimum mitotic cycle time. Proc Natl Acad Sci USA 49:897–902.

Vinogradov AE 2001 Mirrored genome size distributions in monocot and dicot plants. Acta Biotheor 49:43–51.

Wang J, WG Hill, D Charlesworth, B Charlesworth 1999 Dynamics of inbreeding depression due to deleterious mutations in small populations: mutation parameters and inbreeding rate. Genet Res 74: 165–178.

Wang Z, Y Zou, X Li, Q Zhang, L Chen, H Wu, D Su, et al 2006 Cytoplasmic male sterility of rice with boro II cytoplasm is caused by a cytotoxic peptide and is restored by two related *ppr* motif genes via distinct modes of mRNA silencing. Plant Cell 18:676–687.

Weinreich DM, DM Rand 2000 Contrasting patterns of nonneutral evolution in proteins encoded in nuclear and mitochondrial genomes. Genetics 156:385–399.

Whittle CA, MO Johnston 2003 Broad-scale analysis contradicts the theory that generation time affects molecular evolutionary rates in plants. J Mol Evol 56:223–233.

Woodcock G, PG Higgs 1996 Population evolution on a multiplicative single-peak fitness landscape. J Theor Biol 179:61–73.

Wright SI 2003 Effects of recombination rate and mating system on genome evolution and diversity in *Arabidopsis*. PhD thesis. University of Edinburgh.

Wright SI, N Agrawal, TE Bureau 2003a Effects of recombination rate and gene density on transposable element distributions in *Arabidopsis thaliana*. Genome Res 13:1897–1903.

Wright SI, JP Foxe, L DeRose-Wilson, A Kawabe, M Looseley, BS Gaut, D Charlesworth 2006 Testing for effects of recombination rate on nucleotide diversity in natural populations of *Arabidopsis lyrata*. Genetics 174:1421–1430.

Wright SI, G Iorgovan, S Misra, M Mokhtari 2007 Neutral evolution of synonymous base composition in the Brassicaceae. J Mol Evol 64:136–141.

Wright SI, B Lauga, D Charlesworth 2002 Rates and patterns of molecular evolution in inbred and outbred *Arabidopsis*. Mol Biol Evol 19:1407–1420.

——— 2003b Subdivision and haplotype structure in natural populations of *Arabidopsis lyrata*. Mol Ecol 12:1247–1263.

Wright SI, HL Quang, DJ Schoen, TE Bureau 2001 Population dynamics of an *ac*-like transposable element in self- and cross-pollinating *Arabidopsis*. Genetics 158:1279–1288.

Wright SI, DJ Schoen 1999 Transposon dynamics and the breeding system. Genetica 107:139–148.

Wu CI, WH Li 1985 Evidence for higher rates of nucleotide substitution in rodents than in man. Proc Natl Acad Sci USA 82:1741–1745.

Int. J. Plant Sci. 169(1):119–127. 2008.
1058-5893/2008/16901-0010$15.00   DOI: 10.1086/523356

ON THE EVOLUTION OF SELF-FERTILIZATION IN A METAPOPULATION

Daniel J. Schoen and Jeremiah W. Busch

Department of Biology, McGill University, 1205 Avenue Docteur Penfield, Montreal, Quebec H3A 1B1, Canada

The loss of morphological and physiological mechanisms that prevent self-fertilization is perhaps the most common evolutionary trend in the flowering plants. It is generally acknowledged that self-fertilization may often be favored by selection at the individual level, principally by providing reproductive assurance when conditions for vector-mediated pollination are poor and also because mating system modifiers that reduce the rate of outcrossing bias their own transmission. Inbreeding depression is accepted as the principal factor opposing the selection of selfing at the individual level, though this barrier may be transient because of purging of inbreeding load from natural populations. Here we explore the possibility that the selection of selfing may occur not only at the individual level but also at the group level. Accordingly, we model the selection of mating system modifier genes within and among populations and suggest that both levels of selection play a role in the evolution of the mating system. We find that selection among populations can maintain outcrossing through higher extinction rates of selfing groups and through reduced transition rates from outcrossing to selfing.

Keywords: outcrossing, selfing, group selection, individual selection, extinction, evolvability.

Introduction

The mating system is a major determinant of population variation and substructure, and hence, knowledge of a species' mating system is often of primary interest to both evolutionary biologists and breeders (Clegg 1980; Glemin et al. 2006). Among the evolutionary transitions in the higher plants, the loss of mechanisms that enforce outcrossing is probably one of the most prevalent and widespread of all (Stebbins 1974; Barrett 2002; Igic et al. 2008). Evolutionary biologists have long been fascinated by the paradox posed by the frequent occurrence of self-fertilization in the vascular plants on the one hand and by the poor performance of progeny produced through close inbreeding on the other (Darwin 1876; Charlesworth and Charlesworth 1987; Wright et al. 2008).

In the past century, explanations for the evolution of self-fertilization in plants have invoked a variety of factors and levels of selection. A major tension point in the past was the relative importance of group level selection. Geneticists such as Cyril Darlington (1939) and Kenneth Mather (1943) postulated that inbreeding could be favored under conditions in which there is an advantage to maintaining a narrow range of "closely adapted" genotypes, and accordingly, it was imagined that polygenic control of the level of recombination (e.g., obtained by some optimal combination of chromosome number, chiasma frequency, and cross-fertilization) could be the object of group-level selection. Selection was envisioned to fine-tune the recombination level and thereby to preserve coadapted gene complexes, allowing species to persist in their natural environments. This notion was accepted by G. Ledyard Stebbins (1950) as a partial explanation for the evolution of selfing,

though he also noted the importance of individual-level selection, namely, the short-term advantage of autonomous seed set when pollinators are absent, a factor that Darwin (1876) had also emphasized in his work.

Along with George Williams's (1966) cogent and general arguments against the prevalence of group selection, field-based studies of plant mating systems carried out in the middle and latter half of the twentieth century led the way to today's emphasis on individual-level selective factors driving the evolution of selfing. For instance, the work of Herbert Baker (1953, 1955), while recognizing the link between the reproductive system and population-level variability, stressed the importance of uniparental reproduction and autonomous seed set. David Lloyd's (1965, p. 131) classic study of the evolution of self-fertilization in the mustard genus *Leavenworthia* also concluded with an emphasis on individual-level selection (seed set under natural conditions), noting that "the supposed ability of self-fertilized plants to produce more closely adapted offspring has not been a factor in the evolution of self-compatibility" in this genus. Lloyd further promoted the importance of individual-level selection in the evolution of self-fertilization in his classic theoretical analysis of the factors affecting the selection of self-fertilization in plants (Lloyd 1979). This has been one of the most influential articles on the topic because it lucidly demonstrates the several ways by which selfing may evolve by individual-level selection; other similar selection models for selfing published around this time include those of Maynard Smith (1978), Nagylaki (1976), and Wells (1979).

There is now general agreement among evolutionary biologists that individual selection of selfing derives from the inherent transmission bias of a mating system modifier (Fisher 1941) and/or the ability of plants to set seed under conditions of poor pollination (Darwin 1876; Stebbins 1950; Baker 1953). These factors alone can account for the recurrent evolution of selfing, but the advantages they impart may be counteracted by inbreeding depression, the reduced vigor of selfed compared with outcrossed progeny (Charlesworth and Charlesworth 1987). Inbreeding depression as a genetic barrier to the spread of self-fertilization can be mitigated by repeated generations of selection against deleterious recessive mutations in the homozygous state brought about by inbreeding (i.e., purging; Lande and Schemske 1985), although there are also cases in which inbreeding depression levels are not strongly influenced by changes in the mating system (Byers and Waller 1999; Goodwillie et al. 2005). The fact that inbreeding depression generally declines as self-fertilization spreads within a population suggests that there may be additional factors involved in the maintenance of outcrossing.

There are several lines of evidence that point to a role for group selection in the maintenance of outcrossing. For example, phylogenetic analyses of mating system diversity in related taxa often suggest that the transition to self-fertilization is unidirectional (Takebayashi and Morrell 2001; Beck et al. 2006; Igic et al. 2006; but see Ferrer and Good-Avila 2007), that species-rich genera and higher taxonomic groups are rarely composed of a preponderance of self-fertilizing species (Stebbins 1974), and that self-fertilization is often relegated to the tips of phylogenetic trees. In theory, high rates of self-fertilization are associated with reductions in the magnitude of neutral and potentially adaptive genetic variation maintained within populations (Charlesworth and Charlesworth 1995; Charlesworth and Wright 2001; Glemin et al. 2006) that may limit the capacity of selfing populations to respond adaptively to environmental change. A highly homozygous and genetically uniform population may also suffer fitness declines due to relaxed purifying selection (Kimura et al. 1963; Lynch and Gabriel 1990; Schultz and Lynch 1997) and/or the operation of Muller's ratchet (Heller and Maynard Smith 1978).

Although Darlington's and Mather's arguments that group selection favors some optimum level of recombination are now no longer widely accepted, it may nevertheless be worthwhile to reconsider the importance of group-level selection of mating systems, especially in light of the evidence for differences in rates of population extinction and diversification of predominantly selfing versus outcrossing plants. Accordingly, in this article we model mating system evolution in a metapopulation as the means to explore some of the factors that may operate when group-level selection is considered and that may lead to a stable mixture of selfing and outcrossing mating systems within a species. The models we present are admittedly simplistic. Our intent, however, is not to explore the quantitative aspects of mating system selection in a metapopulation. To do so would require detailed consideration of the underlying genetic basis of mating system variation and inbreeding depression, variation in rates of deme extinction, and exploration of the roles of migration among demes and the founding of demes in the metapopulation. Some promising quantitative treatments of the relevant factors have already begun to explore these issues (Ronfort and Couvet 1995; Higgins and Lynch 2001; Pannell and Barrett 2001; Ingvarsson 2002; Theodorou and Couvet 2002). Instead, our focus is on the situation in which individual- and group-level selection for selfing oppose one another. The models we use are simple and heuristic in nature (e.g., Lloyd 1979; Morgan and Schoen 1997) and are intended to highlight some of the factors that may need to be

treated in detail when considering the evolution of selfing in a metapopulation.

Theory and Modeling

Simple Multilevel Selection Model for the Evolution of Selfing

Nunney's (1989) model for the maintenance of sex by group selection can be modified to handle some of the factors that apply to the evolution of self-fertilization. For example, some floral forms promote selfing and reproductive assurance when pollinators are rare but otherwise permit outcrossing (Lloyd and Schoen 1992). As well, while selfing may be associated with the founding of new populations (Baker 1955), inbreeding depression in these newly established demes may elevate extinction probabilities; accumulated load and loss of variation may further contribute to elevated extinction probabilities in selfing demes, as in asexual populations (Schultz and Lynch 1997).

The hierarchical model we develop applies principally to the evolution of selfing in a metapopulation of a single species. While in theory one could extend the approach we use to higher levels of taxonomic organization (e.g., species within a genus), to do so would require consideration of additional factors such as rates of diversification and speciation. To begin, we consider a single species that consists of a large number of populations (demes), some of which may become extinct and whose sites are repopulated (i.e., a metapopulation). Initially, this species is assumed to reproduce by outcrossing alone, but selfing variants may arise by mutation within one or more demes. For simplicity, we will assume that self-pollination has an advantage at the individual selection level, due either to automatic selection (in those populations in which inbreeding depression is not strong enough to present an obstacle to the spread of a selfing variant) or because of reproductive assurance (in those populations in which low pollinator activity limits full seed set). These individual-level advantages will be considered sufficient to favor the selection within demes of a selfing variant if it should arise. Since we are interested in the long-term outcome of mating system evolution, the transition from outcrossing to selfing will be considered instantaneous. Thus, the rate at which selfing variants arise within an outcrossing deme is denoted by $u_{o \rightarrow s}$ and is also assumed to be the rate of transformation of outcrossing to selfing demes—this rate can be thought of as the product of population size and the per-individual rate of mutation from outcrossing to selfing (the rate of occurrence of loss of function mutations at a self-incompatibility locus). Furthermore, outcrossing demes will be assumed to enjoy a reduced rate of extinction compared with selfing demes—this could be due to the lower load of deleterious mutations expected in outcrossers versus the load of mutations in selfers (Schultz and Lynch 1997) or to higher average levels of population genetic variation (Hamrick and Godt 1989; Schoen and Brown 1991) and effective recombination that allows them to better track environmental changes (compared to selfing demes). We denote extinction probabilities for outcrossing and selfing populations as e_o and e_s. The notion that all populations with contrasting mating systems differ in extinction rates by a constant amount, regardless of other biological and environmental factors is, of course, an over-simplification. Moreover, it is likely that a number of generations would be required for the two types of populations to diverge in terms of extinction rates once mating system evolution occurs within a deme. Further exploration of the dynamics of extinction rates is likely to reveal complexities that are not treated in the model.

When a deme becomes extinct, it will be assumed that it is replaced by one of the remaining selfing or outcrossing demes in the metapopulation. The probability that a given deme is the one chosen to recolonize the site formerly held by the extinct deme is initially treated as independent of the mating system. Thus, if O and S represent, respectively, the number of outcrossing and selfing demes in the metapopulation at time t, an outcrossing deme is selected to recolonize an extinct deme with probability $O/(O + S)$, whereas a selfing deme is selected to recolonize with probability $S/(O + S)$; that is, migrants arrive into extinct demes from a "migrant pool" (Slatkin 1977) whose composition reflects the current makeup of the metapopulation with respect to mating system. We denote the number of extinct demes by E. The probability of deme replacement may, however, not be independent of mating system; differential likelihood of recolonization from demes with an alternative mating system will be incorporated into the above model in the sections that follow. Taken together, the basic elements described above lead to a simple compartment model (fig. 1) whose dynamical behavior is captured by three differential equations:

$$\frac{dS}{dt} = Ou_{o \to s} + \frac{SE}{(S + O)} - Se_s,$$

$$\frac{dO}{dt} = \frac{OE}{(S + O)} - O(u_{o \to s} + e_o), \qquad (1)$$

$$\frac{dE}{dt} = Se_s + Oe_o - \frac{SE}{(S + O)} - \frac{OE}{(S + O)}.$$

Setting these equations equal to zero and solving for the equilibrium values of O leads to the following equilibrium proportion of outcrossing demes in the metapopulation:

$$\frac{\hat{O}}{\hat{O} + \hat{S}} = 1 - \frac{u_{o \to s}}{e_s - e_o}. \qquad (2)$$

Thus, once selfing invades the metapopulation by mutation, in order for outcrossing demes to be maintained, it is necessary for $e_s - e_o > u_{o \to s}$. In other words, the extinction probabilities are the main drivers of the equilibrium. For example, suppose one wanted to account for a species in which 90% of the demes are composed of outcrossing plants. If outcrossing demes have zero extinction probability, the maintenance of outcrossing would require that for every selfing variant that arises and successfully invades an outcrossing deme, 10 selfing demes must become extinct. If the percentage of outcrossing demes in the metapopulation is 99%, then 100 selfing demes must become extinct for every successful transition from outcrossing to selfing. As Nunney (1989) pointed out in the case of the evolution of asexuality, excessively large extinction rates such as these seem improbable, and this raises the possibility that outcrossing within the metapopulation could be maintained by other types of group-level selection. One possibility that we explore is that there is selection for a low rate of mu-

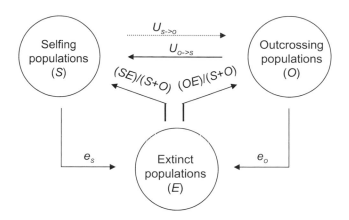

Fig. 1 Simple multilevel selection model of mating system evolution. See text for definitions of variables. The variable $u_{s \to o}$ is assumed to equal 0 in equations (1), (3), and (5).

tational transition from outcrossing to selfing. This will be discussed below after we have introduced a few modifications to the basic model intended to incorporate additional biological features associated with mating system variation.

Colonization Probability Dependent on the Mating System

The simple model presented above does not allow for the possibility that plants from selfing and outcrossing demes may differ in their probability of colonizing sites formerly held by demes that have become extinct. For instance, individuals from selfing demes may be better colonizers due to reproductive assurance (Baker 1955), or alternatively, outcrossing demes may harbor higher levels of genetic variation, which in turn could enhance the likelihood that migrants from them would successfully colonize new sites, for example, under a scenario of spatially varying environmental conditions. Differential colonization can be incorporated into the model by introducing a weighting factor for colonization, c, the relative colonization probability for outcrossers. For simplicity, we will assume that the total number of extinct demes at any given time is not influenced by differences in probability of colonization from selfing and outcrossing demes. This assumption may not hold if, for instance, selfing enhances the probability of founding demes in appropriate habitats never before occupied by the species, such that the overall number of demes comprising the metapopulation increases as the mating system evolves toward increased selfing. Nonetheless, if we are willing to assume that metapopulation size remains constant and that outcrossers have a reduced probability of recolonization compared with selfers ($c < 1$), migrants from selfing demes will fill the void and colonize extinct demes with probability $[S + (1 - c)O]/(S + O)$. This modification gives rise to the modified dynamical equations:

$$\frac{dS}{dt} = Ou_{o \to s} + \frac{[S + (1 - c)O]E}{(S + O)} - Se_s,$$

$$\frac{dO}{dt} = \frac{OcE}{(S + O)} - O(u_{o \to s} + e_o), \qquad (3)$$

$$\frac{dE}{dt} = Se_s + Oe_o - \frac{[S + (1 - c)O]E}{(S + O)} - \frac{OcE}{(S + O)}.$$

Comparison with the simplest model presented above is straightforward. Setting these equations equal to zero and solving for equilibrium values of O leads to

$$\frac{\hat{O}}{\hat{O} + \hat{S}} = \frac{e_o - ce_s + u_{o \to s}}{c(e_o - e_s)}. \qquad (4)$$

If we assume that $e_o = 0$ and that selfers are twice as likely as outcrossers ($c = 0.5$) to colonize extinct demes, then to account for a metapopulation in which 90% of the demes are composed of outcrossers, for every selfing variant that arises and successfully invades an outcrossing deme, 20 selfing demes must become extinct. If selfers are 10 times more likely to colonize empty patches, this ratio rises to 100. On the other hand, if individuals from outcrossing demes are more successful colonizers, the rate of extinction required to account for the same proportion of outcrossers is reduced. These results again raise the question of whether elevated extinction of selfing demes would be sufficient to maintain outcrossing in a metapopulation in which selfing is favored at the within-deme level.

Multilevel Selection of Selfing When There Is Variation in the Rate of Transition from Outcrossing to Selfing

In the models presented above, it has been assumed that the rate of mutational transition from outcrossing to selfing is itself a constant and not subject to evolutionary modification. In this subsection, we note that even though the majority of flowering plants are hermaphroditic and bear perfect flowers, there may nevertheless exist intraspecific variation in the ease in which selfing can evolve from outcrossing. The types of morphological and physiological changes that might allow self-fertilization to evolve vary depending on the floral and physiological characteristics of ancestral species in question. For example, in cases where heterostyly breaks down, a single mutation may be sufficient to produce a plant with a high rate of self-pollination in an otherwise predominantly outcrossing population. This may have been the case, for example, in the evolution of selfing in the boraginaceous species *Amsinckia gloriosa*, in which the long homostylous condition permits high rates of selfing (Johnston and Schoen 1996). Single mutational loss of self-incompatibility could also occur in those self-incompatible (SI) species in which floral morphology permits deposition of self pollen on the stigma but where a functional SI system otherwise prevents selfing from occurring. On the other hand, in cases of more complex floral morphology, several mutations may be needed in order for autonomous selfing to evolve. For example, the simple mutational loss of self-incompatibility in a species with flowers that are highly dichogamous or that exhibit strong spatial separation of anthers and stigmas may be insufficient to permit rapid transition to a predominantly selfing form. Additional change in the timing and/or spatial separation of stamens and stigmas may be required, and these may require evolution at other loci.

Variation in the rate of transition from outcrossing to selfing within a single species may be incorporated into the simple model outlined above by initially allowing a range of transition rates to exist among the different outcrossing demes in the metapopulation. Let there be K total classes of outcrossing demes that differ in their transition rates to selfing and where the ith class has transition rate $u_{oi \to s}$ (fig. 2). This is comparable to Nunney's (1989) model, in which transition rates from sexuality to asexuality are allowed to vary among species. Nunney examined the consequences of such variation via simulation, but it is also possible to examine such a model numerically. Modifying equations (1) to incorporate variation in transition from outcrossing to selfing, the dynamical equations become

$$\frac{dS}{dt} = \sum_{i=1}^{K} O_i u_{oi \to s} + \frac{SE}{\left(S + \sum_{i=1}^{K} O_i\right)} - Se_s,$$

$$\frac{dO_i}{dt} = \frac{O_i E}{\left(S + \sum_{i=1}^{K} O_i\right)} - O(u_{oi \to s} + e_{oi}), \qquad (5)$$

$$\frac{dE}{dt} = Se_s + \sum_{i=1}^{K} O_j e_{oi} - \frac{SE}{\left(S + \sum_{i=1}^{K} O_i\right)} - \frac{\sum_{i=1}^{K} O_i E}{\left(S + \sum_{i=1}^{K} O_i\right)}.$$

In this model, the outcrossing deme class with the lowest rate of transition from outcrossing to selfing is selected and will eventually replace the other deme classes (fig. 3). Thus, modification of the simple model to allow variation in the rate of the outcrossing to selfing transition shows that, in theory, there can be interdemic selection for those demes with the lowest rate of transition from outcrossing to selfing.

Interaction between the Transition to Selfing and the Probability of Extinction

The modeling exercises above pose the question of what maintains the outcrossing state in hermaphroditic plants when individual selection favors the evolution of selfing. We have seen one possibility is that group-level selection in a metapopulation may bring about the evolution of reduced transition rates from outcrossing to selfing. This result is a specific example of the more general finding that group-level selection, if sufficiently strong to counteract individual selection, may favor the suppression of "evolutionary pathologies" (traits that threaten the survival of the group, e.g., imprudent predation, habitat overuse) by reducing the evolvability of such traits (Altenberg 2005). However, there may be other factors at play. In the model just considered, the transition rates from outcrossing to selfing were allowed to vary among demes, but the extinction rates of all classes of outcrossing demes were treated as independent of other demographical characteristics. This need not be the case, and in fact, there may be good reason to expect that the extinction rates of demes are not independent of rates of transition from outcrossing to selfing. To see why this may be so, it is useful to consider the various modes of selfing that have been described by Lloyd (1979) and Lloyd and Schoen (1992). Certain modes of selfing (e.g., delayed selfing) involve autonomous self-pollination and may allow full seed set even when pollinators are in low abundance. In these instances, we expect the deme's selfing rate to be inversely related to pollinator activity; that is, when there are high levels of pollinator activity in such demes, plants will be predominant outcrossers, yet if pollinators largely fail or are lost from these demes, the mating system would make the "immediate" transition from outcrossing to predominant selfing.

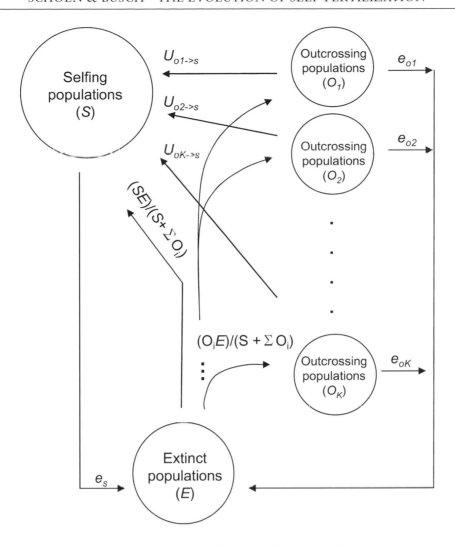

Fig. 2 Selection model of mating system evolution when outcrossing demes differ in rates of transition to selfing and in extinction rates.

It is interesting to speculate that extinction rates of outcrossing demes that possess this potential for "background" (or backup) autonomous selfing may, in fact, be lower than rates of those that do not possess such mechanisms; that is, their ability to persist during periods of pollinator failure could provide a buffer against demographic instability. Such a relationship between extinction rates of outcrossing demes and transition rates from outcrossing to selfing might look like the function graphed in figure 4. In this case, demes with low but nonzero transition rates would be favored, and the model suggests that some proportion of selfing demes will persist in the metapopulation; that is, the ability to evolve selfing from outcrossing will be retained. A prediction of this model, therefore, is that selfing species should often be composed of plants with the ability to self-pollinate autonomously. In fact, although autonomous selfing can be favored by individual selection alone (Lloyd 1979; Lloyd and Schoen 1992), multilevel selection models outlined above provide yet another explanation as to why autonomous selfing may be a common mode of selfing in plants.

Discussion

We have motivated our simple models with the suggestion that selection against selfing at the group level may oppose selection for selfing that occurs within demes and thereby help to account for the preponderance of outcrossing among the flowering plants. These models make a variety of assumptions about the effects of self-fertilization on population fitness, extinction probabilities, and the establishment of newly selfing demes by migration and colonization. In the remainder of the article, we discuss some of the empirical evidence (or lack thereof) necessary to address the role of multilevel selection in the evolution of self-fertilization, along with some qualitative predictions that follow from these models.

Is Self-Fertilization Associated with Elevated Rates of Extinction in Nature?

Inbreeding is expected to inhibit the decay of linkage disequilibrium and result in reduced effective population size com-

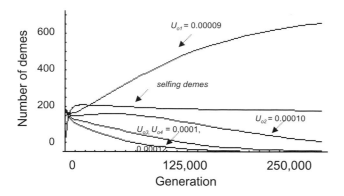

Fig. 3 Numerical iteration of equations (5) in the case where there are four classes of outcrossing demes with transition rates to selfing ranging from $u_{oi \longrightarrow s} = 0.9 \times 10^{-4}$ to 1.2×10^{-4}. In this example, the extinction rates of outcrossing and selfing demes are $e_{oi} = 1.0 \times 10^{-4}$ (for all deme classes $i = 1, 4$), and $e_s = 5.0 \times 10^{-4}$. Initial numbers of demes in each of the K classes is 250. Demes with high transition rates are lost, and there is selection for the demes with the lowest transition rate from outcrossing to selfing ($u_{o1 \longrightarrow s} = 0.9 \times 10^{-4}$) leading to an equilibrium metapopulation consisting of ca. 775 outcrossing and 225 selfing demes. Iterations of equations (5) were carried out using Mathematica, version 3.0.

pared with outbreeding (Pollak 1987; Charlesworth 2003). These effects on the genetic characteristics of the population should result in an elevated rate of accumulation of deleterious mutations, because of selective interference between loci under selection (Hill and Robertson 1966; Betancourt and Presgraves 2002) and/or because of a decrease in the efficacy of purifying selection (Kimura et al. 1963). Since all natural populations are finite and experience some degree of inbreeding, there has been interest in testing the idea that restricted outcrossing leads to reduction in population fitness (Frankham 2005). In general, evidence from a large number of plant species is consistent with a role for inbreeding in reducing effective population size and endangering population fitness (Paland and Schmid 2003; Tallmon et al. 2004). Empirical evidence also suggests that self-fertilization, the most severe form of inbreeding, is the single most important factor reducing the effective size and effective recombination rate of natural populations (Glemin et al. 2006). For instance, in the model plant species *Arabidopsis thaliana* (in which populations have selfing rate of ca. 99%), analyses of sequence variation have shown a slight increase in the ratio of nonsynonymous to synonymous substitutions and the accumulation of transposable elements (Abbott and Gomes 1989; Wright et al. 2001; Bustamante et al. 2002). Both an elevated rate of protein evolution and an increased density of selfish DNA are consistent with relaxed purifying selection against mildly deleterious mutations in highly inbred genomes.

Our models suggest that the extinction rate of selfing populations must be very high ($e_s \gg e_o$) to alone account for the predominance of outcrossing in a metapopulation. Moreover, if there is a low degree of genetic structure ($F_{ST} \sim 0$) among demes in the metapopulation, selection against self-fertilization at the population level should be relatively weak compared to selection at the individual level. That being said, the evolution of

self-fertilization could increase the genetic structure of a metapopulation (Ingvarsson 2002; Charlesworth 2003) and thereby elevate the level of among-population variation on which group selection may act (Wade and Griesemer 1998). The importance of population structure has been debated in theoretical discussions on the role of group selection in the maintenance of sexual reproduction, with most believing that the opportunity for group selection is limited (Williams 1966; Maynard Smith 1978). It seems, however, that selection among highly inbred populations could, in theory, be effective, so it is important to determine whether this level of selection can limit or stabilize the spread of self-fertilization in naturally occurring metapopulations (sensu Stevens et al. 1995). Application of the Price equation (Price 1972) to examine the contribution of individual- and group-level selection to the evolution of selfing may help to clarify this issue (e.g., see Frank 1987).

Even if selfing populations do not experience mutation accumulation rapidly enough to counter selection of selfing at the individual level, the types of habitats often colonized by plants with uniparental reproduction could be associated with unpredictable or fluctuating conditions (Levin 1975; Lloyd 1980), and this could lead to an association between selfing and elevated extinction probabilities. For example, there are some selfing taxa (e.g., *Clarkia* and *Leavenworthia*) in which the evolution of selfing occurs in populations at range margins (Lloyd 1965; Moore and Lewis 1965), although this is by no means a generality. In order to examine the effects of selfing on population viability and longevity, studies should examine the consequences of a change in the rate of self-fertilization on population fitness (Busch 2006). Furthermore, if we are to understand the potential role of among-population selection in the maintenance of outcrossing in flowering plants, we will have to address the relationship between rates of self-fertilization and extinction in natural environments. While studying extinction rates among populations with variable mating systems may be difficult in nature, this approach has shown that inbreeding drives patch extinction in a butterfly metapopulation (Saccheri et al. 1998).

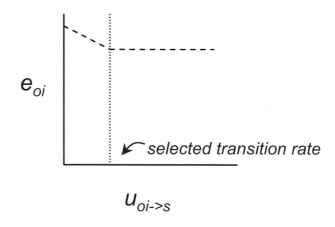

Fig. 4 Possible functional relationship between extinction rates of outcrossing demes and transition rates from outcrossing to selfing in the case where selfing occurs autonomously and provides reproductive assurance.

Does Self-Fertilization Influence Rates of Colonization in Nature?

Colonization is likely to be a powerful factor in the evolution of self-fertilization at the metapopulation level (Holsinger 1986). The question naturally arises as to whether the spread of alleles for self-fertilization realistically occurs as characterized in the models. In nature, the evolution of self-fertilization within a local deme is generally followed by declines in allocation to male function (Cruden 1977; Charnov 1982). As a result, one might expect demes with the highest rates of self-fertilization to produce smaller total pollen pools and consequently export less pollen throughout the metapopulation. On the other hand, selfing demes may export more alleles via seed migration. It is not immediately obvious whether the effects of selfing on the joint rates of pollen and seed migration will be positive or negative.

Self-fertilization is likely to be associated with a colonization benefit because a single selfing seed can found a new population, as discussed by Baker (1955). "Baker's law" was originally stated with respect to the notion of a colonization filter favoring the establishment of species with self-compatible mating systems on oceanic islands, although it also probably applies to any species in which populations exhibit turnover and are connected to one another by migration (Pannell and Barrett 1998; Pannell and Dorken 2006). It seems probable that whenever there is competition among outcrossing and selfing plants to colonize empty patches, selection for selfing via this filter will be quite strong ($c \ll 1$) in a metapopulation. If this factor were operating in nature, one might expect that strong metapopulation structure (high rates of extinction/colonization and relatively low migration) would support plant species with relatively high rates of self-fertilization or at least the ability to produce seed autonomously (Pannell and Barrett 1998). The fact that the nature of metapopulation structure has been shown to influence selection for uniparental reproduction in groups that ancestrally have separate sexes further suggests that selection among demes has the potential to influence the spread of alleles causing self-fertilization (Pannell 1997; Obbard et al. 2006).

How Can Natural Selection Reduce Transition Rates to Self-Fertilization?

Our final model incorporates variation among populations in their propensity to make the transition from outcrossing to self-fertilization. Variation at this level can occur through two primary mechanisms: (1) populations differ in the mutation rate at loci controlling selfing and/or (2) populations differ in the complexity of the underlying control features that maintain outcrossing because of underlying constraints on floral morphology or physiology. In the first case, populations that have a high probability of recruiting mutations for selfing would presumably undergo high rates of population extinction due to mutation accumulation at other loci that contribute to fitness. This hypothesis predicts that outcrossing groups will have experienced selection for reduced mutation rates at loci controlling self-fertilization. That being said, it seems unlikely that

selfing groups will have higher rates of mutation at the subset of loci controlling gamete transfer alone, because selection against deleterious mutation should favor the evolution of lower genomewide mutation rates (Kondrashov 1995). A recent study found no evidence for differences in the rates of deleterious mutation between closely related outcrossing and selfing species, although this work did not specifically address rates of mutation at the loci controlling self-fertilization (Schoen 2005).

The genetic architecture of traits maintaining outcrossing can presumably influence the tempo of the transition from outcrossing to selfing. In cases where only a single mutation is necessary to allow outcrossing plants to become selfing, these mutations may rapidly increase in frequency until they equilibrate at a frequency determined by the conflict between the fertility benefit they provide and inbreeding depression (Lloyd 1979; Charlesworth 1988). This is thought to be the case for mutations causing self-fertility in species with genetically controlled self-incompatibility systems (Marshall and Abbott 1984; Fenster and Barrett 1994; Royo et al. 1994; Tsukamoto et al. 2003). In species that are self-compatible yet have morphological adaptations that prevent selfing, the genetic control of outcrossing is likely to be polygenic, as suggested by biometric and quantitative trait loci studies of the genetic basis of floral traits underlying mating system evolution in a variety of taxonomic groups (Shore and Barrett 1990; Holtsford and Ellstrand 1992; Fenster and Ritland 1994; Lin and Ritland 1997; Georgiady et al. 2002; Fishman and Stratton 2004; Goodwillie et al. 2006). In these species, it may take many generations of selection to assemble a multilocus genotype capable of both self-pollination and self-fertilization. In such a situation, there can be selection against the most inbred (and therefore least fit) genotypes, with recurrent outcrossing restoring fitness to a subset of the population (Latta and Ritland 1993).

The question remains, however, as to whether selection among demes has modulated the genetic architecture of outcrossing such that it is sufficiently complex so as to maintain widespread outcrossing, as outlined in the model we present. Indeed, an important question to ask, arising from the perspective of evolutionary transitions, is simply which factors, genetic, physiological, or developmental, determine the probability that a selfing form can arise rapidly in a population of outcrossing plants. Comparative studies may help to answer this question, but in the end, a firm understanding of floral development is likely to be more informative.

Acknowledgments

We thank Spencer Barrett, Pierre-Olivier Cheptou, Josh Kohn, and two anonymous reviewers for comments on earlier versions of the manuscript. D. J. Schoen acknowledges continuing support from the Natural Sciences and Engineering Research Council of Canada (NSERC). J. W. Busch was supported by a Tomlinson Fellowship from McGill University and by the NSERC.

Literature Cited

Abbott RJ, MF Gomes 1989 Population genetic structure and outcrossing rate of *Arabidopsis thaliana* (L.) Heynh. Heredity 62:411–418.

Altenberg L 2005 Evolvability suppression to stabilize far-sighted adaptations. Artif Life 11:407–426.

Baker HG 1953 Race formation and reproductive method in flowering plants. Symp Soc Exp Biol 7:114–145.

——— 1955 Self-compatibility and establishment after "long-distance" dispersal. Evolution 9:347–348.

Barrett SCH 2002 The evolution of plant sexual diversity. Nat Rev Genet 3:274–284.

Beck JB, IA Al-Shehbaz, BA Schaal 2006 Leavenworthia (Brassicaceae) revisited: testing classic systematic and mating system hypotheses. Syst Bot 31:151–159.

Betancourt AJ, DC Presgraves 2002 Linkage limits the power of natural selection in Drosophila. Proc Natl Acad Sci USA 99:13616–13620.

Busch JW 2006 Heterosis in an isolated, effectively small, and self-fertilizing population of the flowering plant Leavenworthia alabamica. Evolution 60:184–191.

Bustamante CD, R Nielsen, SA Sawyer, KM Olsen, MD Purugganan, DL Hartl 2002 The cost of inbreeding in Arabidopsis. Nature 416:531–534.

Byers D, DM Waller 1999 Do plant populations purge their genetic load? effects of population size and mating history on inbreeding depression. Annu Rev Ecol Syst 30:479–513.

Charlesworth D 1988 Evolution of homomorphic sporophytic self-incompatibility. Heredity 60:445–453.

——— 2003 Effects of inbreeding on the genetic diversity of populations. Philos Trans R Soc B 358:1051–1070.

Charlesworth D, B Charlesworth 1987 Inbreeding depression and its evolutionary consequences. Annu Rev Ecol Syst 18:237–268.

——— 1995 Quantitative genetics in plants: the effect of the breeding system on genetic variability. Evolution 49:911–920.

Charlesworth D, SI Wright 2001 Breeding systems and genome evolution. Curr Opin Genet Dev 11:685–690.

Charnov EL 1982 On sex allocation and selfing in higher plants. Evol Ecol 1:30–36.

Clegg MT 1980 Measuring plant mating systems. BioScience 30:814–818.

Cruden RW 1977 Pollen-ovule ratios: conservative indicator of breeding systems in flowering plants. Evolution 31:32–46.

Darlington CD 1939 The evolution of genetic systems. Cambridge University Press, Cambridge. 149 pp.

Darwin C 1876 The effects of cross and self-fertilisation in the vegetable kingdom. J Murray, London. 488 pp.

Fenster CB, SCH Barrett 1994 Inheritance of mating-system modifier genes in Eichhornia paniculata (Pontederiaceae). Heredity 72:433–445.

Fenster CB, K Ritland 1994 Quantitative genetics of mating system divergence in the yellow monkeyflower species complex. Heredity 73:422–435.

Ferrer MM, SV Good-Avila 2007 Macrophylogenetic analyses of the gain and loss of self-incompatibility in the Asteraceae. New Phytol 173:401–414.

Fisher RA 1941 Average excess and average effect of a gene substitution. Ann Eugen 11:53–63.

Fishman L, DA Stratton 2004 The genetics of floral divergence and postzygotic barriers between outcrossing and selfing populations of Arenaria uniflora (Caryophyllaceae). Evolution 58:296–307.

Frank SA 1987 Individual and population sex allocation patterns. Theor Popul Biol 31:47–74.

Frankham R 2005 Genetics and extinction. Biol Conserv 126:131–140.

Georgiady MS, RW Whitkus, EM Lord 2002 Genetic analysis of traits distinguishing outcrossing and self-pollinating forms of currant tomato, Lycopersicon pimpinellifolium (Jusl.) Mill. Genetics 161:333–344.

Glemin S, E Bazin, D Charlesworth 2006 Impact of mating systems on patterns of sequence polymorphism in flowering plants. Proc R Soc B 273:3011–3019.

Goodwillie C, S Kalisz, CG Eckert 2005 The evolutionary enigma of mixed mating in plants. Annu Rev Ecol Syst 36:47–79.

Goodwillie C, C Ritland, K Ritland 2006 The genetic basis of floral traits associated with mating system evolution in Leptosiphon (Polemoniaceae): an analysis of quantitative trait loci. Evolution 60:491–504.

Hamrick JL, MJ Godt 1989 Allozyme diversity in plant species. Pages 43–63 in AHD Brown, MT Clegg, AL Kahler BS Weir, eds. Plant population genetics, breeding and germplasm resources. Sinauer, Sunderland, MA.

Heller R, J Maynard Smith 1978 Does Muller's ratchet work with selfing? Genet Res 32:289–293.

Higgins K, M Lynch 2001 Metapopulation extinction caused by mutation accumulation. Proc Natl Acad Sci USA 98:2928–2933.

Hill WG, A Robertson 1966 The effect of linkage on limits to artificial selection. Genet Res 8:269–294.

Holsinger KE 1986 Dispersal and plant mating systems: the evolution of self-fertilization in subdivided populations. Evolution 40:405–413.

Holtsford TP, NC Ellstrand 1992 Genetic and environmental variation in floral traits affecting outcrossing rate in Clarkia tembloriensis (Onagraceae). Evolution 46:216–225.

Igic B, L Bohs, JR Kohn 2006 Ancient polymorphism reveals unidirectional breeding system shifts. Proc Natl Acad Sci USA 103:1359–1363.

Igic B, R Lande, JR Kohn 2008 Loss of self-incompatibility and its evolutionary consequences. Int J Plant Sci 169:93–104.

Ingvarsson PK 2002 A metapopulation perspective on genetic diversity and differentiation in partially self-fertilizing plants. Evolution 56:2368–2373.

Johnston MO, DJ Schoen 1996 Correlated evolution of self-fertilization and inbreeding depression: an experimental study of nine populations of Amsinckia (Boraginaceae). Evolution 50:1478–1491.

Kimura M, T Maruyama, JF Crow 1963 The mutation load in small populations. Genetics 48:1303–1312.

Kondrashov AS 1995 Modifiers of mutation-selection balance: general approach and the evolution of mutation rates. Genet Res 66:53–69.

Lande R, DW Schemske 1985 The evolution of self-fertilization and inbreeding depression in plants I. Genetic models. Evolution 39:24–40.

Latta R, K Ritland 1993 Models for the evolution of selfing under alternative modes of inheritance. Heredity 71:1–10.

Levin DA 1975 Pest pressure and recombination systems in plants. Am Nat 109:437–457.

Lin J-Z, K Ritland 1997 Quantitative trait loci differentiating the outbreeding Mimulus guttatus from the inbreeding M. platycalyx. Genetics 146:1115–1121.

Lloyd DG 1965 Evolution of self-compatibility and racial differentiation in Leavenworthia (Cruciferae). Contrib Gray Herb Harv Univ 195:3–134.

——— 1979 Some reproductive factors affecting the selection of self-fertilization in plants. Am Nat 13:67–79.

——— 1980 Demographic factors and mating patterns in angiosperms. Pages 67–88 in OT Solbrig, ed. Demography and evolution in plant populations. Blackwell, Oxford.

Lloyd DG, DJ Schoen 1992 Self- and cross-fertilization in plants. 1. Functional dimensions. Int J Plant Sci 153:358–369.

Lynch M, W Gabriel 1990 Mutation load and the survival of small populations. Evolution 44:1725–1737.

Marshall DF, RJ Abbott 1984 Polymorphism for outcrossing frequency at the ray floret locus in Senecio vulgaris L. 2. Confirmation. Heredity 52:331–336.

Mather K 1943 Polygenic inheritance and natural selection. Biol Rev 18:32–64.

Maynard Smith J 1978 The evolution of sex. Cambridge University Press, Cambridge. 222 pp.

Moore DM, H Lewis 1965 The evolution of self-pollination in Clarkia xantiana. Evolution 19:104–114.

Morgan MT, DJ Schoen 1997 The role of theory in an emerging new plant reproductive biology. Trends Ecol Evol 12:231–234.

Nagylaki T 1976 A model for the evolution of self-fertilization and vegetative reproduction. J Theor Biol 58:55–58.

Nunney L 1989 The maintenance of sex by group selection. Evolution 43:245–257.

Obbard DJ, SA Harris, JR Pannell 2006 Sexual systems and population genetic structure in an annual plant: testing the metapopulations model. Am Nat 167:354–366.

Paland S, B Schmid 2003 Population size and the nature of genetic load in Gentianella germanica. Evolution 57:2242–2251.

Pannell J 1997 The maintenance of gynodioecy and androdioecy in a metapopulation. Evolution 51:10–20.

Pannell JR, SCH Barrett 1998 Baker's law revisited: reproductive assurance in a metapopulation. Evolution 52:657–668.

——— 2001 Effects of population size and metapopulation dynamics on a mating-system polymorphism. Theor Pop Biol 59:145–155.

Pannell JR, ME Dorken 2006 Colonisation as a common denominator in plant metapopulations and range expansions: effects on genetic diversity and sexual systems. Land Ecol 21:837–848.

Pollak E 1987 On the theory of partially inbreeding finite populations. I. Partial selfing. Genetics 117:353–360.

Price GR 1972 Extension of covariance selection mathematics. Ann Hum Genet 35:485–490.

Ronfort J, D Couvet 1995 A stochastic model of selection on selfing rates in structured populations. Genet Res 65:209–222.

Royo J, C Kunz, Y Kowyama, M Anderson, AE Clark, E Newbigin 1994 Loss of a histidine residue at the active site of S-locus ribonuclease is associated with self-compatibility in Lycopersicon peruvianum. Proc Natl Acad Sci USA 91:6511–6514.

Saccheri I, M Kuussaari, M Kankare, P Vikman, W Fortelius, I Hanski 1998 Inbreeding and extinction in a butterfly metapopulation. Nature 392:491–494.

Schoen DJ 2005 Deleterious mutation in related species of the plant genus Amsinckia with contrasting mating systems. Evolution 59:2370–2377.

Schoen DJ, AHD Brown 1991 Interspecific variation in population gene diversity and effective population size correlates with the mating system in plants. Proc Natl Acad Sci USA 88:4494–4497.

Schultz ST, M Lynch 1997 Deleterious mutation and extinction: effects of variable mutational effects, synergistic epistasis, beneficial mutations, and degree of outcrossing. Evolution 51:1363–1371.

Shore JS, SCH Barrett 1990 Quantitative genetics of floral characters in homostylous Turnera ulmifolia var. angustifolia Wild. (Turneraceae). Heredity 64:105–112.

Slatkin M 1977 Gene flow and genetic drift in a species subject to frequent local extinctions. Theor Pop Biol 12:253–262.

Stebbins GL 1950 Variation and evolution in plants. Columbia University Press, New York. 643 pp.

——— 1974 Flowering plants: evolution above the species level. Belknap, Cambridge, MA. 399 pp.

Stevens L, CJ Goodnight, S Kalisz 1995 Multilevel selection in natural populations of Impatiens capensis. Am Nat 145:513–526.

Takebayashi N, PL Morrell 2001 Is self-fertilization an evolutionary dead end? revisiting an old hypothesis with genetic theories and a macroevolutionary approach. Am J Bot 88:1143–1150.

Tallmon DA, G Luikart, RS Waples 2004 The alluring simplicity and complex reality of genetic rescue. Trends Ecol Evol 19:489–496.

Theodorou K, D Couvet 2002 Inbreeding depression and heterosis in a subdivided population: influence of the mating system. Genet Res 80:107–116.

Tsukamoto T, T Ando, K Takahashi, T Omori, H Watanabe, H Kokubun, E Marchesi, T-H Kao 2003 Breakdown of self-incompatibility in a natural population of Petunia axillaris caused by loss of pollen function. Plant Physiol 131:1903–1912.

Wade MJ, JR Griesemer 1998 Populational heritability: empirical studies of evolution in metapopulations. Am Nat 151:135–147.

Wells H 1979 Self-fertilization: advantageous or deleterious? Evolution 33:252–255.

Williams GC 1966 Adaptation and natural selection. Princeton University Press, Princeton, NJ. 307 pp.

Wright SI, QH Le, DJ Schoen, TE Bureau 2001 Population dynamics of an Ac-like transposable element in self- and cross-pollinating Arabidopsis. Genetics 158:1279–1288.

Wright SI, RW Ness, JP Foxe, SCH Barrett 2008 Genomic consequences of outcrossing and selfing in plants. Int J Plant Sci 169:105–118.

Int. J. Plant Sci. 169(1):129–139. 2008.
1058-5893/2008/16901-0011$15.00   DOI: 10.1086/523360

GENDER VARIATION AND TRANSITIONS BETWEEN SEXUAL SYSTEMS IN *MERCURIALIS ANNUA* (EUPHORBIACEAE)

John R. Pannell, Marcel E. Dorken, Benoit Pujol, and Regina Berjano

Department of Plant Sciences, University of Oxford, South Parks Road, Oxford OX1 3RB, United Kingdom

Evolutionary transitions between hermaphroditism and dioecy have occurred numerous times in the land plants. We briefly review the factors thought to be responsible for these transitions, and we provide a synthesis of what has been learned from recent studies of the annual herb *Mercurialis annua*, in which dioecy (males and females), monoecy (functional hermaphrodites), and androdioecy (males and hermaphrodites) occur in different parts of its geographic range. Previous research on *M. annua* has revealed the importance of genome duplication and hybridization in the origin of much of the observed variation. Here we show, however, that spatial transitions in the sexual system also occur within the same ploidy level. In particular, we present an analysis, using flow cytometry data, of ploidy variation across a previously unstudied transition between hermaphroditism and androdioecy, in which we find that the sexual-system transition is uncoupled from the shift in ploidy levels. We review recent research that shows that such transitions between sexual systems in *M. annua* are consistent with differential selection at the regional level for reproductive assurance during colonization. We also present new experimental data that highlight both the importance of the resource status of plants and that of their local mating context in regulating gender strategies and sex ratios. The studies reviewed and the new results presented emphasize the role that shifts in the ecological and genetic context of plant populations may play in causing transitions between sexual systems.

Keywords: androdioecy, dioecy, gynodioecy, hermaphroditism, mating system, monoecy.

Introduction

By far, the majority of flowering plants are hermaphroditic (Yampolsky and Yampolsky 1922; Sakai and Weller 1999), but transitions between hermaphroditism and dioecy have been frequent (Renner and Ricklefs 1995; Weiblen et al. 2000; Vamosi et al. 2003; Case et al. 2008). Most research has been directed toward understanding shifts from hermaphroditism toward dioecy, with a range of hypothesized evolutionary paths invoked (reviewed by Charlesworth [1999]; Webb [1999]). Indeed, Darwin (1877) set the stage for much of this work by noting the substantial advantages of hermaphroditism and asking why "hermaphrodite plants should ever have been rendered dioecious" (p. 279). In contrast, we might similarly ask why males and females, which may enjoy advantages of gender specialization, should ever be replaced by hermaphrodites. Although probably less frequent than transitions from hermaphroditism to dioecy, the breakdown of dioecy toward hermaphroditism is known to have occurred in both plants and animals (Desfeux et al. 1996; Wolf et al. 2001; Kiontke et al. 2004; Weeks et al. 2006), and a growing body of ideas exists on when such shifts might occur (Charnov et al. 1976; Charnov 1982; Maurice and Fleming 1995; Wolf and Takebayashi 2004). Nevertheless, the empiri-

cal foundation for understanding transitions from dioecy to hermaphroditism remains weak.

From a theoretical point of view, the breakdown of dioecy requires the invasion and spread in a population of either pollen-producing females or seed-producing males. Initially, the spread of hermaphrodites in a dioecious populations will yield "trioecy," a sexual system in which all three gender classes are maintained together, but the conditions under which such a gender trimorphism can be maintained evolutionarily appear to be rather limited (Maurice and Fleming 1995; Wolf and Takebayashi 2004). More likely is the rapid displacement of either the females or the males from the population, with the invading hermaphrodites maintained in an androdioecious or a gynodioecious population, respectively. Although gynodioecy is understood to be an important step in the evolution of dioecy from hermaphroditism, it is not known to have played a major role in the breakdown of dioecy. In contrast, while androdioecy is exceedingly rare in absolute terms (Darwin 1877; Charlesworth 1984), almost all the known androdioecious species, both in animals and in plants, appear to have evolved from a dioecious rather than a hermaphroditic ancestor (Pannell 2002; Weeks et al. 2006).

Transitions from dioecy to hermaphroditism by way of androdioecy have been studied in only two plant species in much detail. In the wind-pollinated North American perennial herb *Datisca glomerata*, hermaphrodites appear to be modified (pollen-producing) females with the same sex determination as females in its dioecious sister species *Datisca cannabina* (Wolf et al. 2001). In the European herb *Mercurialis annua* (fig. 1), which is also wind-pollinated, androdioecy

Fig. 1 Male (*A*), female (*B*), and monoecious (*C*) individuals of *Mercurialis annua*. The male flowers on the male and the monoecious individuals have the same morphology, but male flowers of male plants are held on erect peduncles. Also, the female flowers of both female and monoecious plants have the same morphology and placement in the leaf axils (in the photographs, female flowers have already set fruit). Males of diploid and hexaploid populations have the same morphology. Monoecious plants are effectively females that produce staminate flowers around their pistillate flowers.

is found only in polyploid populations of a complex in which dioecy is clearly the ancestral trait (Durand 1963; Durand and Durand 1992; Pannell 1997*d*; Pannell et al. 2004). *Mercurialis annua* is unusual in the extraordinary diversity of sexual systems it displays across its geographic range; this variation makes it a useful study system in which to address questions concerning evolutionary transitions between sexual systems and the origins and maintenance of androdioecy.

In this article, we present new data that advance our understanding of transitions in the sexual system of *M. annua* in two ways. First, we ask how closely the spatial transition between monoecy and androdioecy in a hitherto poorly studied part of the species' range on the Atlantic coast of Morocco corresponds to a shift between tetraploidy and hexaploidy. This question is important because elsewhere in the distributional range of the species complex, spatial (and perhaps evolutionary) transitions in the sexual system and ploidy levels are confounded; instances in which sexual-system transitions are uncoupled from those in ploidy provide particularly fertile ground for invoking the maintenance of different sexual systems by natural selection (rather than as a result of the historical distribution of ploidy levels). Second, we present results of an experiment that asks how the mating system is affected by

plastic responses to resource availability in the relative allocation to male and female functions. We address both these issues within the context of a synthesis of recent work on *M. annua*. We begin by considering the phylogenetic and phylogeographic history of the species complex in Europe. We then review empirical tests of a hypothesis that invokes differential extinction-colonization dynamics in a metapopulation to explain the maintenance of different sexual systems. Finally, we consider the potential importance of phenotypic plasticity in sex allocation in regulating the selection of combined versus separate sexes. Overall, we highlight the importance of both shifts in the genetic system associated with polyploidization and hybridization and shifts in the ecological and demographic context of selection for transitions between sexual systems.

Transitions in the Sexual System: The Role of Polyploidy and Hybridization

Mercurialis annua belongs to a small European genus in the Euphorbiaceae that comprises mainly dioecious woody or herbaceous perennials. Phylogenetic reconstruction of the genus by Krahenbuhl et al. (2002), based on ITS sequence analysis, indicated that dioecy and perenniality are ancestral in the genus and that monoecy has evolved on at least two independent occasions, one of which was in the annual polyploid complex *M. annua*. This complex comprises ploidy levels ranging from diploid through at least 12-ploid, with dioecy confined to the diploids and monoecy found only in the polyploid lineages. Here, the tetraploids, octaploids, and higher ploidy levels are exclusively monoecious, but the hexaploids show remarkable variation in their sexual systems, with populations ranging from monoecy through androdioecy to subdioecy in various parts of their range (Durand 1963; Durand and Durand 1992).

Early studies of morphology and meiotic pairing behavior suggested that *M. annua* was an autopolyploid series and that polyploidization had precipitated the evolution of monoecy (Durand 1963). However, more recent analysis identified at least two divergent ITS paralogues in hexaploid populations of *M. annua*, only one of which occurred in diploid and tetraploid individuals (Obbard et al. 2006*b*). The other ITS sequence, absent in diploids and tetraploids, was found in *Mercurialis huetii*, a diploid sister species to *M. annua*, which has the same number of chromosomes and is also annual and dioecious. Obbard et al. (2006*b*) interpreted these results as evidence for hybridization between tetraploid *M. annua* and *M. huetii*, yielding triploids and followed by polyploidization to produce the hexaploid lineage (fig. 2). This hypothesis requires further testing with more exhaustive sampling of genotypes and loci, but preliminary results from microsatellite loci, some of which amplify only in hexaploid *M. annua* and *M. huetii*, are consistent with the ITS sequence analysis (P. Rymer, H. Stone, G. Korbecka, and J. R. Pannell, unpublished data).

Polyploidy represents a dramatic shift in the genetic system of a lineage. Not only may it alter the expression of inbreeding depression, potentially allowing the spread of self-fertilization in a population (Ronfort 1999; Husband et al. 2008), but it may also alter a broad range of phenotypic traits that may allow a lineage to occupy new and different habitats; i.e., it can cause ecological shifts that may also favor a transition in the sexual system of a population (Stebbins 1950; Pannell et al. 2004). Pannell et al. (2004) reviewed the interactions that are expected to occur between ploidy and sexual-system evolution. Their main conclusion was that the complexities involved make general predictions difficult. For example, on the one hand, polyploidization may allow self-fertile hermaphroditism to replace dioecy by causing a reduction in inbreeding depression (Lande and Schemske 1985; Pannell et al. 2004; Husband et al. 2008). On the other hand, genome duplication can cause the breakdown of self-incompatibility in hermaphroditic populations, allowing dioecy to evolve as an alternative outcrossing mechanism (Miller and Venable 2000; but see Mable 2004; Pannell et al. 2004). Thus, it can be tempting to invoke polyploidization as a cause of transitions both from hermaphroditism to dioecy and vice versa. In the case of *M. annua*, it would seem not only that polyploidization initiated the shift from dioecy to hermaphrodism but also that it has not precluded the maintenance of males in hexaploid populations of *M. annua* or the existence of dioecy in the newly discovered tetraploid species *Mercurialis canariensis* (Obbard et al. 2005).

Phylogeography and Regional Transitions in the Sexual System

Spatial transitions in the sexual system of *Mercurialis annua* in Europe (fig. 3) correspond, to an important extent (though not exclusively; see below), with transitions in the ploidy level. Fully dioecious populations are exclusively diploid; these are widespread throughout Europe. In contrast, populations containing hermaphrodites (with or without males) are polyploids; these are largely restricted to the western Mediterranean Basin and northwestern Africa. Tetraploids occur south of Rabat on the Atlantic coast of Morocco, and these meet with hexaploid populations to the north. Tetraploids are hermaphroditic, whereas the hexaploids, which are very widespread in northwestern Morocco and around the coast of the Iberian Peninsula, are variously hermaphroditic or androdioecious. Hexaploid populations meet the diploids at two contact zones in northeastern and northwestern Spain (Durand 1963; Obbard et al. 2006*b*).

Patterns of allelic richness and genetic diversity at several isozyme loci suggest that the Spanish diploid-hexaploid transitions are secondary contact zones, with diploids having expanded across Europe from an eastern Mediterranean refugium and the hexaploids having moved north along the coasts of the Iberian Peninsula from southern Spain or North Africa (Obbard et al. 2006*b*). This expansion of the geographic range of diploid *M. annua* appears to be continuing apace, with surveys suggesting that the diploids have displaced the hexaploids by some 80 and 200 km within about four decades in northeastern and northwestern Spain, respectively, apparently both as a result of diploid superiority in their growth (Buggs and Pannell 2007) and through the ability of diploids to swamp the monoecious hexaploids with the large amounts of pollen that diploid males produce (Buggs and Pannell 2006).

The higher pollen production of diploid males over hermaphrodites provides a plausible explanation for the rapid displacement of hermaphroditic populations by the dioecious lineage. However, what should we predict for a contact zone between dioecy and androdioecy, where males co-occur with the hermaphrodites? This is a pertinent question because, should the dioecious lineage continue its advance down the coasts of the Iberian Peninsula, it will soon encounter androdioecious populations farther south. Dorken and Pannell (2007) addressed this question, using spatially explicit computer simulations. They expected that the presence of males in androdioecious hexaploid populations might slow the diploid advance by competing with the diploid males to sire outcrossed progeny. However, because range expansion requires recurrent successful colonization (Baker 1955; Pannell and Barrett 1998), the occurrence of males with monoecious individuals actually diluted one of the benefits that monoecy has over dioecy: the advantage of reproductive assurance during colonization. Dorken and Pannell's (2007) simulations therefore predict that the diploid advance might accelerate when androdioecious regions are encountered, and they recall the complexities that can result when selection acts on plant reproductive strategies at both the population and metapopulation levels (Barrett and Pannell 1999; also see next section).

The transitions between dioecy and hermaphroditism in northern Spain, described above, are fully confounded by

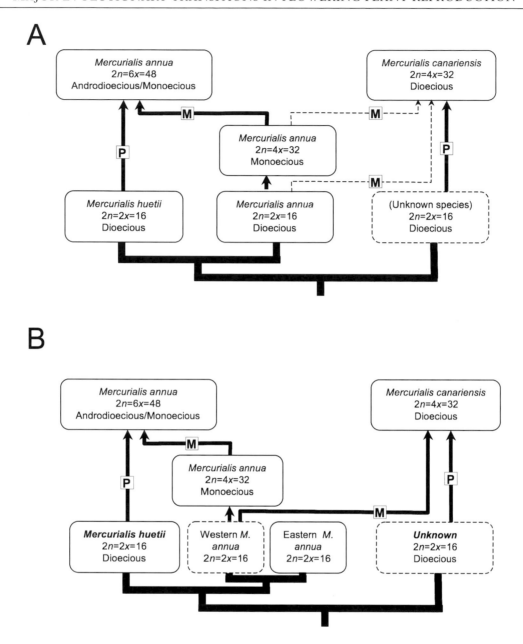

Fig. 2 Hypotheses for the relationships between the annual lineages of *Mercurialis*. Heavy lines indicate the phylogenetic relationships between diploid species, and thin arrows show polyploidization or hybridization events; *M* indicates proposed maternal parentage, and *P* indicates proposed paternal parentage. *A*, Diploid *M. annua* proposed as the parent of polyploid *M. annua*. The heterogeneous ITS types present in hexaploid *M. annua* show that it has an allopolyploid origin through hybridization between *M. annua* and *M. huetii*, and the hexaploid chromosome complement is consistent with hybridization between a tetraploid and a diploid lineage, followed by chromosome doubling. ITS data also show *M. canariensis* to be allopolyploid in origin, probably a hybrid between *M. annua* s.l. and an unknown taxon (dashed box). Chloroplast sequence similarity to *M. annua* s.l. suggests that *M. annua* s.l. was the maternal parent, and chromosome numbers are consistent with both parents being diploid. *B*, Given the greater similarity between ITS sequences from *M. canariensis* and tetraploid and hexaploid *M. annua*, the diploid progenitor of these taxa may have be an earlier western lineage of *M. annua*, divergent from the lineage that has recently expanded from the east. From Obbard et al. (2006*a*).

shifts between diploidy and hexaploidy. However, transitions also occur between androdioecy and hermaphroditism elsewhere in the Iberian Peninsula and in Morocco. In the Iberian Peninsula, these transitions occur within the hexaploid lineage, but in Morocco, the transition occurs along the Atlantic coast in a region broadly coincident with a transition between hexaploidy in the north and tetraploidy in the south

(Durand 1963). To determine whether these two transitions coincide precisely, we used flow cytometry to assay the ploidy level of 22 populations at intervals along the Atlantic coast of Morocco (fig. 4). For the flow cytometry measurements, we used leaf material from seeds grown in the greenhouse. Material was prepared using the "LB01" method of Dolezel et al. (1989) and *Lycopersicon esculentum* cv. Gardener's Delight

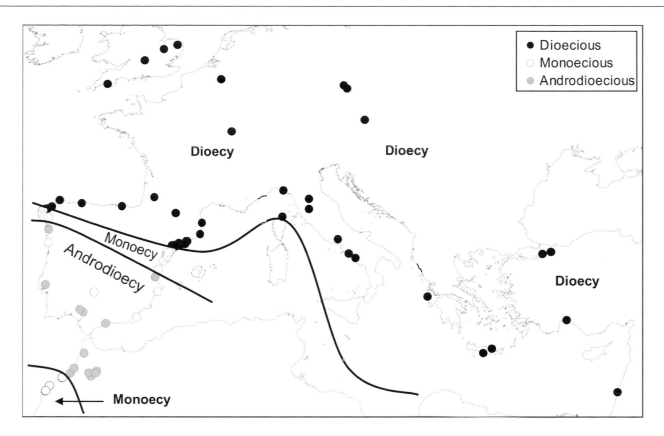

Fig. 3 Distribution of *Mercurialis annua* in Europe and around the Mediterranean Basin. In the north and east of this range, *M. annua* is dioecious and diploid, whereas in Iberia and North Africa, it is monoecious (and androdioecious) and polyploid. Circles indicate the locations of seed collections used to estimate patterns of diversity by Obbard et al. (2006*b*). Regions marked "Dioecy," "Monoecy," and "Androdioecy" denote zones occupied by the corresponding sexual systems (see "Phylogeography and Regional Transitions in the Sexual System"). Modified from Obbard et al. (2006*b*).

leaf material as a standard. Each sample was assessed on a Becton Dickinson FACScan flow cytometer and analyzed using Becton Dickinson CellQuest software (BD BioSciences, Franklin Lakes, NJ).

Our results indicate that the transition between tetraploidy and hexaploidy in Morocco occurs some 100 km south of the monoecy-androdioecy transition, which is located at Rabat; males are found only in populations north of this point (fig. 4). (Note that two tetraploid populations were found north of the major ploidy transition within the hexaploid zone. Such populations are likely to be the result of earth-moving roadworks, although they might also be remnant populations following a hexaploid advance south; see Buggs and Pannell 2006.) The geographic shift in the sexual system is thus uncoupled from the shift in ploidy in Morocco, and it occurs within the hexaploid lineage just as it does in the other monoecy-androdioecy transitions in the Iberian Peninsula. This poses the question of why the sexual system should change over geographical space, given that populations on both sides of these transitions have the same genetic system and appear to exchange genes (Obbard et al. 2006*b*).

Mechanisms of Transitions among Sexual Systems: The Role of Metapopulation Dynamics

To explain the maintenance of hermaphroditism versus dioecy or androdioecy in different regions of the range of hexa-

ploid *Mercurialis annua*, Pannell (1997*c*) hypothesized a metapopulation model that invokes differential selection for reproductive assurance during colonization. Because only self-fertile hermaphrodites, and not males, can colonize unoccupied habitat on their own, we should expect extinction-colonization dynamics to reduce the frequency of males at the metapopulation (i.e., regional) level. At the same time, hermaphrodites with female-biased sex allocation should be favored by selection during colonization over those with more equal sex allocation because their populations will grow more quickly (note that selection for female-biased sex allocation at the metapopulation level can be seen equivalently in terms of local mate competition during selfing or in terms of deme-level selection; see Frank 1986). As populations grow denser after colonization, the selective advantage of maleness increases, both because of the availability of large numbers of ovules to fertilize and because of the low numbers of pollen grains produced by hermaphrodites with which they must compete. Thus, whereas males are disfavored during colonization, they may be favored as migrants (or, probably much more rarely, as new mutants) into established demes. Under this model, whether males are found at a regional level depends largely on the balance between these two opposing selective forces (Pannell 1997*a*; see also Pannell 2001).

The metapopulation model makes a number of predictions that have recently been tested. One demographic prediction is that, if the regional absence of males is the result of rapid

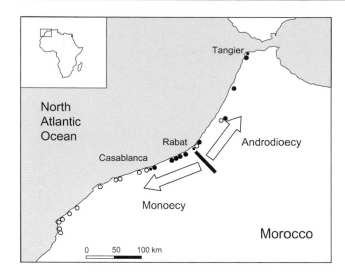

Fig. 4 Map showing the locations of 22 populations of *Mercuralis annua* for which the ploidy level was determined along the Atlantic coast of Morocco. Open and filled circles represent tetraploid and hexaploid populations, respectively. The solid line near Rabat indicates the point of transition between monoecious populations (to the south) and androdioecious populations (to the north). Two tetraploid populations were found within the area occupied principally by hexaploid populations.

population turnover compared with regions in which males occur, then populations in monoecious regions ought to be smaller than those in androdioecious regions (because they are, on average, younger). Another prediction is that rates of habitat occupancy should be lower in monoecious than in androdioecious regions if extinction-colonization rates are higher in the former (Gaston et al. 2000; Freckleton et al. 2005). Patterns of occupancy and abundance observed across several clines in the sexual system of *M. annua* in Spain were consistent with both these predictions (Eppley and Pannell 2007*b*).

From a population genetics perspective, the metapopulation model predicts that monoecious populations should have lower genetic variation and should be more strongly differentiated from one another than androdioecious populations. This is because monoecious populations are expected to display genetic signatures of more recent colonization bottlenecks that subsequent migration has not had time to erase (e.g., Slatkin 1977; Wade and McCauley 1988; Pannell and Charlesworth 2000). To test these predictions, Obbard et al. (2006*b*) measured genetic diversity in hexaploid monoecious and androdioecious populations across their range in the Iberian Peninsula and North Africa and in dioecious populations across their European range. They found that, even after the effects of the hypothesized range expansion had been accounted for (see "Phylogeography and Regional Transitions in the Sexual System"), hexaploid androdioecious populations were strikingly more diverse than their monoecious counterparts, as expected. Moreover, whereas pairwise genetic differentiation among hexaploid monoecious populations was highly variable and often large, pairs of androdioecious populations were invariably very similar, as were dioecious

populations. These patterns suggest a history of homogenizing gene flow among androdioecious populations and among dioecious populations (though not between androdioecious and dioecious populations, owing to their different ploidy levels; Obbard et al. 2006*b*). Although populations of *M. annua* with separate sexes might be linked by more gene flow because they contain males that disperse more pollen than hermaphrodites, such an explanation would not account for the low diversity of monoecious populations occurring within androdioecious regions (Obbard et al. 2006*b*).

A key assumption of the metapopulation model is that selfing rates are density dependent: hermaphrodites self-fertilize their progeny when mates are absent, but opportunities for outcrossing increase when populations grow. Eppley and Pannell (2007*a*) tested this assumption by estimating selfing rates in populations growing at different densities. Selfing rates were high when individuals were more than ca. 30 cm apart but quickly dropped in denser stands, as predicted by the model. Eppley and Pannell (2007*a*) used their results to predict the threshold density below which males should be excluded from a population and above which their maintenance was assured. This threshold density was much lower than commonly measured in purely monoecious populations, indicating that the presence of males in a population is indeed migration limited (Eppley and Pannell 2007*a*).

Evolution of Androdioecy: Sex Allocation and Inflorescence Structure

We have seen that dense populations of *Mercurialis annua* are largely outcrossing, easing the potential invasion of males into hermaphroditic populations. But how do males surpass the twofold threshold in pollen production required for their maintenance at frequencies greater than zero (Lloyd 1975; Charlesworth and Charlesworth 1978; Charlesworth 1984)? The metapopulation hypothesis suggests one mechanism by invoking selection for female-biased sex allocation in the hermaphrodites. Thus, if hermaphrodites allocate a proportion $x < 0.5$ of their reproductive resources to their male function, then female-sterile individuals that divert all their resources otherwise invested in seeds to pollen production will be able to disperse $1/x > 2$ units of pollen relative to the pollen produced by hermaphrodites. Full "compensation" in resource allocation between male and female functions would therefore be sufficient for male invasion as long as the hermaphrodites are female biased in their sex allocation.

Another way in which males might cross the twofold invasion threshold is by dispersing the pollen they produce better than hermaphrodites. Assume again that hermaphrodites allocate a proportion x of their resources to pollen production and that males allocate a proportion 1.0. Assume further that pollen grains dispersed by males are γ times more likely to find a receptive stigma than those dispersed by hermaphrodites. Then males should enjoy a relative siring success of γ/x. Clearly, if $\gamma > 1.0$, then males might invade a population of hermaphrodites even if $x = 0.5$, i.e., the population is unbiased in its sex allocation. Because we expect the sex allocation of outcrossing hermaphrodites to be ca. 0.5, if the male and female fitness gain curves are not very different

(Lloyd and Bawa 1984), this reasoning shows that the condition $\gamma > 1.0$ ought to be sufficient for the evolution of maleness fairly generally. Eppley and Pannell (2007*a*) tested this idea and estimated a value of $\gamma = 1.6$.

That males are so much better than hermaphrodites at dispersing their pollen is almost certainly due to the differences between the two morphs in their inflorescence architectures. Although monoecious plants have a female morphology, with male flowers held around a subsessile female flower in the leaf axils, males disperse their pollen from flowers held above the plant on long, erect peduncles. Inflorescences such as those of *M. annua* males are widely found in wind-pollinated herbs and are expected to increase the siring success of pollen by allowing it to travel farther after release (Levin and Kerster 1974; Niklas 1985; Burd and Allen 1988). Such adaptations almost certainly give rise to an accelerating male fitness gain curve, and they should thus stabilize the maintenance of separate sexes (Charnov et al. 1976). It is poorly understood whether the evolution of such inflorescences is more likely to precede, and thus to precipitate, a transition from hermaphroditism to dioecy or to follow the evolution of dioecy and wind pollination. However, transitions between wind pollination and dioecy often go hand-in-hand (Renner and Ricklefs 1995; Wallander 2001; Friedman and Barrett 2008), and they may coincide with secondary sexual adaptations such as those found in *M. annua* (see Weller et al. 1998, 2006; Karrenberg et al. 2002; Friedman and Harder 2004; Golonka et al. 2005).

Maintenance and Breakdown of Androdioecy: The Role of Phenotypic Plasticity

Sex expression is notoriously labile in plants, with males and, less often, females of dioecious populations frequently producing flowers of the opposite sex (Lloyd and Bawa 1984; Korpelainen 1998). In gynodioecious and androdioecious species, the hermaphrodites too may vary in their sex allocation in response to environmental cues (reviewed by Delph and Wolf [2005]). Whereas the implications and functional significance of complete gender switches remain poorly understood, more subtle expressions of sexual lability may have important consequences for sexual-system evolution, e.g., by conferring upon a lineage an ability to self-fertilize after long-distance dispersal (Baker and Cox 1984). Plasticity in the sex allocation of hermaphrodites is probably driven by variation in the marginal cost of resources needed for reproduction or in changes to the shapes of fitness gain curves (Lloyd and Bawa 1984; Klinkhamer et al. 1997), and it has important implications for sexual-system evolution through its effect on the maintenance of males or females.

The phenotypic plasticity in sex allocation of *Mercurialis annua* hermaphrodites should have important implications for the frequency of males that can be maintained in androdioecious populations. Indeed, along with metapopulation dynamics, it might be a further factor underlying the large among-population variation observed in male frequencies of the species. As argued by Delph (2003), plasticity in sex allocation by hermaphrodites can change the relative fertility of unisexual and hermaphrodite plants (see also Delph and Lloyd 1991). Because the relative pollen or seed fertility of unisexuals and hermaphrodites regulates their equilibrium frequencies (Lloyd 1976), such plasticity can cause variation in unisexual frequencies across ecological gradients (Delph 1990; Delph and Lloyd 1991; Asikainen and Mutikainen 2003; Barr 2004). For example, males always allocate all of their reproductive resources to pollen production (i.e., $x_m = 1$). If hermaphrodites change the proportion of resources allocated to pollen in response to an ecological gradient (i.e., x_h varies across the gradient), then this also changes the relative pollen fertilities of males and hermaphrodites (i.e., $r = x_m/x_h$). Half of the progeny sired by males are, on average, male, via the segregation of dominant male-determining alleles (Pannell 1997*b*). Thus, differences in r across environmental gradients will affect the relative frequency of male-determining pollen in the population and the frequency of males at equilibrium. Interestingly, such a scenario would concur with Darwin's (1877) reasoning that unfavorable environmental conditions can lead to an increased separation of the sexes and follows similar patterns shown by other species (reviewed by Delph and Wolf [2005]).

In a previous experiment, M. E. Dorken and J. R. Pannell (unpublished manuscript) found that plant density governs evolutionary trajectories in male frequencies by affecting the magnitude of r. Under high plant densities, r was more than three times higher than under low densities (and see Pannell 1997*c*). In the next generation, male frequencies were 38% higher among the progeny of plants grown under high densities (M. E. Dorken and J. R. Pannell, unpublished manuscript). Density can affect male siring success in two ways. First, density may affect the local availability of resources by increasing competition among plants. Second, because pollen is dispersed locally (Eppley and Pannell 2007*a*), density directly affects male siring ability. We attempted to dissect these confounded effects by controlling plant density and manipulating resource availability. Does resource availability affect the magnitude of r? If so, does variation in resource availability regulate male frequencies in *M. annua*?

To address these questions, we mixed seeds from five nearby sites in Morocco, each with high frequencies of males (i.e., between 25% and 47%). We grew plants from seed in 12 3×3-m raised beds in standardized, low-nutrient soil mixtures at the Wytham Field Lab between July and October 2004. We added slow-release nutrient fertilizer pellets (Osmocote, Scotts, Marysville, OH) at the recommended dosage to half of the beds, using a randomized block design with three blocks. On average, there were 566 ± 97 (SD) plants in the high-resource plots and 566 ± 42 (SD) plants in the low-resource plots. We maintained high male frequencies in each plot by transplanting males from additional similar plots not included in this analysis here (average male frequency in high-resource plots = 52.8% ± 1.4% [SE], in low-resource plots = 52.5% ± 8.2% [SE]). At the end of the experiment, we harvested a standardized sample of 20 hermaphrodite and 10 male plants from each plot by sampling individuals at regular intervals along four evenly spaced transects. Following Pannell (1997*c*), we calculated hermaphrodite allocation to male function (r) as the aboveground proportion of biomass allocated to pollen of hermaphrodites (π_h) relative to that of males (π_m; i.e., $r = \pi_m/\pi_h$). Thus, lower values of r indicate higher hermaphrodite allocation to pollen.

We found that hermaphrodites grown under high-resource conditions (average $r = 11.2 \pm 0.9$ [SE]) had nearly double the allocation to pollen of plants grown under low-resource conditions (average $r = 21.6 \pm 2.3$ [SE]; fig. 5a; two-tailed t-test with equal variances: $t = -4.28$, df $= 10$, $P < 0.005$). At the end of September, we harvested all hermaphrodites in

a)

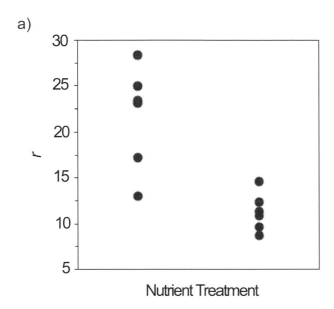

b)

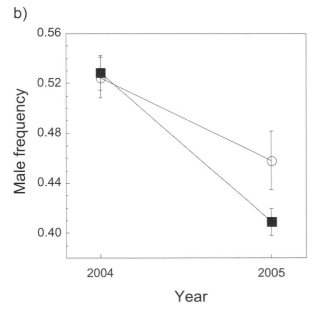

Fig. 5 Phenotypic plasticity in hermaphrodite sex allocation affects evolutionary trajectories in male frequencies. *a*, The relative pollen production (*r*) of males, compared with that of hermaphrodites, differs between low- and high-resource conditions. *b*, Because the pollen fertility of males relative to that of hermaphrodites regulates the frequency of males, different values of *r* between low- (circles) and high-resource conditions (squares) lead to divergence in male frequencies in the next generation. In *b*, error bars represent standard errors.

a plot and collected their seed in bulk, and we sowed an average of 1500 seeds (estimated by weighing out standard amounts of seeds) back into the plot in June 2005. Because hermaphrodites had higher allocation to pollen in high-resource conditions, the proportion of progeny sired by males should have been lower than that under low-resource conditions. Moreover, because maleness is governed by the segregation of a dominant Mendelian allele (Pannell 1997b), we predicted that the frequency of males among the progeny of plants grown under high-resource conditions should be lower than that among those from low-resource conditions. Our results are consistent with this prediction (fig. 5). The average frequency of males from high-resource plots was 40.9% ± 1.1% (SE; average number of plants per plot $= 720 \pm 74$ [SE]), compared with 45.8% ± 1.6% (SE) for low-resource plots (average number of plants per plot $= 501 \pm 83$ [SE]; one-tailed t-test with equal variances: $t = -1.92$, df $= 10$, $P < 0.05$).

Our results indicate that plasticity in sex allocation regulates the frequency of unisexual plants in gender-dimorphic populations, supporting the prediction made by Delph (2003). Even though our experiment was conducted across a single generation, we found considerable changes in male frequencies in response to environmental conditions. Specifically, we found that male frequencies not only can respond to density, which regulates male siring ability (as predicted by Eppley and Pannell 2007a), but also can respond directly to variation in r imposed by variation in the resource status of hermaphrodites. These results demonstrate the importance of phenotypic plasticity in hermaphrodite sex allocation for regulating male frequencies in androdioecious *M. annua*.

Genetics of Sex Determination and Sex Allocation in *Mercurialis annua*

The genetic basis of sex expression in dioecious *Mercurialis annua* was studied intensively by Durand and co-workers (Louis and Durand 1978; Dauphin-Guerin et al. 1980; Durand and Durand 1991a, 1991b). On the basis of crossing experiments using a small sample of original genotypes, they concluded that sex was determined by epistatic interactions between alleles segregating at three independently segregating loci. The fact that family sex ratios are often strongly male or female biased, despite a 1 : 1 sex ratio at the population level (J. R. Pannell, personal observations), would seem to confirm that more than one locus is involved, but results of recent genetic analysis based on sequence-characterized amplified-region (SCAR) markers were consistent with a single locus (Khadka et al. 2002). Pannell (1997b) concluded that maleness in an androdioecious population was determined by the presence of a single dominant allele, but his analysis also pointed to the importance of a plastic component of sex expression. Clearly, much remains to be learned about sex determination in *M. annua* and about the evolution of sex determination in the complex through its sexual-system transitions.

It is clear from the studies reviewed above that sex allocation in hermaphrodites of *M. annua* is also highly variable, albeit as a continuous trait. Much of this variation is evidently due to phenotypic plasticity, but comparisons between

Moroccan and Spanish populations suggest an important genetic component, too (J. R. Pannell, R. Berjano, and S. M. Eppley, unpublished manuscript); however, the architecture of this genetic variation in *M. annua* is unknown. If a more complete picture of past evolutionary transitions between sexual systems in the species complex is to emerge, we need further information about the quantitative genetics of sex allocation.

Two specific questions are currently the focus of attention of our research, and both relate to the so-called breeder's equation, $R = h^2 S$, which relates the potential response of a population to selection, R, to the product of the selection coefficient, S, and the narrow-sense heritability of the trait, h^2. Because $h^2 = V_A/V_P$, i.e., the fraction of total phenotypic variance that is due to additive genetic variance, the potential for response to selection on sex allocation will depend on both the absolute amount of additive genetic variance, V_A, and the size of V_P, which will be inflated by phenotypic plasticity (Falconer and Mackay 1996). The first question thus concerns how phenotypic plasticity in sex allocation in *M. annua* will affect responses to selection on pollen production. This question is of general importance for our understanding of transitions from gynodioecy and androdioecy to dioecy (Delph and Wolf 2005).

The second question concerns the size of V_A: how much additive genetic variance in sex allocation is present in populations of *M. annua* with different sexual systems, and how might this vary geographically? The geographic perspective on this question is important because range expansions, which are known to have occurred in *M. annua* (see "Phylogeography and Regional Transitions in the Sexual System"), typically involve repeated population bottlenecks that are expected to reduce quantitative genetic variation (Lande 1992). Indeed, range expansion in *M. annua* was inferred on the basis of reduced genetic variation at isozyme loci with distance from putative refugia (Obbard et al. 2006b). What we do not know is whether the hypothesized range expansions affected genetic variation in sex allocation, which is under frequency-dependent selection, differently. Current work is addressing this question through an assessment of the response to selection on pollen production in hermaphrodites sampled from populations at different latitudes in Spain and North Africa.

Conclusions

Mercurialis annua displays unusual variation in its sex expression at several levels in the genealogical hierarchy, including within and among genotypes, among populations in putative metapopulations, among regions in different parts of Europe, and between related species in a clade represented by lineages with reticulate phylogenetic relationships. This variation reflects a complex history of divergence, migration, and gene flow, which have all contributed to transitions between sexual systems. Research on *M. annua* has thrown light onto a number of themes in sexual-system evolution. These include the importance of ploidy, hybridization, and range expansions in regulating sexual-system variation and thus setting the stage on which natural selection then acts; the importance of phenotypic plasticity underlying much of the variation observed between individuals and its responsibility in regulating details of sex ratio evolution; the role of context-dependent mating in species characterized by demographic fluctuations and possible metapopulation dynamics; the adaptive value of sexual specialization for pollen dispersal under wind pollination; and the potential role of androdioecy in evolutionary paths between dioecy and hermaphroditism, a pathway about which very little was hitherto known. Future work aims to characterize the demographic processes hypothesized to have played a role in driving the observed sexual-system transitions at regional and local-patch scales as well as to describe the quantitative genetic architecture of sex determination and sex allocation across the phylogenetic and geographic range of the species complex.

Acknowledgments

We thank Spencer Barrett for inviting us to contribute to this special issue of the journal, Anne Sakai and two anonymous reviewers for useful comments on the manuscript, and the Wellcome Trust and F. M. Platt for flow cytometry facilities. J. R. Pannell and B. Pujol were supported by grants from the Natural Environment Research Council (NERC) of the United Kingdom; M. E. Dorken was supported jointly by NERC and the Natural Sciences and Engineering Research Council of Canada; and R. Berjano was supported by a Spanish grant of the Ministerio de Educación y Ciencia.

Literature Cited

Asikainen E, P Mutikainen 2003 Female frequency and relative fitness of females and hermaphrodites in gynodioecious *Geranium sylvaticum* (Geraniaceae). Am J Bot 90:226–234.

Baker HG 1955 Self-compatibility and establishment after "long-distance" dispersal. Evolution 9:347–348.

Baker HG, PA Cox 1984 Further thoughts on dioecism and islands. Ann Mo Bot Gard 71:244–253.

Barr CM 2004 Soil moisture and sex ratio in a plant with nuclear-cytoplasmic sex inheritance. Proc R Soc B 271:1935–1939.

Barrett SCH, JR Pannell 1999 Metapopulation dynamics and mating-system evolution in plants. Pages 74–100 *in* P Hollingsworth, R Bateman, R Gornall, eds. Molecular systematics and plant evolution. Chapman & Hall, London.

Buggs RJA, JR Pannell 2006 Rapid displacement of a monoecious plant lineage is due to pollen swamping by a dioecious relative. Curr Biol 16:996–1000.

——— 2007 Ecological differentiation and diploid superiority across a moving ploidy contact zone. Evolution 61:125–140.

Burd M, TFH Allen 1988 Sexual allocation strategy in wind-pollinated plants. Evolution 42:403–407.

Case AL, SW Graham, TD Macfarlane, SCH Barrett 2008 A phylogenetic study of evolutionary transitions in sexual systems in Australasian *Wurmbea* (Colchicaceae). Int J Plant Sci 169:141–156.

Charlesworth D 1984 Androdioecy and the evolution of dioecy. Biol J Linn Soc 23:333–348.

——— 1999 Theories of the evolution of dioecy. Pages 33–60 *in* MA Geber, TE Dawson, LF Delph, eds. Gender and sexual dimorphism in flowering plants. Springer, Heidelberg.

Charlesworth D, B Charlesworth 1978 A model for the evolution of dioecy and gynodioecy. Am Nat 112:975–997.

Charnov EL 1982 The theory of sex allocation. Princeton University Press, Princeton, NJ.

Charnov EL, J Maynard Smith, JJ Bull 1976 Why be an hermaphrodite? Nature 263:125–126.

Darwin C 1877 The different forms of flowers on plants of the same species. Appleton, New York.

Dauphin-Guerin B, G Teller, B Durand 1980 Different endogenous cytokinins between male and female *Mercurialis annua* L. Planta 148:124–129.

Delph LF 1990 Sex-ratio variation in the gynodioecious shrub *Hebe strictissima* (Scrophulariaceae). Evolution 44:134–142.

——— 2003 Sexual dimorphism in gender plasticity and its consequences for breeding system evolution. Evol Dev 5:34–39.

Delph LF, DG Lloyd 1991 Environmental and genetic control of gender in the dimorphic shrub *Hebe subalpina*. Evolution 45:1957–1964.

Delph LF, DE Wolf 2005 Evolutionary consequences of gender plasticity in genetically dimorphic breeding systems. New Phytol 166:119–128.

Desfeux C, S Maurice, JP Henry, B Lejeune, PH Gouyon 1996 Evolution of reproductive systems in the genus *Silene*. Proc R Soc B 263:409–414.

Dolezel J, P Binarova, S Lucretti 1989 Analysis of nuclear DNA content in plant cells by flow cytometry. Biol Plant 31:113–120.

Dorken ME, JR Pannell 2007 The maintenance of hybrid zones across a disturbance gradient. Heredity 99:89–101.

Durand B 1963 Le complexe *Mercurialis annua* L. *s.l.*: une étude biosystématique. Ann Sci Nat Bot 12:579–736.

Durand B, R Durand 1991*a* Male sterility and restored fertility in annual mercuries, relations with sex differentiation. Plant Sci 80:107–118.

——— 1991*b* Sex determination and reproductive organ differentiation in *Mercurialis*. Plant Sci 80:49–66.

——— 1992 Dioecy, monoecy, polyploidy and speciation in annual mercuries. Bull Soc Bot Fr Lett Bot 139:377–399.

Eppley SM, JR Pannell 2007*a* Density-dependent self-fertilization and male versus hermaphrodite siring success in an androdioecious plant. Evolution 61:2349–2359.

——— 2007*b* Sexual systems and measures of occupancy and abundance in an annual plant: testing the metapopulation model. Am Nat 169:20–28.

Falconer DS, TFC Mackay 1996 Introduction to quantitative genetics. Longman, Essex.

Frank SA 1986 Hierarchical selection theory and sex ratios. 1. General solutions for structured populations. Theor Popul Biol 29:312–342.

Freckleton RP, JA Gill, D Noble, AR Watkinson 2005 Large-scale population dynamics, abundance-occupancy relationships and the scaling from local to regional population size. J Anim Ecol 74:353–364. (Erratum, 75:314.)

Friedman J, SCH Barrett 2008 A phylogenetic analysis of the evolution of wind pollination in the angiosperms. Int J Plant Sci 169:49–58.

Friedman J, LD Harder 2004 Inflorescence architecture and wind pollination in six grass species. Funct Ecol 18:851–860.

Gaston KJ, TM Blackburn, JJD Greenwood, RD Gregory, RM Quinn, JH Lawton 2000 Abundance-occupancy relationships. J Appl Ecol 37:39–59.

Golonka AM, AK Sakai, SG Weller 2005 Wind pollination, sexual dimorphism, and changes in floral traits of *Schiedea* (Caryophyllaceae). Am J Bot 92:1492–1502.

Husband BC, B Ozimec, SL Martin, L Pollock 2008 Mating consequences of polyploid evolution in flowering plants: current trends and insights from synthetic polyploids. Int J Plant Sci 169:195–206.

Karrenberg S, J Kollmann, PJ Edwards 2002 Pollen vectors and inflorescence morphology in four species of *Salix*. Plant Syst Evol 235:181–188.

Khadka DK, A Nejidat, M Tal, A Golan-Goldhirsh 2002 DNA markers for sex: molecular evidence for gender dimorphism in dioecious *Mercurialis annua* L. Mol Breed 9:251–257.

Kiontke K, NP Gavin, Y Raynes, C Roehrig, F Piano, DHA Fitch 2004 *Caenorhabditis* phylogeny predicts convergence of hermaphroditism and extensive intron loss. Proc Natl Acad Sci USA 101:9003–9008.

Klinkhamer PGL, TJ de Jong, H Metz 1997 Sex and size in cosexual plants. Trends Ecol Evol 12:260–265.

Korpelainen H 1998 Labile sex expression in plants. Biol Rev Camb Philos Soc 73:157–180.

Krahenbuhl M, YM Yuan, P Kupfer 2002 Chromosome and breeding system evolution of the genus *Mercurialis* (Euphorbiaceae): implications of ITS molecular phylogeny. Plant Syst Evol 234:155–170.

Lande R 1992 Neutral theory of quantitative genetic variance in an island model with local extinction and colonization. Evolution 46:381–389.

Lande R, DW Schemske 1985 The evolution of self-fertilization and inbreeding depression in plants. I. Genetic models. Evolution 39:24–40.

Levin DA, HW Kerster 1974 Gene flow in seed plants. Evol Biol 7:139–220.

Lloyd DG 1975 The maintenance of gynodioecy and androdioecy in angiosperms. Genetica 45:325–339.

——— 1976 The transmission of genes via pollen and ovules in gynodioecious angiosperms. Theor Popul Biol 9:299–316.

Lloyd DG, KS Bawa 1984 Modification of the gender of seed plants in varying conditions. Evol Biol 17:255–338.

Louis J-P, B Durand 1978 Studies with the dioecious angiosperm *Mercurialis annua* L. (2n=16): correlation between genic and cytoplasmic male sterility, sex segregation and feminising hormones (cytokinins). Mol Gen Genet 165:309–322.

Mable BK 2004 Polyploidy and self-compatibility: is there an association? New Phytol 162:803–811.

Maurice S, TH Fleming 1995 The effect of pollen limitation on plant reproductive systems and the maintenance of sexual polymorphisms. Oikos 74:55–60.

Miller JS, DL Venable 2000 Polyploidy and the evolution of gender dimorphism in plants. Science 289:2335–2338.

Niklas KJ 1985 The aerodynamics of wind-pollination. Bot Rev 51:328–386.

Obbard DJ, SA Harris, RJA Buggs, JR Pannell 2006*a* Hybridization, polyploidy, and the evolution of sexual systems in *Mercurialis* (Euphorbiaceae). Evolution 60:1801–1815.

Obbard DJ, SA Harris, JR Pannell 2006*b* Sexual systems and population genetic structure in an annual plant: testing the metapopulation model. Am Nat 167:354–366.

Obbard DJ, JR Pannell, SA Harris 2005 A new species of *Mercurialis* (Euphorbiaceae). Kew Bull 61:99–106.

Pannell J 1997*a* The maintenance of gynodioecy and androdioecy in a metapopulation. Evolution 51:10–20.

——— 1997*b* Mixed genetic and environmental sex determination in an androdioecious population of *Mercurialis annua*. Heredity 78:50–56.

——— 1997*c* Variation in sex ratios and sex allocation in androdioecious *Mercurialis annua*. J Ecol 85:57–69.

——— 1997*d* Widespread functional androdioecy in *Mercurialis annua* L. (Euphorbiaceae). Biol J Linn Soc 61:95–116.

Pannell JR 2001 A hypothesis for the evolution of androdioecy: the joint influence of reproductive assurance and local mate competition in a metapopulation. Evol Ecol 14:195–211.

——— 2002 The evolution and maintenance of androdioecy. Annu Rev Ecol Syst 33:397–425.

Pannell JR, SCH Barrett 1998 Baker's law revisited: reproductive assurance in a metapopulation. Evolution 52:657–668.

Pannell JR, B Charlesworth 2000 Effects of metapopulation processes on measures of genetic diversity. Philos Trans R Soc B 355:1851–1864.

Pannell JR, DJ Obbard, RJA Buggs 2004 Polyploidy and the sexual system: what can we learn from *Mercurialis annua*? Biol J Linn Soc 82:547–560.

Renner SS, RE Ricklefs 1995 Dioecy and its correlates in the flowering plants. Am J Bot 82:596–606.

Ronfort J 1999 The mutation load under tetrasomic inheritance and its consequences for the evolution of the selfing rate in autotetraploid species. Genet Res 74:31–42.

Sakai AK, SG Weller 1999 Gender and sexual dimorphism in flowering plants: a review of terminology, biogeographic patterns, ecological correlates, and phylogenetic approaches. Pages 1–31 *in* MA Geber, TE Dawson, LF Delph, eds. Gender and sexual dimorphism in flowering plants. Springer, Heidelberg.

Slatkin M 1977 Gene flow and genetic drift in a species subject to frequent local extinction. Theor Popul Biol 12:253–262.

Stebbins GL 1950 Variation and evolution in plants. Columbia University Press, New York.

Vamosi JC, SP Otto, SCH Barrett 2003 Phylogenetic analysis of the ecological correlates of dioecy in angiosperms. J Evol Biol 16:1006–1018.

Wade MJ, DE McCauley 1988 Extinction and recolonization: their effects on the genetic differentiation of local populations. Evolution 42:995–1005.

Wallander E 2001 Evolution of wind pollination in *Fraxinus* (Oleaceae): an ecophylogenetic approach. PhD thesis. Botanical Institute, Göteborg University.

Webb CJ 1999 Empirical studies: evolution and maintenance of dimorphic breeding systems. Pages 61–95 *in* MA Geber, TE Dawson, LF Delph, eds. Gender and sexual dimorphism in flowering plants. Springer, Heidelberg.

Weeks SC, C Benvenuto, SK Reed 2006 When males and hermaphrodites coexist: a review of androdioecy in animals. Integr Comp Biol 46:449–464.

Weiblen GD, RK Oyama, MJ Donoghue 2000 Phylogenetic analysis of dioecy in monocotyledons. Am Nat 155:46–58.

Weller SG, AK Sakai, TM Culley, DR Campbell, AK Dunbar-Wallis 2006 Predicting the pathway to wind pollination: heritabilities and genetic correlations of inflorescence traits associated with wind pollination in *Schiedea salicaria* (Caryophyllaceae). J Evol Biol 19:331–342.

Weller SG, AK Sakai, AE Rankin, A Golonka, B Kutcher, KE Ashby 1998 Dioecy and the evolution of pollination systems in *Schiedea* and *Alsinidendron* (Caryophyllaceae: Alsinoideae) in the Hawaiian Islands. Am J Bot 85:1377–1388.

Wolf DE, JA Satkoski, K White, LH Rieseberg 2001 Sex determination in the androdioecious plant *Datisca glomerata* and its dioecious sister species *D. cannabina*. Genetics 159:1243–1257.

Wolf DE, N Takebayashi 2004 Pollen limitation and the evolution of androdioecy from dioecy. Am Nat 163:122–137.

Yampolsky C, H Yampolsky 1922 Distribution of sex forms in the phanerogamic flora. Bibl Gen 3:4–62.

Int. J. Plant Sci. 169(1):141–156. 2008.
1058-5893/2008/16901-0012$15.00   DOI: 10.1086/523368

A PHYLOGENETIC STUDY OF EVOLUTIONARY TRANSITIONS IN SEXUAL SYSTEMS IN AUSTRALASIAN *WURMBEA* (COLCHICACEAE)

Andrea L. Case,* Sean W. Graham,[†] Terry D. Macfarlane,[‡] and Spencer C. H. Barrett[§]

*Department of Biological Sciences, Kent State University, Kent, Ohio 44242, U.S.A.; †University of British Columbia Botanical Garden and Centre for Plant Research, Faculty of Land and Food Systems, and Department of Botany, University of British Columbia, Vancouver, British Columbia V6T 1Z4, Canada; ‡Manjimup Research Centre, Department of Environment and Conservation, Brain Street, Manjimup, Western Australia 6258, Australia; and §Department of Ecology and Evolutionary Biology, University of Toronto, 25 Willcocks Street, Toronto, Ontario M5S 3B2, Canada

Using phylogenies to make sound inferences about character evolution depends on a variety of factors, including tree uncertainty, taxon sampling, and the degree of evolutionary lability in the character of interest. We explore the effect of these and other sources of ambiguity for maximum likelihood (ML)–based inferences of sexual-system evolution in *Wurmbea*, a small genus of geophytic monocots from the Southern Hemisphere. We reconstructed *Wurmbea* phylogeny using four noncontiguous regions (ca. 5.5 kb) of the plastid genome across a broad sampling of taxa, and we confirm that the genus is divided into two well-supported clades, each defined by its geography (Africa vs. Australasia) and variation in sexual system (i.e., uniformly monomorphic vs. sexually variable, respectively). We demonstrate that the predominantly Australian clade includes the sexually monomorphic species *Iphigenia novae-zelandiae*. We observe treewide uncertainty in the state of all ancestral nodes, and therefore all state transitions, when all taxa in *Wurmbea* are considered. We demonstrate that this is primarily a consequence of interspersion of terminals with gender dimorphism vs. monomorphism throughout the Australasian clade, rather than tree uncertainty or the presence of very short internal branches. We accounted for tree uncertainty by randomly sampling alternative resolutions of branches that are poorly supported by ML bootstrap analysis, effectively interpreting these as soft polytomies. Under the assumption that well-supported aspects of our gene tree accurately depict organismal phylogeny, there is a marked evolutionary lability in the sexual systems of Australasian *Wurmbea*. A more problematic issue is that our results contradict the monophyly of two sexually polymorphic Australian species, *Wurmbea dioica* and *Wurmbea biglandulosa*. If this reflects paraphyly at the species level, lateral gene transfer, or failed coalescence, then the interpretations of character transitions will need to be adjusted. Our analysis provides an example of the impediments to linking macroevolutionary pattern with microevolutionary processes for evolutionarily labile traits in recently evolved plant groups that possess a high degree of variation in sexual characters.

Keywords: ancestral-state reconstruction, cosexuality, dioecy, gynodioecy, noncoding DNA, paraphyletic species, gender strategies, polyphyly, sexual-system evolution, species concepts, subdioecy.

Introduction

Evolutionary transitions between hermaphroditism (gender monomorphism) and separate sexes (gender dimorphism) have occurred frequently in the history of flowering plants (e.g., Weiblen et al. 2000; Vamosi et al. 2003). Understanding the evolutionary history of sexual systems and the mechanisms responsible for such transitions is thus an important issue in plant evolutionary biology (reviewed in Geber et al. 1999). The conditions for the evolution and maintenance of gender dimorphism have been the subject of much theoretical and empirical research, aimed primarily at explaining the sporadic but taxonomically widespread occurrence of this condition in the angiosperms (reviewed in Thomson and Brunet 1990;

Renner and Ricklefs 1995; Sakai and Weller 1999; Heilbuth 2000). Phylogenies can provide us with windows into the past history of organisms, enabling investigation of the long-term significance of evolutionary transitions and ecological processes that otherwise can be observed only on local or instantaneous time frames. For example, because phylogenies span multiple speciation events, they can be used to identify convergent traits that arose under similar selective regimes in distantly related taxa, to develop chronological sequences of character transitions by reconstructing states at ancestral nodes (e.g., Kohn et al. 1996), and to measure cross-taxon correlations among characters of interest (e.g., Friedman and Barrett 2008; Sargent and Vamosi 2008).

A number of studies have employed phylogenies to investigate evolutionary transitions in plant reproductive characters (e.g., Weller and Sakai 1999 and references therein; Weiblen et al. 2000; Renner and Won 2001; Krahenbuhl et al. 2002; Miller 2002; Vamosi et al. 2003; Graham and Barrett 2004; Givnish et al. 2005; Gleiser and Verdú 2005; Levin and Miller 2005). These studies are primarily aimed at estimating the number of times particular traits arose and/or the ecological or morphological context in which traits may have evolved, which provide indirect evidence of the proximate mechanisms responsible for phenotypic change and the adaptive significance of trait transitions. Although character correlations and estimates of transition rates do not necessarily require precise localization of character state changes on a phylogenetic tree, it is often useful to know the local direction of change (e.g., on which terminal or near-terminal branches particular transitions occurred). Pinpointing transitions on particular branches can guide further sampling efforts and subsequent microevolutionary studies that delve into the local-scale processes responsible for observed macroevolutionary patterns. In practice, mapping traits onto trees with precision is not necessarily straightforward, particularly for characters that show a high degree of evolutionary lability and/or intraspecific polymorphism (Weller and Sakai 1999). In this article, we describe a phylogenetic analysis of transitions in the sexual systems of the Australasian members of *Wurmbea* (Liliales: Colchicaceae), a widespread genus that displays considerable variation in gender strategies. In doing so, we address some of the challenges associated with inferences of historical transitions in variable characters.

Recent advances have been made in developing and implementing model-based methods for studying character evolution using phylogenies (e.g., Pagel 1994, 1999; Cunningham et al. 1998; Mooers and Schluter 1999; Huelsenbeck et al. 2003; Pagel et al. 2004; Ronquist 2004; Maddison and Maddison 2005). This has led to more nuanced approaches than maximum parsimony (MP), which accommodate uncertainty in how ancestral states are reconstructed on a given tree and, increasingly, underlying uncertainty about the phylogenetic tree itself. MP estimates of character evolution are still widely used but could be misleading because they do not take account of available branch-length information (e.g., multiple changes along a branch are not permitted in MP reconstructions). Maximum likelihood (ML) estimates of character states and their transitions offer the advantage that they are explicitly model based and can incorporate branch-length information (Pagel 1994, 1999). A Bayesian approach that stochastically maps character evolution (Nielson 2002; Huelsenbeck et al. 2003; Pagel et al. 2004; Bollback 2006) can accommodate uncertainty in assignment of ancestral states, in addition to uncertainty in the tree itself, using Markov chain Monte Carlo (MCMC) samples from the posterior distribution of trees, substitution model parameters, and character ancestral states (Bollback 2006). Stochastic character mapping requires that the prior probability of the rate parameter of character state change be assigned, although usually this will not be known. In this method, tree uncertainty can be accounted for by using post-burn-in trees sampled by MrBayes using MCMC (Huelsenbeck and Ronquist 2001). The MCMC tree output has also been used to account for tree uncertainty in MP and ML ancestral-state reconstructions in recent studies (e.g., Miadlikowska and Lutzoni 2004; Lewis and Lewis 2005; Jones and Martin 2006; Vanderpoorten and Goffinet 2006; Galley and Linder 2007). However, MCMC samples may lead to inflated estimates of clade confidence (e.g., Suzuki et al. 2002; Cummings et al. 2003; Erixon et al. 2003; Simmons et al. 2004), potentially biasing reconstructions of ancestral states.

The most common method of accounting for tree uncertainty in phylogenetic analysis is the bootstrap (Felsenstein 1985). Bootstrapping provides a nonparametric estimate of the effect of sampling error on tree topology and is performed by resampling the characters used to infer the tree (Felsenstein 1985). Using the bootstrap profile of trees for character mapping (Jones and Martin 2006) arguably provides a more conservative approach for taking account of tree uncertainty than using an MCMC tree sample. Instead of using the raw bootstrap profile for character mapping in our study of *Wurmbea*, we retained branches that were moderately to strongly supported by ML bootstrap analysis and randomly sampled possible resolutions of poorly supported branches (i.e., we considered branches with <70% support to represent soft polytomies). These trees were used to perform ML and MP reconstructions of the ancestral states of sexual systems in *Wurmbea*. It is not our intent to compare the efficacy of different character mapping approaches, although recent empirical studies suggest that reconstructions based on MP, ML, and stochastic mappings may be quite closely related (e.g., Lewis and Lewis 2005; Jones and Martin 2006; Vanderpoorten and Goffinet 2006; Leschen and Buckley 2007).

The diversity of gender strategies within and among species of *Wurmbea*—ca. 48 species of perennial, geophytic herbs (Nordenstam 1978, 1986, 1998; Macfarlane 1980, 1986, 1987; Bates 1995, 2007; Macfarlane and van Leeuwen 1996)—has drawn much recent attention (Barrett 1992; Vaughton and Ramsey 1998, 2002, 2003, 2004; Barrett et al. 1999; Case and Barrett 2001, 2004a, 2004b; Jones and Burd 2001; Ramsey and Vaughton 2001, 2002; Barrett and Case 2006; Ramsey et al. 2006a, 2006b). Most of this work has been conducted on *Wurmbea dioica* and *Wurmbea biglandulosa*, two wide-ranging polymorphic species that exhibit considerable intraspecific variation in sexual systems, including hermaphroditism, gynodioecy, subdioecy and dioecy (Barrett 1992; Vaughton and Ramsey 2002; Case and Barrett 2004a). Biogeographical surveys and experimental studies have investigated the role of resource availability and pollination biology in the evolution and maintenance of gender dimorphism in these taxa (Case and Barrett 2001, 2004a, 2004b; Vaughton and Ramsey 2004). However, evaluating these selective hypotheses in an ahistorical context could be misleading, not least because of the potential problems of species circumscription that are common in morphologically diverse groups (e.g., Funk and Omland 2003).

Although *Wurmbea* has a relatively even African-Australian distribution (Goldblatt 1978; Nordenstam 1978; Conran 1985), the types of sexual system represented in the two continental regions are strikingly different. Gender dimorphism is present only among the Australian species, while African *Wurmbea* and all other members of Colchicaceae are uniformly monomorphic for gender (Nordenstam 1978; Macfarlane 1980; Dahlgren et al. 1985). *Wurmbea* probably originated in southern Africa and arrived in western Australia via long-distance dispersal (Nordenstam 1978; Barrett 1992; Vinnersten and Bremer 2001). The nested position of *Wurmbea* in Colchicaceae, along with the probable monophyly of the Australian taxa (Vinnersten and Reeves 2003), tends to support the idea that gender dimorphism is evolutionarily derived in *Wurmbea* and that one or more origins of dimorphism followed its arrival and establishment in

Australia (Nordenstam 1978; Barrett 1992). Within Australia, gender and sexual dimorphism are more prevalent in eastern states, with a larger proportion of gender-dimorphic taxa and greater sexual dimorphism there than in western Australia. This may indicate region-specific selection for gender and sexual dimorphism.

Biogeographical signals evident in our phylogenetic analysis may inform hypotheses for variation in sexual systems as *Wurmbea* spread across the continent and perhaps beyond. We were also interested in investigating whether an additional eastward dispersal event from Australia to New Zealand occurred in *Wurmbea*. *Iphigenia novae-zelandiae* is the sole New Zealand member of Colchicaceae, and it may be a taxonomically misplaced member of *Wurmbea* (Moore and Edgar 1970; T. D. Macfarlane, personal observation). Plants of *I. novae-zelandiae* are small and have few leaves and tiny, solitary, and "poorly-formed" flowers (Moore and Edgar 1970); firm decisions on its generic position based on morphology alone are difficult because of its floral instability and reduced stature. The high degree of developmental instability of flowers may indicate that *I. novae-zelandiae* is self-compatible and predominantly autogamous (e.g., Barrett 1985). Although no mating system information is available for *I. novae-zelandiae*, some *Wurmbea* species are self-compatible (Vaughton and Ramsey 1998, 2003; Case 2000; Ramsey and Vaughton 2002). Baker's Law (Baker 1967) predicts an increased likelihood of successful establishment after long-distance dispersal by plants capable of autonomous selfing. Thus, an additional goal of our study was to use molecular data to determine whether *I. novae-zelandiae* is indeed a species of *Wurmbea* whose New Zealand distribution resulted from a more recent long-distance dispersal event from Australia.

We used phylogenetic evidence from four noncontiguous regions of the plastid genome to investigate evolutionary transitions between monomorphic and dimorphic sexual systems in Australian species of *Wurmbea*, the biogeographical structure of gender dimorphism within Australia, and the phylogenetic status of *I. novae-zelandiae*. We encountered several possible sources of uncertainty in our inferences of character evolution. These included pronounced interspersion of monomorphic and dimorphic taxa on inferred trees, suggesting substantial evolutionary lability in character shifts; tree uncertainty (reflected in several polytomies and short branches with limited bootstrap support); and evidence of the nonmonophyly of at least two of the species that are polymorphic for sexual system. We explored the implications of these sources of uncertainty on the inferred evolutionary history of gender variation in the Australian *Wurmbea*. Our exploration provides general insights into the challenges associated with the reconstruction of evolutionary labile characters on phylogenetic trees, particularly for groups with a recent origin or those that include classically described species that may not be monophyletic.

Methods

Taxon Sampling and Outgroup Selection

Wurmbea has been the subject of recent taxonomic study; species delimitations are generally well accepted (Nordenstam 1978, 1986; Macfarlane 1980, 1986, 1987; Bates 1995, 2007; Macfarlane and van Leeuwen 1996). Of the ca. 48 species of *Wurmbea*, 18 occur in Africa and 30 in Australia. Our sampling focused on the latter because the former are exclusively sexually monomorphic. We include 16 Australian species and three species from the Cape Province of South Africa. The Australian sample includes representatives of all five dimorphic species of *Wurmbea* (see appendix). According to the most recent circumscription (Bates 2007), the widespread species *Wurmbea dioica* consists of four subspecies (*alba*, *dioica*, *brevifolia*, and *lacunaria*), and the species *Wurmbea biglandulosa* consists of two (*biglandulosa* and *flindersica*). *Wurmbea dioica* subspecies are uniformly dimorphic for gender, except in western Australia, where monomorphic and dimorphic populations of *W. dioica* ssp. *alba* co-occur (Macfarlane 1980; Barrett 1992; Case and Barrett 2001, 2004a, 2004b). Both subspecies of *W. biglandulosa* have monomorphic populations, whereas ssp. *biglandulosa* also contains dimorphic populations (Vaughton and Ramsey 2002). Our analysis included one population of each of the following: *W. biglandulosa* ssp. *flindersica*, both subspecies of *Wurmbea latifolia*, and two subspecies of *W. dioica*. We included both monomorphic and dimorphic populations of the two polymorphic species, one population of each sexual system in *W. dioica* ssp. *alba*, and two populations of each sexual system in *W. biglandulosa* ssp. *biglandulosa*.

We sampled multiple putative outgroups representing three of the four tribes of the Colchicaceae: *Onixotis punctata* and *Onixotis stricta* (Anguillarieae); *Baeometra uniflora* (Baeometreae); *Iphigenia indica*, *Iphigenia novae-zelandiae*, and *Gloriosa superba* (Iphigenieae); and *Colchicum* (Colchiceae). Among these outgroups, the nearest relatives to *Wurmbea* are predicted to belong to *Onixotis* because *Wurmbea* and *Onixotis*, together with South African *Neodredgea*, make up tribe Anguillarieae (Nordenstam 1978, 1986). Vinnersten and Manning (2007) recognize *Onixotis* and *Neodredgea* under *Wurmbea* (see also Vinnersten 2003), but for the purposes of this article we follow the narrower circumscription. We also included *Burchardia multiflora*, a species of previously uncertain taxonomic position that is now well supported as a member of Colchicaceae (Chase et al. 1995; Rudall et al. 2000; Vinnersten and Bremer 2001; see also Nordenstam 1998).

Molecular Data

We extracted DNA from frozen or silica-gel-dried leaf tissue using either CTAB (Doyle and Doyle 1987) or Qiagen DNEasy isolation kits (Qiagen, Valencia, CA). We sequenced four segments of the large single-copy region of the chloroplast genome possessing a range of evolutionary rates (e.g., Graham et al. 2006): a cluster of four photosystem II genes (*psbB*, *psbT*, *psbN*, and *psbH*; hereafter *psbBTNH*), two tRNA genes with an intervening spacer region and a single intron (*trnL-UAA* and *trnF-GAA*; hereafter *trnLF*), the intergenic spacer region between *atpB* and *rbcL* (hereafter *atpB-rbcL*), and the 3′ end of the coding region of NADH dehydrogenase, subunit F (hereafter *ndhF*). These regions were amplified and sequenced using primers described by Taberlet et al. (1991) for *trnLF*, Graham and Olmstead (2000) for *psbBTNH*, Mannen et al. (1994) and Savolainen et al. (1994) for *atpB-rbcL*, and Olmstead and Sweere (1994) and Graham et al. (1998) for *ndhF*. Two of the internal sequencing primers of *atp*B-*rbc*L did not work for *Wurmbea*, so we designed two additional primers for sequencing (IGS4A = 5′-AATTGTGA-GTAAATGTGTTTAT; IGS4B = 5′-GATTCATTATTTCGATC-TTACC). A few regions could not be recovered for individual taxa, as is noted in the appendix.

We conducted PCR amplifications, using standard methods and profiles (e.g., Graham and Olmstead 2000). PCR products were purified with QiaQuick PCR purification columns (Qiagen) following manufacturer instructions; for cycle sequencing we used primers located internally to those used for amplification. For each gene region, we compiled and base-called contiguous sequence fragments using Sequencher (Gene Codes, Ann Arbor, MI). We used Clustal W (Thompson et al. 1994) to generate initial sequence alignments and manually adjusted alignments using Se-Al 1.0 (Rambaut 1998), following criteria described by Graham et al. (2000). Alignment gaps were coded as missing data. The aligned matrix is available on request from the first author.

Phylogenetic Inference Methods

We inferred trees with ML, using PhyML, version 2.4.4 (Guindon and Gascuel 2003), and MP, using PAUP*, version 4.0b10 (Swofford 2002), and we used Mesquite, version 1.12 (Maddison and Maddison 2005) for ML reconstructions of ancestral character states. The optimal model of DNA sequence evolution, chosen using hierarchical likelihood ratio tests (hLRTs) in ModelTest, version 3.06 (Posada and Crandall 1998), is TVM + Γ + I (see ModelTest documentation for model details). We performed heuristic searches in PhyML while determining optimal model parameters from the data but otherwise using default settings. We assessed branch support using 100 bootstrap replicates (Felsenstein 1985), using 10 random-addition replicates for each bootstrap replicate for the MP analysis.

Reconstructions of Sexual-System Evolution

We performed ML optimization of sexual systems on the best ML tree to trace their evolutionary history, using Mesquite (equivalent to the "global" option in Pagel's Discrete). We coded sexual system as a binary character (monomorphic vs. dimorphic) rather than assigning dimorphic taxa as gynodioecious, subdioecious, or dioecious, because extensive surveys indicate a substantial degree of gender variation among populations of all dimorphic taxa (e.g., Barrett 1992; Barrett et al. 1999; Vaughton and Ramsey 2002). Branch lengths were estimated from the data using PAUP*. The best ML tree contained four polytomies, which we interpret as "soft," reflecting a lack of evidence for resolving nodes (Maddison 1997). Because Mesquite cannot fully accommodate zero-length branches in ML ancestral-state reconstructions, we randomly resolved these four polytomies using the "create trees" tool in MacClade, version 4.03 (Maddison and Maddison 2001) and assigned them very short nonzero branch lengths (0.000001 substitutions per site; see also Ferrer and Good-Avila 2007). We refer to this tree as the "resolved best tree."

We estimated the overall likelihoods of one-rate and two-rate (asymmetric) models of evolution; the one-rate model (the Mk1 model in Mesquite, equivalent to the Markov model of Lewis [2001]) was the better-fitting one based on likelihood-

ratio tests (Pagel 1999). We considered ancestral states of individual nodes with a difference in log-likelihood scores greater than 2.0 to be statistically significant (Edwards 1972). Preliminary reconstructions indicated numerous transitions in sexual system in *Wurmbea* and uncertain reconstructions (monomorphic vs. dimorphic) for all nodes subtending *Wurmbea*. Although dimorphism is restricted to *Wurmbea*, the rest of Colchicaceae (and possibly related families) are uniformly sexually monomorphic. It is therefore quite possible that the one-rate model or its rate parameter does not apply consistently across the whole family. Most implementations of ML-based inference of ancestral states assume that a single, consistent evolutionary model applies across the tree under consideration. Mooers and Schluter (1999) describe a method for taking account of a shift in the evolutionary model on a specific branch. However, it is not clear how to pinpoint where (or whether) shifts in the evolutionary model or model parameters occurred. Simply placing a rate shift between an ingroup and outgroup is an unwarranted assumption. In the absence of knowledge of whether or where a rate shift occurred, we conducted separate ML reconstructions on *Wurmbea* alone and on Australasian *Wurmbea* alone (an approach suggested by W. Maddison, personal communication). We assessed both cases because it is not clear whether we should include the monomorphic African species in ML ancestral-state reconstructions; like the outgroups, the African species of *Wurmbea* are sexually monomorphic.

We used an additional approach to account for uncertainty in tree topology (phylogenetic uncertainty) and its effect on our character reconstructions. We used MacClade to collapse four branches with <70% ML bootstrap support in the best ML tree (fig. 1), taking 70% support as an indication of moderately well-supported branches (e.g., Graham et al. 1998). We generated a random sample of >10,000 resolutions of the eight resulting polytomies (the original four plus the four additional collapsed branches). We then used the "condense trees" function in PAUP* to remove duplicate trees, leaving 10,000 unique topologies for character reconstructions. After recalculating ML branch lengths for all trees using PAUP*, we assigned very short nonzero branch lengths (0.000001 substitutions per site), as we did for the resolved best tree. It is unlikely that this small departure from the ML branch lengths has a substantive effect on our overall conclusions from character reconstructions (see "Results").

Because zero-length or very short branches with poor support may make it difficult to make robust ancestral-state reconstructions, we repeated our reconstructions on trees with all branches set to unit length (e.g., Mooers and Schluter 1999) and also on trees with all zero-length branches assigned the median length estimated within the *Wurmbea* clade (0.001445 substitutions per site). The former ignores branch-length information altogether and reflects solely the number of speciation events separating species (Pagel 1994; Mooers et al. 1999); the latter does not reflect actual evolution on the tree but permits us to assess how robust inferences of ancestral states at individual nodes would be, had there been a more even timing of speciation events.

We used the "trace character over trees" function in Mesquite to determine how many of the 10,000 topologies had significant ancestral-state reconstructions for the nodes considered in the

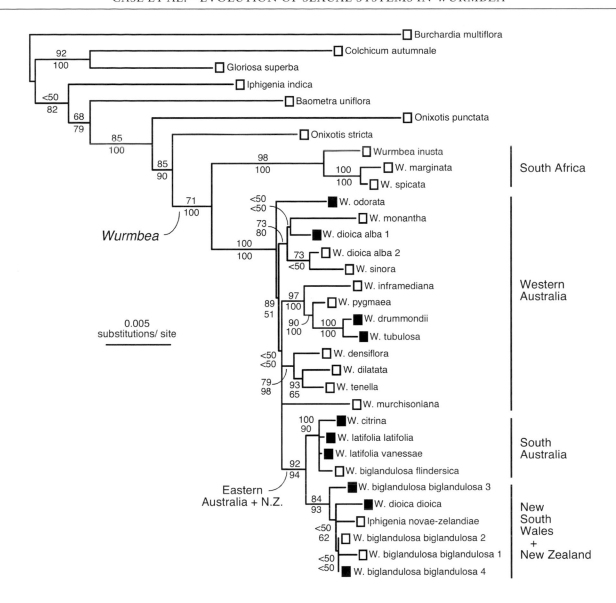

Fig. 1 Phylogenetic relationships in *Wurmbea* and relatives based on four disjunct plastid genome regions. Maximum likelihood (ML) analysis yields a single tree ($-\ln L = 13,084.47$); branch lengths are ML estimates. Bootstrap values are noted near branches (ML above, maximum parsimony below). Boxes at the ends of terminal branches indicate the sexual system of each sampled taxon: white indicates monomorphic, black indicates dimorphic. Multiple populations sampled within *Wurmbea dioica* ssp. *alba* and *Wurmbea biglandulosa* ssp. *biglandulosa* are distinguished using numbers.

resolved best tree, also noting the number of trees in which each node appeared. In addition, we used MP to estimate the minimum total number of transitions between monomorphism and dimorphism in either direction. We estimated total parsimony changes (Fitch 1971; Hartigan 1973) using the summary table option of the "state changes and stasis" tool in MacClade.

The best ML tree indicates that at least two species of *Wurmbea* are not monophyletic (see "Species Nonmonophyly"). We evaluated whether enforcing the monophyly of each species results in a significant increase in tree score by running heuristic ML searches in PAUP* with topological constraints enforced. The constraints enforced the monophyly of all (or subsets of) the populations for these species. We evaluated the significance of difference in scores of the shortest unconstrained and constrained trees, using

the Shimodaira-Hasegawa test (Shimodaira and Hasegawa 1999), with resampling estimated log likelihoods (RELL) bootstrap, using ML estimates of TVM $+ \Gamma + I$ parameters as determined on an MP tree (base frequencies were estimated empirically) and otherwise using default settings. All tree score comparisons were made simultaneously (see Goldman et al. 2000).

Results

Data Characteristics

The total data set, including all outgroups, consists of 5337 aligned nucleotides (233 are parsimony informative), excluding a short (<15 bp) homopolymer region in the *trn*LF intergenic spacer region. The data set for all *Wurmbea* samples contains 131 informative sites, with 79 informative sites for the Australia/New Zealand samples. For the full taxon set, the four chloroplast regions consist of 2090 (*psb*BTNH), 1262 (*trn*LF), 857 (*atp*B-*rbc*L), and 1128 bp (*ndh*F), which contribute 24%, 30%, 17%, and 28% of the informative sites, respectively. A 41-bp inversion in the *trn*LF intergenic spacer of one taxon (*Wurmbea inframediana*; see table 3 in Graham et al. 2000) was reinverted before alignment, following Graham and Olmstead (2000).

Phylogenetic Relationships in Wurmbea *and Relatives*

We rooted the best ML tree such that *Burchardia multiflora* is sister group to all other sampled Colchicaceae (Vinnersten and Reeves 2003). *Wurmbea* is then nested within a grade of *Onixotis*, a South African genus (fig. 1). *Iphigenia novae-zelandiae* is supported as a misclassified member of *Wurmbea* because of its nested position within a well-supported clade of eastern Australian members of the genus. *Wurmbea* is therefore monophyletic only if it is considered to include *I. novae-zelandiae*. A formal combination has not yet been made, but for simplicity in the rest of this article we will refer to *Wurmbea* as if it includes *I. novae-zelandiae*. There is a strongly supported basal division of *Wurmbea* into an African and an Australia/New Zealand clade. *Wurmbea odorata* is sister to the rest of the Australasian taxa, with weak to moderate support (fig. 1). A broadly eastern Australian clade of *Wurmbea* is also well supported and contains all taxa sampled from South Australia (which constitute a well-supported subclade), plus all taxa from New South Wales and New Zealand (another well-supported subclade). Relationships among four distinct *Wurmbea* lineages in western Australia are unresolved and poorly supported: (1) *W. murchisoniana*; (2) *W. inframediana* + *W. pygmaea* + *W. drummondii* + *W. tubulosa*; (3) *W. densiflora* + *W. dilatata* + *W. tenella*; and (4) *W. monantha* + *W. sinora* + *W. dioica* ssp. *alba*. Clades 2 and 3 are well supported; clade 4 is moderately supported.

Inference of Ancestral States

The resolved best tree shown in figure 2 includes all nodes in the best ML tree and one possible set of resolutions of the four polytomies in the best ML tree (fig. 1). No internal nodes on the resolved best ML tree for *Wurmbea* have significantly supported ancestral-state reconstructions (fig. 2). In large part, this appears to be because the 10 dimorphic taxa (terminals) in our taxonomic sample are widely interspersed among monomorphic Australian taxa. All but one of the major well-supported subclades within eastern and western Australia (see above) contain both dimorphic and monomorphic taxa. Deletion of the African clade of *Wurmbea* had no effect on the lack of significant reconstructions (data not shown).

Logically, the interspersion of dimorphic and monomorphic taxa must require multiple origins and/or losses of gender dimorphism. Mesquite does not provide an estimate of the ML number of transitions, but lower bounds to the number of transitions can be estimated using MP. The MP estimate should be close to the minimum possible number of transitions, because it permits at most one state change per branch to explain the observed distribution of character states; the true number of changes may be higher. The MP estimate of the range of total transitions across the 10,000 alternative resolutions of the weakest parts of the tree (as revealed by ML bootstrap analysis) is five to eight. This spans zero to eight gains versus zero to eight losses of gender dimorphism. The same ranges are obtained for reconstructions made across *Wurmbea* as a whole and for Australasian *Wurmbea* alone. This indicates that there were frequent evolutionary transitions in the sexual systems of *Wurmbea*.

As stated above, the resolved best tree has no significantly supported ancestral-state reconstructions (fig. 2). We explored whether any nodes on the resolved best tree might have significant ancestral reconstructions, given a large sampling of alternate resolutions of eight nodes—the four original polytomous nodes, plus the additional four nodes with relatively weak (<70%) ML bootstrap support, which were collapsed for this analysis. Below, we describe the outcome of these analyses. As a rule of thumb, we consider an assignment of an ancestral state in these reconstructions noteworthy if two conditions are met: (1) the node is present in all of the sampled alternate resolutions (indicated by "10,000" in the second column of table 1) and (2) it is assigned a significant ancestral state in the majority (>50%) of these 10,000 resolutions (indicated by values >50% in cols. A–F of table 1; see Miadlikowska and Lutzoni 2004).

First, we address the tree resolutions where the collapsed branches are resolved and assigned very short lengths (cols. A and D in table 1). Of all nodes that are strongly supported in the best ML tree (i.e., present in all 10,000 resolutions; second column in table 1), significant inferences of ancestral state were assigned in less than 5% of the sampled resolutions (all values for these nodes are ≤4.1 in cols. A and D of table 1). In only one case, which represents a random resolution of a polytomy in the best tree (node q; fig. 2; table 1), is there unambiguous assignment of an ancestral state in more than 50% of cases (underlined numbers in col. A of table 1); however, this node is present in only 1126 of the 10,000 sampled resolutions (second column in table 1), and this level of significant ancestral-state assignments is seen only in trees that include African *Wurmbea* (cf. cols. A and D in table 1).

When branch-length information is ignored (i.e., all branches are assumed to be equal; cols. B and E in table 1), the proportion of resolutions with significant ancestral states is generally higher than when inferred branch lengths are considered (cf. cols. A and B, D and E in table 1). However, the improvement is modest; reconstructions are still ambiguous in most cases. Only three strongly supported nodes (i, w, and y; fig. 2; table 1) are significantly assigned a state in >50% of the sampled resolutions (monomorphism in all three cases; underlined numbers in col. B in table 1). Of these, only i is unambiguous when Australian *Wurmbea* are considered alone (underlined numbers in col. E in table 1).

When zero-length branches in the 10,000 resolutions of the best tree are artificially inflated to the median branch length, the significance of ancestral-state reconstructions is also generally higher (cf. cols. A vs. C and D vs. F in table 1). However, short branch lengths cannot be the primary explanation for treewide uncertainty, because after branch-length inflation, most tree resolutions still do not result in significant assignments (i.e., most values in cols. C and F in table 1 are

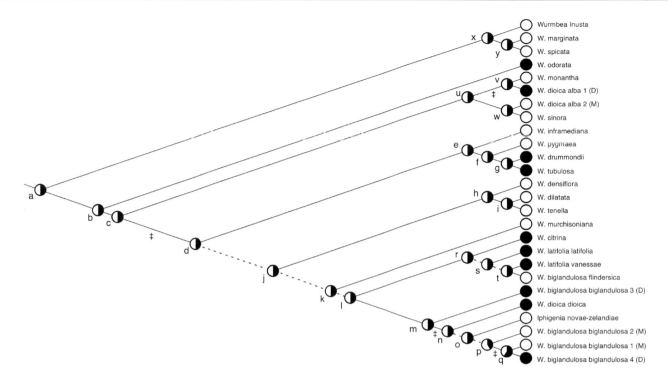

Fig. 2   Maximum likelihood ancestral-state reconstruction of sexual-system evolution on the shortest tree shown in fig. 1, with four polytomies arbitrarily resolved (the "resolved best" tree; see text). Branch lengths are assumed to be proportional to inferred DNA changes (fig. 1) but are presented as nonproportional lengths. Dashed lines denote the branches that collapse to polytomies in fig. 1; daggers indicate branches with <70% ML bootstrap support. The character reconstruction is limited to the genus *Wurmbea* (including *Iphigenia novae-zelandiae*). State inferences at nodes are represented as proportional likelihoods, shown as pie charts. The marginal probability reconstruction with model Mk1 in Mesquite has a rate of 760,745 ($-\ln L = 18.03$). Gender dimorphism is indicated in black, gender monomorphism in white. No nodes have significant ancestral-state assignments. Multiple populations sampled within *Wurmbea dioica* ssp. *alba* and *Wurmbea biglandulosa* ssp. *biglandulosa* are distinguished using numbers and the letters *M* and *D* (for monomorphic vs. dimorphic populations).

low) most of the time (i.e., most nodes have low values in cols. C and F in table 1).

A further indication of uncertainty is evident for several nodes (e.g., c, m, and r) that have significant reconstructions of *both* sexual-system states in at least some tree resolutions (occurrences of both M and D in cols. A–F in table 1). None of these findings is substantially affected by inclusion or exclusion of African *Wurmbea* from the reconstructions (cf. cols. A–C vs. D–F in table 1), although the proportion of nodes with high levels of significance is generally higher with these taxa included (cols. A–C in table 1).

Note that in order to pinpoint a transition between sexual systems with confidence, we would need to detect a change in sexual states at consecutive nodes in which each has a significant inference of ancestral state. Although we detected significant reconstructions in >50% of the sampled resolutions in a few instances (see underlined values in table 1), these particular branches are all terminal or near-terminal, and most do not indicate transitions between monomorphism and dimorphism because the inferred ancestral state in each case is identical to the states of the terminals it subtends (nodes g, i, w, y; see fig. 2). Nodes q and t provide evidence of transitions in some reconstructions, predicting a recent reversal from D to M in *Wurmbea biglandulosa* ssp. *biglandulosa* (population M1) and *W. biglandulosa* ssp. *flindersica*, respectively (fig. 2; table 1).

Species Nonmonophyly

Within the ingroup, we found evidence for nonmonophyly of two species containing both monomorphic and dimorphic populations (*Wurmbea dioica* and *W. biglandulosa*). When the monophyly of *W. dioica* as a whole is enforced, the score of the constrained tree is significantly higher than that of the best unconstrained tree (difference in $-\ln L = 111.92$; $P < 0.001$); however, when the two populations of *W. dioica* ssp. *alba* sampled here are constrained to form a small clade, this does not result in a significant increase in tree length (difference in $-\ln L = 4.03$; $P = 0.785$). *Wurmbea biglandulosa* is also not monophyletic in our analyses, but forcing the four sampled populations of *W. biglandulosa* ssp. *biglandulosa* together as a clade did not result in a significant increase in tree length (difference in $-\ln L = 5.56$; $P = 0.81$). *Wurmbea biglandulosa* ssp. *flindersica* is well supported as the sister group of *Wurmbea citrina* and *Wurmbea latifolia*, and constraining the monophyly of *W. biglandulosa* as a whole resulted in significantly longer trees (difference in $-\ln L = 85.9$; $P < 0.001$).

Discussion

We address the implications of our data for understanding sexual-system evolution in *Wurmbea*. Several strong biogeographic patterns are evident, and our reconstructions indicate

Table 1

Percentage of Nodes with Significant Maximum Likelihood (ML) Assignments of Ancestral States in 10,000 Resolutions of Eight Polytomies in the "Resolved Best" ML Tree

Node in figure 2	No. trees with node	All *Wurmbea* species Shortest branches = .000001 (A)	All branches equal (B)	Shortest branches = median length (C)	Australasian *Wurmbea* Shortest branches = .000001 (D)	All branches equal (E)	Shortest branches = median length (F)
a	10,000				na	na	na
b	10,000			M: .02	D: .1		
c	10,000		M: .8	M: .2		M: .01; D: .02	M: .1
d	1154		M: 5.0	M: .9		M: .7	M: .2
e	10,000		M: 12.3	M: 4.0		M: 5.7	M: 1.4
f	10,000			M: 3.0			M: 1.2
g	10,000	D: 2.1	D: .6	D: 86.2	D: .2		D: 59.6
h	10,000	M: .1	M: 47.8	M: 30.8		M: 25.3	M: 13.6
i	10,000	M: .3	M: 77.9	M: 49.3		M: 50.6	M: 21.8
j	489		M: 5.7	M: .8		M: 2.7	M: .4
k	506		M: 6.5	M: 1.6		M: 2.4	M: 1.0
l	1092		M: 1.7	M: 1.2; D: 1.3		M: .5	M: .2; D: .4
m	10,000	M: .7; D: 1.2	M: 6.5; D: .4	M: 10.5; D: 7.2	M: .3; D: .8	M: 3.7; D: .3	M: 7.8; D: 5.3
n	1099	D: .3	D: 2.5	M: .7; D: 7.7		D: 1.8	M: .3; D: 6.7
o	500		M: .2	M: .8; D: 14.4			D: 7.4
p	473		M: .4	M: 1.3			
q	1126	D: 57.2		M: .1	D: 29.3		
r	10,000	D: 4.1	D: .9	M: .03; D: 38.3	D: 3.6	D: .8	D: 26.4
s	2002	D: 3.6	D: 1.6	D: 58.8	D: 3.4	D: 1.2	D: 36.1
t	1911	D: 4.0		D: 63.9	D: 3.5		D: 34.7
u	10,000		M: 4.9	M: .7		M: 1.6	M: .1
v	3288		M: .2				
w	10,000	M: .4	M: 57.7	M: 43.9	M: .03	M: 34.3	M: 19.0
x	10,000		M: 20.3	M: 2.0	na	na	na
y	10,000	M: 3.0	M: 75.5	M: 88.8	na	na	na

Note. The eight resolved polytomies comprise the four original polytomies plus four branches with <70% bootstrap support collapsed post hoc (fig. 2). Lowercase letters correspond to nodes present in this tree, which shows one possible resolution of the four original polytomies seen in figure 1. Numbers in the second column indicate the number of resolutions in which a node was observed. The last six columns (A–F) represent reconstructions for each of two taxon sets (with analyses limited either to *Wurmbea* as a whole [A–C] or to Australasian *Wurmbea* alone [D–F]) but considering branch lengths in three different ways. The percentage of cases with a particular state at a node is noted in columns A–F (M = monomorphic; D = dimorphic); underlined numbers indicate cases for which at least 50% of the resolved trees had a significant ancestral state. Analyses were performed with zero-length branches assigned a very short length (A, D), with all branches assigned equal lengths (B, E), or with all zero-length branches assigned the median branch length within *Wurmbea* (C, F). na = not applicable.

that sexual systems in the Australian clade are evolutionarily labile. However, we were unable to pinpoint where sexual-system transitions occurred with any statistical confidence. We address the source of this uncertainty and discuss how the nonmonophyly of individual species may further complicate attempts to make inferences about the evolution of gender strategies using phylogenies. The inference-related problems that we encountered are unlikely to be restricted to *Wurmbea* and may be commonplace (but poorly recognized) in morphologically variable groups with a recent origin or in those that contain wide-ranging polymorphic taxa with evolutionarily labile traits.

Geography and Phylogeny

Species of *Wurmbea* tended to be geographically clustered within well-supported clades and subclades. The existence of African and Australasian clades received strong statistical support, as did an eastern Australian/New Zealand subclade. The relationships among western Australian subclades and of these to the eastern Australian/New Zealand subclade were poorly supported. However, there was weak support for the eastern subclade being nested in a western Australian grade. This arrangement, if correct, is consistent with migration and diversification eastward from the coast of western Australia. The Australasian clade of *Wurmbea* is nested in a larger African grade that includes the African clade of *Wurmbea* (mostly from the Cape Province of South Africa), a grade of *Onixotis* (two of three species in the genus are sampled here, all from the Cape Province), and *Baeometra uniflora* (the sole species in this genus, native to the Cape Province but collected in western Australia). Vinnersten and Manning (2007) recognize *Onixotis* and *Neodredgea* as part of *Wurmbea*; the former genus is paraphyletic with respect to *Wurmbea*, and the latter is monotypic and sister to *Onixotis* + *Wurmbea* s.s. (Vinnersten and Reeves 2003). Arguably, *Baeometra* might also be recognized under *Wurmbea* in a broader sense (*Wurmbea* would have taxonomic priority in this situation) because it is monotypic and sister to all of these taxa (see Backlund and Bremer 1998 for a rationale). Regardless of the classification used, the taxa that we accept as *Wurmbea* are clearly nested in a grade of South African taxa. Thus, our results (and those of Vinnersten and colleagues) support a single origin of Australasian *Wurmbea*, following a relatively recent long-distance dispersal event from southern Africa.

Paradoxically, such long-distance dispersal is inconsistent with several features of colchicoid life history. Seed shadows are probably very local in Colchicaceae; most plants have a diminutive stature, and seeds are generally shaken from dry dehiscent capsules. Moreover, aside from the possibility of rafting, fruits and seeds are unlikely to be dispersed by water, and many are ant dispersed (Nordenstam 1978). Although some species of *Wurmbea* have the capacity for limited clonal reproduction, daughter corms remain enclosed within the tunics covering the parent corm and apparently do not disperse readily (Nordenstam 1978; Macfarlane 1980; A. L. Case, personal observation). Hence, the mechanism(s) by which long-distance dispersal was achieved in *Wurmbea*, and in Colchicaceae as a whole, remains an enigmatic problem for biogeographical research (Vinnersten and Bremer 2001). However, this predicted long-distance dispersal was clearly not a unique event in *Wurmbea* biogeographic history, as exemplified by our finding that *Iphigenia novae-zelandiae* dispersed from within the Australian *Wurmbea* clade to the South Island of New Zealand.

Dynamic Evolutionary History of Sexual Systems in Wurmbea

Previous phylogenetic analyses of the evolution of gender strategies have been aimed at identifying independent transitions to evaluate the selective mechanisms responsible for gender dimorphism (e.g., Hart 1985; Wagner et al. 1995; Weller et al. 1995; Soltis et al. 1996; Sakai et al. 1997; Sakai and Weller 1999; Renner and Won 2001; Levin and Miller 2005). These findings, in conjunction with population-level studies, can support hypotheses that specific ecological factors favor transitions from gender monomorphism to dimorphism (reviewed in Ashman 2006; Barrett and Case 2006). If our plastid-based estimate of phylogeny represents species relationships accurately, then there were numerous transitions in sexual system in the Australian clade of *Wurmbea*. This means that the high degree of sexual variation in this clade mirrors high evolutionarily lability of sexual systems. This makes evaluating ecological correlates of dimorphic sexual systems more difficult within a phylogenetic framework, although uncertainty in ancestral states is not necessarily problematic when among-character correlations are assessed using ML analysis (Pagel 1999). Our analyses indicate that there were a minimum of five changes in sexual system, but they do not indicate the balance or direction of these changes. For example, we cannot distinguish whether dimorphism evolved repeatedly within Australia or just once after dispersal to Australia, followed by repeated reversions to monomorphism. Moreover, the possibility that an unknown dimorphic taxon was dispersed from Africa, although it might be considered

less parsimonious (and less probable because it would usually require dispersal of both sexes), is not altogether unlikely, as indicated in the ML reconstructions (see the root node of the Australian taxa in figs. 2, 3).

Reversals from gender dimorphism to monomorphism are not commonly reported and are traditionally thought of as being exceedingly unlikely to occur, particularly in the case of dioecy, which has previously been thought generally irreversible (Bull and Charnov 1985). Bull and Charnov's arguments for the irreversibility of dioecy are probably appropriate for animal systems, but sex expression in flowering plants is subject to very different constraints (e.g., modular body plans, sexual inconstancy). Gender dimorphism has arisen independently from cosexual ancestors in almost half of flowering plant families (Renner and Ricklefs 1995), suggesting that it is not necessarily difficult to evolve. Although there are a few notable exceptions (e.g., *Cotula*, Lloyd 1975; *Fuchsia*, Sytsma et al. 1991; *Mercurialis*, Obbard et al. 2006; *Sagittaria*, Dorken et al. 2002; Cucurbitaceae, Zhang et al. 2006), reversals to cosexuality from gender dimorphism are still typically presumed to be less likely in plants (note, however, that many gender-dimorphic groups that include dioecious species exhibit a continuum of sex expression between unisexuality and hermaphroditism; Delph and Wolf 2005; Ehlers and Bataillon 2007). In the few cases (i.e., nodes) where we detected a high proportion of significant ancestral-state assignments in *Wurmbea*, our inferences of sexual-system transitions indicated reversals from dimorphism to monomorphism. These cases all included a reversal in *Wurmbea biglandulosa* ssp. *flindersica* and often also included reversals in *W. biglandulosa* ssp. *biglandulosa* and *I. novae-zelandiae*. This is not surprising, given the extensive sexual inconstancy within *Wurmbea* species (Barrett et al. 1999; Ramsey and Vaughton 2001; Barrett and Case 2006). It would be interesting to know what features of Australian *Wurmbea* biology make the clade so unusually sexually labile. Comparative microevolutionary studies would be particularly informative, as would broadscale comparative analyses of the ecology and biology of African versus Australian species.

One possible reversal to monomorphism may be associated with the long-distance dispersal event that yielded *I. novae-zelandiae*. Losses of dimorphism have been associated with long-distance dispersal in other plant groups in the Southern Hemisphere (Baker and Cox 1984; Sytsma et al. 1991). Because *I. novae-zelandiae* is nested within a local clade that includes some dimorphic taxa (*Wurmbea dioica* ssp. *dioica* and *W. biglandulosa*), a dimorphic common ancestor could have given rise to it via (or after) dispersal. A scenario in which a polleniferous plant with sex inconstancy from a gender-dimorphic species could have established a monomorphic population is consistent with ecological evidence from several species. First, there is ample evidence of labile sex expression in *Wurmbea*; individual plants with sex inconstancy reproduce as hermaphrodites in some years but as males in others (Barrett et al. 1999; Ramsey and Vaughton 2001). A single founding individual would need to reproduce as a hermaphrodite to establish a monomorphic population. Second, this type of dispersal event would also require self-compatibility for successful establishment. Self-compatibility has been reported in several *Wurmbea* species (Barrett 1992; Ramsey and Vaughton 2002;

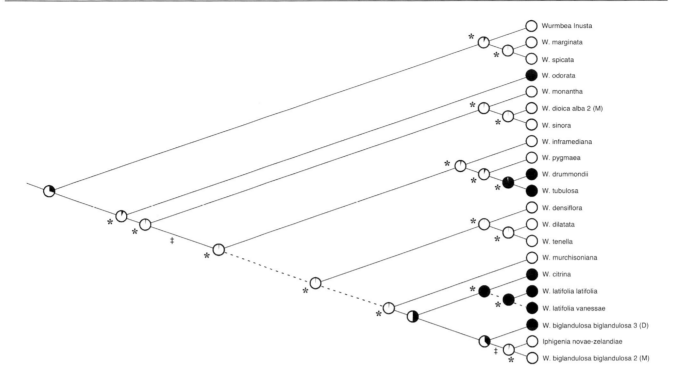

Fig. 3 Sensitivity of maximum likelihood ancestral-state reconstructions of sexual-system evolution to taxon sampling. Five population subsamples were deleted relative to fig. 2 (i.e., *Wurmbea dioica* ssp. *alba* [D], *Wurmbea dioica* ssp. *dioica*, *Wurmbea biglandulosa* ssp. *flindersica*, and *W. biglandulosa* ssp. *biglandulosa* 1 [M] and 4 [D]; see text for details). The character reconstruction is limited to the genus *Wurmbea* (including *Iphigenia novae-zelandiae*); branch lengths are assumed to be proportional to inferred DNA changes (fig. 1) but are presented as nonproportional lengths. Dashed lines denote the branches that collapse to polytomies in fig. 1; daggers indicate branches with <70% ML bootstrap support. State inferences at nodes are represented as proportional likelihoods, shown as pie charts. The marginal probability reconstruction with model Mk1 in Mesquite has a rate of 102.5 ($-\ln L = 12.13$). Gender dimorphism is indicated in black, gender monomorphism in white. The majority of internal nodes (17 out of 20) have significant ancestral-state assignments, indicated by asterisks. Multiple populations sampled within *Wurmbea dioica* ssp. *alba* and *Wurmbea biglandulosa* ssp. *biglandulosa* are distinguished using numbers and the letters *M* and *D* (for monomorphic vs. dimorphic populations).

Vaughton and Ramsey 2003), and the biology of *I. novae-zelandiae* suggests that it is predominantly selfing. Autonomous selfing would aid in the establishment of this lineage after dispersal from Australia (Baker 1967), particularly in the context of the relatively depauperate pollinator fauna of New Zealand (Webb and Kelly 1993).

We have shown that the largest source of uncertainty in the reconstruction of sexual systems in *Wurmbea* is the intermingling of monomorphic and dimorphic taxa across the phylogenetic tree. This interspersion contributes substantially more to mapping uncertainty than does the presence of zero- or near-zero-length branches (or otherwise poorly supported branches; see table 1). Studies of groups with a high degree of gender variation are especially likely to result in ambiguous ancestral-state reconstructions, while reconstructions from only a few dimorphic taxa, where there is less interspersion of sexual systems (e.g., *Lycium*; Levin and Miller 2005), should be more straightforward. In the same way that rapid speciation can obscure phylogenetic signal for determining species relationships, the complex distribution of sexual systems across our *Wurmbea* phylogeny obscures the localization of specific character transitions on the tree.

Taxon Sampling, Short Branches, Uncertain Species Boundaries, and Their Effects on Inferences of Character-State Transitions

Some of the dispersion of dimorphic sexual systems across the tree reflects the dispersed pattern of population terminals of *W. dioica* and *W. biglandulosa*. One or more of the population terminals in these taxa may represent distinct (cryptic) species, which may provide an explanation for at least part of the scattered distribution of *W. dioica* populations on the plastid phylogeny (see Vanderpoorten and Goffinet 2006 for a potentially similar situation in mosses). As currently circumscribed, *W. dioica* has the widest distribution of Australian *Wurmbea* species and a high degree of morphological variation, particularly with regard to floral traits and sex expression. Distinct morphological characters acknowledged in taxonomic treatments of each recognized subspecies of *W. dioica* (see Macfarlane 1980, 1986; Bates 1995) lend credence to the idea that at least some of these represent distinct species, either closely related or not. Indeed, during the preparation of this article, one Australian taxon was elevated to species status (*Wurmbea citrina*) from its previous designation as a subspecies of *W. dioica* (Bates 2007). Trees with *W. citrina* and *W.*

dioica ssp. *dioica* enforced as a sister taxa are significantly longer than our best tree (difference in $-\ln L = 62.0$; $P < 0.025$), supporting this proposal. In contrast to W. *dioica*, there are few obvious morphological or ecological traits in W. *biglandulosa* ssp. *biglandulosa* that distinguish the different populations or sexual morphs, particularly in the northern part of the species range, where our samples were collected (M. Ramsey, personal communication).

It is possible that only one of the population terminals in the case of W. *dioica* or W. *biglandulosa* is the correct placement for either species. For example, chloroplast capture events or failed coalescence might explain at least some of their dispersion across the phylogeny. Different terminals for individual populations of W. *dioica* and W. *biglandulosa* are more closely related to geographically proximal species than to each other, which is consistent with local introgression. Alternatively, these geographic patterns may reflect common descent from geographically proximal ancestors in which at least one member of each descendant species pair retains a more plesiomorphic floral condition and ends up being misclassified (this situation differs from the plesiospecies concept of Olmstead [1995]; in our case the error may be associated with taxonomic definitions that focus on plesiomorphic reproductive characters, whereas Olmstead's potentially genetically cohesive plesiospecies gives rise to isolated apospecies). At present, we cannot distinguish among these scenarios. However, chloroplast capture or failed coalescence may provide the correct explanation for at least some of the dispersion of dimorphic terminals across the tree.

It is worth considering what we might have reconstructed had we included a more limited sampling of populations, as many phylogenetic studies of reproductive transitions involve restricted taxon sampling. We therefore experimented with character reconstructions on trees with reduced taxon sampling, which may incidentally delete misplaced terminals. Because the main goal of our study was to reconstruct sexual-system evolution, we maintained a reasonable density of sampling, including both monomorphic and dimorphic taxa from across Australia in our taxon-deletion experiments. We subsampled taxa from the tree shown in figure 2 and traced sexual-system transitions in Mesquite, using the same procedure as for the full taxon sample, noting changes in the significance of ancestral states at each node. These taxonomic subsamplings could easily have been the samples collected for our study, if there had been a slightly different emphasis to our field collections.

We found that it can take very few terminal deletions to dramatically increase the number of significantly reconstructed ancestral states across the tree, regardless of the short internal branches observed here. For example, in figure 3 we illustrate a taxon subsample that leaves some level of interspersion and two nonmonophyletic taxa. The number of nodes inferred to have a significant ancestral-state change varies substantially—from no ancestral nodes in figure 2 to 85% significant nodes in figure 3 (i.e., 17 of 20 internal nodes). The estimate for the character transition rate for the trace shown in figure 3 (rate = 102.5) is substantially lower than that of the resolved best tree shown in figure 2 (rate = 760,745).

We experimented with different taxon samplings and found that deletion of just one branch has a large effect on the significance of ancestral-state reconstructions. Specifically,

when population D4 or M2 of W. *biglandulosa* ssp. *biglandulosa* is removed from consideration, the number of significant nodes jumps from zero to seven or nine, respectively (these significant nodes are all near-terminal; data not shown). The greater uncertainty in ancestral-state reconstructions when both populations are included is a consequence of the populations having different sexual-system states (dimorphic vs. monomorphic) with essentially no evolutionary distance between them (fig. 1). Because they have different states, a state transition must be inferred; because there is no detectable distance between them, this also implies near-instantaneous change, requiring a substantially higher treewide transition rate with both included (760,745) than with either sample excluded (357.4 or 354.9).

The existence of closely related populations of W. *biglandulosa* with alternate sexual systems contributes to the difficulty in inferring ancestral states. However, it should be emphasized that the overall uncertainty in ancestral-state reconstructions here (i.e., very few nodes with significant ancestral-state reconstructions; table 1) is not a function of high or low numerical values for the overall character transition rate under different treatments of branch length. For example, the resolved best tree (fig. 2) has a transition rate of 248.0 when zero-length branches are replaced with branches of the median length and a rate of only 0.393 when all branches are set to equal length, and yet most reconstructions are still equivocal for all three branch-length scenarios (see table 1). Furthermore, when population M2 of W. *biglandulosa* is deleted and all branch lengths are treated as having unit length, the transition rate falls to 0.4086 from 357.4, but only two nodes are inferred to have significant states (vs. nine when the ML branch lengths are considered). Although branch length affects character reconstructions in a complex way, the biggest source of uncertainty in our reconstructions is not the overall ML rate or the very short branches subtending some taxa, but rather the interspersion of taxa with different character states on the tree (fig. 2; table 1). This interspersion requires frequent transitions in any explanation of character evolution.

Whatever the cause of the dispersion of W. *dioica* and W. *biglandulosa* terminals across the tree, it serves to underline the dramatic effect that the observed interspersion has on character reconstructions. Undoubtedly, both speciation and extinction contribute to the observed patterns of taxa and character states across any phylogeny. The effect of extinction is expected to be comparable to that of incomplete taxonomic sampling, and there is some evidence that dioecious plant lineages have a higher probability of extinction than their hermaphroditic counterparts (Vamosi and Otto 2002). This is difficult to deal with in practice because extinct taxa are usually not observable (unless very good fossil data are available).

Ancestral-State Reconstruction When Species Boundaries Are Unclear

In the analyses described above, we treated each population as if it were a valid terminal for phylogenetic inference. Phylogenetic studies often include at most one or two exemplar specimens per species and tend not to be designed to test the (implicit) assumption that species boundaries are well-defined or readily definable (e.g., Funk and Omland 2003). In our case, we sampled four and five populations in *W. dioica* and *W. biglandulosa*, respectively, and found clear evidence for the nonmonophyly of both species. This may reflect failure of ancestral polymorphisms to coalesce within a species, chloroplast capture events, or the existence of cryptic species, as described above, or even true species-level paraphyly. These alternate hypotheses cannot be distinguished using the current evidence, and, in general, it is not clear how to deal with these contrasting possibilities in phylogenetic studies of character transitions. Approaches exist for estimating species phylogeny when different terminals from a given species are interspersed on a gene tree, but these work well only if the explanation can be ascribed to a single cause (failed coalescence, for example; Maddison and Knowles 2006). It is clear that whatever the source(s) of real or apparent nonmonophyly is for *W. dioica* and *W. biglandulosa*, there is a considerable need for population genetic, demographic, and taxonomic work to determine the limits and degree of leakiness in species boundaries in these taxa.

The observed uncertainty in species definitions for *W. dioica* and *W. biglandulosa* is problematic for interpreting character reconstructions and for linking these to results from microevolutionary or ecological studies (e.g., Barrett 1992; Vaughton and Ramsey 1998, 2002, 2003, 2004; Case and Barrett 2001, 2004a, 2004b; Ramsey and Vaughton 2001, 2002). Our sensitivity analysis using altered taxon sampling indicates that if we accept that there are multiple independent terminals within each "species," the level of taxon sampling, which is often strongly affected by sampling logistics in a study, can have a profound effect on ancestral-state reconstruction. This should be of concern for any study in which character reconstructions use only one or a few population samples from widespread, highly polymorphic species. We suggest that this would be a fertile area of investigation for other studies of character evolution in recently evolved organisms. The potential error levels in inferences of ancestral state associated with poor taxon sampling may be greater than those associated with tree uncertainty or other sources of analytical error (e.g., use of MP vs. ML in ancestral-state reconstructions). It is clear that all relevant species should be included in a study, but a strong case can also be made that funding agencies should permit (i.e., support) active field collection of multiple population samples for each species. This would at least provide insights into whether exemplar-based sampling of species (the use of one or a few samples to represent a species) may be providing a biased view of the underlying phylogeny or of reconstructions of character transitions.

The inferred lability of sexual systems in *Wurmbea* is in large part a function of the interspersion of dimorphic terminals of *W. dioica* and *W. biglandulosa* among monomorphic taxa. If the observed placement of different populations of *W. dioica* is a consequence of plastid capture or failed coalescence, this may substantially mislead inference of ancestral

states. It is possible that further sampling within Australian *Wurmbea* will reveal additional problems with species definitions. However, including additional unsampled species is unlikely to overturn our main finding from the current gene tree that sexual-system transitions in *Wurmbea* are frequent, because, except for the two remaining species of *W. dioica*, the remaining unsampled taxa appear to be all monomorphic for gender. Wherever these species are located in a better-sampled phylogeny, the various dimorphic terminals will still be substantially interspersed among monomorphic taxa.

It is worth considering how often failure to subsample from across a species' geographic range may miss nonmonophyly of individual species. Two reviews of this phenomenon suggest that this may be a widespread phenomenon (Crisp and Chandler 1996; Funk and Omland 2003), and it seems especially probable in classically defined species that are geographically widespread and that have a high degree of phenotypic variation. Ironically, these kinds of groups, which provide multiple inferences of convergence for macroevolutionary studies (and therefore are most likely to cause problems in ancestral-state reconstructions as a result of character lability), are also those more likely to have been used in microevolutionary studies concerning the mechanisms responsible for evolutionary transitions. In comparisons involving recently evolved taxa, taking full account of the complexity of species boundaries is likely to be a critical problem for evolutionary biologists to resolve before using phylogenies to make accurate inferences about character transitions.

Acknowledgments

We gratefully acknowledge the cooperation and support of the Western Australian Department of Environment and Conservation, the State Herbarium of South Australia, and Commonwealth Scientific and Industrial Research Organization, Canberra, for collections of Australian taxa. We thank R. Bates, M. Chase, B. Hall, L. Jesson, P. Johnson, D. Jones, P. Linder, M. Ramsey, K. Shepherd, G. Vaughton, and C. Wilkins for leaf material or DNA; A. Calladine, W. Cole, P. Lorch, H. Stace, and M. Waycott for field assistance; V. Biron, A. Brown, D. Cherniawsky, W. Case, A. Gilpin, P. Murray, H. O'Brien, M. Waycott, and J. Willis for lab assistance and support; and L. Delph, J. Eckenwalder, A. Schwarzbach, J. Thomson, M. Ramsey, and anonymous reviewers for critical comments on earlier drafts of the manuscript. W. Maddison provided advice on ancestral-state reconstructions using maximum likelihood. This research was supported by grants from the Natural Science and Engineering Research Council of Canada to S. Barrett and S. Graham and by graduate fellowships from the Ontario Government and the University of Toronto to A. Case.

Appendix

Table A1

Sexual System and Collection Localities of *Wurmbea* Species and Outgroups

Taxon (tribe)	Geographic range	Sexual system	Collection locality	Voucher	GenBank accession numbers			
					*ndh*F	*atp*B-*rbc*L	*trn*LF	*psb*BTNH
Wurmbea (Anguillareae), Australia:								
W. biglandulosa ssp. biglandulosa	NSW, Vic, ACT	M (1)	Rocky River, NSW	MR1	EU044643	EU044623	EU044698	EU044683
W. biglandulosa ssp. biglandulosa	NSW, Vic, ACT	M (2)	Mt. Yarrowyck, NSW	MR2	EU044644	EU044618	EU044699	EU044681
W. biglandulosa ssp. biglandulosa	NSW, Vic, ACT	D (3)	Warrumbungle, NSW	MR3	EU044645	EU044621	EU044700	EU044680
W. biglandulosa ssp. biglandulosa	NSW, Vic, ACT	D (4)	Gooniwigall, NSW	MR4	EU044646	EU044620		EU044682
W. biglandulosa ssp. flindersica	SA	M	Mt. Remarkable, SA	AC9	EU044642	EU044598	EU044701	EU044667
W. citrine	SA	D	Quorn, SA	AC34	EU044639	EU044596	EU044695	EU044664
W. dioica ssp. alba	WA	M	Brookton, WA	AC53	EU044633	EU044601	EU044705	EU044671
W. dioica ssp. alba	WA	D	Eneabba, WA	AC38	EU044634	EU044600	EU044704	EU044670
W. dioica ssp. dioica	SA, Vic, NSW, ACT	D	Burrenjuck Dam, NSW	AC13	EU044647	EU044594	EU044693	EU044662
W. densiflora	WA	M	Murchison, WA	AC80	EU044632	EU044610	EU044711	EU044677
W. dilatata	WA	M	Kalbarri, WA	AC3	EU044637	EU044608	EU044712	EU044678
W. drummondii	WA	D	Duck Pool, WA	AC17	EU044636	EU044606	EU044709	EU044675
W. inframediana	WA	M	Overlander, WA	AC75	EU044638	EU044605	EU044708	EU044674
W. latifolia ssp. latifolia	SA	D	Barunga Gap, SA	AC25	EU044640	EU044597	EU044696	EU044665
W. latifolia ssp. vanessae	SA	D	Williamston, SA	AC29	EU044641	EU044599	EU044697	EU044666
W. monantha	WA	M	Windy Harbour, WA	AC22	EU044652	EU044602	EU044702	EU044668
W. murchisoniana	WA	M	Murchison, WA	AC2	EU044635	EU044604	EU044703	EU044669
W. odorata	WA	D	Coral Bay, WA	AC68	EU044649	EU044611	EU044707	EU044673
W. pygmaea	WA	M	Helena Valley, WA	AC77	AF547012	AY699143	EU044714	AY465584
W. sinora	WA, SA	M	Venus Bay, SA	AC28	EU044650	EU044603	EU044706	EU044672
W. tenella	WA	M	Murchison, WA	AC78	EU044631	EU044609	EU044713	EU044679
W. tubulosa	WA	D	Guranu, WA	AC72	EU044651	EU044607	EU044710	EU044676
Wurmbea (Anguillareae), Africa:								
W. inusta	South Africa	M	Strand-Gordons Bay, South Africa	LKJ1		EU044614	EU044692	EU044661
W. marginata	South Africa	M	Darling, South Africa	LKJ2		EU044612	EU044690	EU044659
W. spicata var. spicata	South Africa	M	Little Lions Head, South Africa	LKJ3		EU044613	EU044691	EU044660
Outgroups (tribe or family):								
Baeometra uniflora (Baeometreae)	South Africa	M	Lesmurdie, WA	AC24	EU044628	EU044622	EU044687	EU044656
Burchardia multiflora	WA	M	Helena Valley, WA	AC81	EU044625	EU044616	EU044684	EU044653
Gloriosa superba (Iphigenieae)	Africa, Asia	M	Cultivated, Toronto, ON	AC82	EU044626	EU044617	EU044686	EU044655
Iphigenia indica (Iphigenieae)	Asia, WA, NT, Qld	M	India	MWC1028			EU044685	EU044654
Iphigenia novae-zelandiae (Iphigenieae)	NZ	M	Southland, NZ	AC83	EU044648	EU044595	EU044694	EU044663
Onixotis punctata (Anguillarieae)	South Africa	M	Dutoitskloof, South Africa	LKJ4	EU044629	EU044619	EU044688	EU044657
Onixotis stricta (Anguillarieae)	South Africa	M	Cultivated, Canberra, ACT	AC16	EU044630	EU044615	EU044689	EU044658
Colchicum autumnale (Colchicaceae)	...	M	Cultivated, Toronto, ON	AC84	EU044627	EU044624		

Note. Sexual systems are either monomorphic (M) or dimorphic (D) for gender. Locality and range abbreviations are Western Australia (WA), South Australia (SA), New South Wales (NSW), Victoria (Vic), Australian Capital Territory (ACT), Northern Territory (NT), Queensland (Qld), and New Zealand (NZ). Vouchers for all WA species are deposited in the Western Australian Herbarium, South Perth (PERTH); the voucher for *Iphigenia indica* is lodged at the Royal Botanic Gardens, Kew (K); all other vouchers are in the Vascular Plant Herbarium, Royal Ontario Museum, Toronto (TRT).

Literature Cited

Ashman T-L 2006 The evolution of separate sexes: a focus on the ecological context. Pages 204–222 in LD Harder, SCH Barrett, eds. The ecology and evolution of flowers. Oxford University Press, Oxford.

Backlund A, K Bremer 1998 To be or not to be: principles of classification and monotypic families. Taxon 47:391–400.

Baker HG 1967 Support for Baker's Law—as a rule. Evolution 21:853–856.

Baker HG, PA Cox 1984 Further thoughts on islands and dioecism. Ann Mo Bot Gard 71:244–253.

Barrett SCH 1985 Floral trimorphism and monomorphism in continental and island populations of *Eichhornia paniculata* (Pontederiaceae). Biol J Linn Soc 25:41–60.

——— 1992 Gender variation and the evolution of dioecy in *Wurmbea dioica* (Liliaceae). J Evol Biol 5:423–444.

Barrett SCH, AL Case 2006 The ecology and evolution of gender strategies in plants: the example of Australian *Wurmbea* (Colchicaceae). Aust J Bot 54:417–433.

Barrett SCH, AL Case, GB Peters 1999 Gender modification and resource allocation in subdioecious *Wurmbea dioica* (Colchicaceae). J Ecol 87:123–137.

Bates RJ 1995 The species of *Wurmbea* (Liliaceae) in South Australia. J Adelaide Bot Gard 16:33–53.

——— 2007 A review of South Australian *Wurmbea* (Colchicaceae ~ Liliaceae): keys, new taxa and combinations, and notes. J Adelaide Bot Gard 21:75–81.

Bollback JP 2006 SIMMAP: stochastic character mapping of discrete traits on phylogenies. BMC Bioinformatics 7:88.

Bull JJ, EL Charnov 1985 On irreversible evolution. Evolution 39:1149–1155.

Case AL 2000 The evolution of combined versus separate sexes in *Wurmbea* (Colchicaceae). PhD diss. University of Toronto.

Case AL, SCH Barrett 2001 Ecological differentiation of combined and separate sexes of *Wurmbea dioica* (Colchicaceae) in sympatry. Ecology 82:2601–2616.

——— 2004a Environmental stress and the evolution of dioecy: *Wurmbea dioica* (Colchicaceae) in Western Australia. Evol Ecol 18:145–164.

——— 2004b Floral biology of gender monomorphism and dimorphism in *Wurmbea dioica* (Colchicaceae) in Western Australia. Int J Plant Sci 165:289–301.

Chase MW, MR Duvall, HG Hills, JG Conran, AV Cox, LE Eguiarte, J Hartwell, et al 1995 Molecular phylogenetics of Lilianae. Pages 109–137 in PJ Rudall, PJ Cribb, DF Cutler, CJ Humphries, eds. Monocotyledons: systematics and evolution. Royal Botanic Gardens, Kew.

Conran JG 1985 Family distributions in the Liliiflorae and their biogeographical implications. J Biogeogr 22:1023.

Crisp MD, GT Chandler 1996 Paraphyletic species. Telopea 6:813–844.

Cummings MP, SA Handley, DS Myers, DL Reed, A Rokas, K Winka 2003 Comparing bootstrap and posterior probability values in the four-taxon case. Syst Biol 52:477–487.

Cunningham CW, KE Omland, TH Oakley 1998 Reconstructing ancestral character states: a critical reappraisal. Trends Ecol Evol 9:361–366.

Dahlgren RMT, HT Clifford, PF Yeo 1985 The families of the monocotyledons: structure, evolution, and taxonomy. Springer, New York.

Delph LF, DE Wolf 2005 Evolutionary consequence of gender plasticity in genetically dimorphic breeding systems. New Phytol 166:119–128.

Dorken ME, J Friedman, SCH Barrett 2002 The evolution and maintenance of monoecy and dioecy in *Sagittaria latifolia* (Alismataceae). Evolution 56:31–41.

Doyle JJ, JL Doyle 1987 A rapid DNA isolation procedure from small quantities of fresh leaf tissues. Phytochem Bull 19:11–15.

Edwards AWF 1972 Likelihood. Cambridge University Press, Cambridge.

Ehlers BK, T Bataillon 2007 "Inconstant males" and the maintenance of labile sex expression in subdioecious plants. New Phytol 174:194–211.

Erixon P, B Svennblad, T Britton, B Oxelman 2003 Reliability of Bayesian posterior probabilities and bootstrap frequencies in phylogenetics. Syst Biol 52:665–673.

Felsenstein J 1985 Confidence limits on phylogenies: an approach using the bootstrap. Evolution 39:783–791.

Ferrer MM, SV Good-Avila 2007 Macrophylogenetic analyses of the gain and loss of self-incompatibility in the Asteraceae. New Phytol 173:401–414.

Fitch WM 1971 Towards defining the course of evolution: minimum change for a specific tree topology. Syst Zool 20:406.

Friedman J, SCH Barrett 2008 A phylogenetic analysis of the evolution of wind pollination in the angiosperms. Int J Plant Sci 169:49–58.

Funk DJ, KE Omland 2003 Species-level paraphyly and polyphyly: frequency, causes, and consequences with insights from animal mitochondrial DNA. Annu Rev Ecol Evol Syst 34:397–423.

Galley C, HP Linder 2007 The phylogeny of the *Penaschistis* clade (Danthonioideae, Poaceae) based on chloroplast DNA, and the evolution and loss of complex characters. Evolution 61:864–884.

Geber MA, TE Dawson, LF Delph, eds 1999 Gender and sexual dimorphism in flowering plants. Springer, New York.

Givnish TJ, JC Pires, SW Graham, MA McPherson, LM Prince, TB Patterson, HS Rai, et al 2005 Repeated evolution of net venation and fleshy fruits among monocots in shaded habitats confirms a priori predictions: evidence from an *ndh*F phylogeny. Proc R Soc B 272:1481–1490.

Gleiser G, M Verdú 2005 Repeated evolution of dioecy from androdioecy in *Acer*. New Phytol 165:633–640.

Goldblatt P 1978 An analysis of the flora of Southern Africa: its characteristics, relationships, and origins. Ann Mo Bot Gard 65:369–436.

Goldman N, JP Anderson, AG Rodrigo 2000 Likelihood-based tests of topologies in phylogenetics. Syst Biol 49:652–670.

Graham SW, SCH Barrett 2004 Phylogenetic reconstruction of the evolution of stylar polymorphisms in *Narcissus* (Amaryllidaceae). Am J Bot 91:1007–1021.

Graham SW, JR Kohn, BR Morton, JE Eckenwalder, SCH Barrett 1998 Phylogenetic congruence and discordance among one morphological and three molecular data sets from Pontederiaceae. Syst Biol 47:545–567.

Graham SW, RG Olmstead 2000 Utility of 17 chloroplast genes for inferring the phylogeny of the basal angiosperms. Am J Bot 87:1712–1730.

Graham SW, PA Reeves, ACE Burns, RG Olmstead 2000 Microstructural changes in noncoding chloroplast DNA: interpretation, evolution, and utility of indels and inversions in basal angiosperm phylogenetic inference. Int J Plant Sci 161(suppl):S83–S96.

Graham SW, JM Zgurski, MA McPherson, DM Cherniawsky, JM Saarela, EFC Horne, SY Smith, et al 2006 Robust inference of monocot deep phylogeny using an expanded multigene plastid data set. Pages 3–20 in JT Columbus, EA Friar, JM Porter, LM Prince, MG Simpson, eds. Monocots: comparative biology and evolution (excluding Poales). Rancho Santa Ana Botanic Garden, Claremont, CA.

Guindon S, O Gascuel 2003 A simple, fast, and accurate algorithm to estimate large phylogenies by maximum likelihood. Syst Biol 52:696–704.

Hart JA 1985 Evolution of dioecism in *Lepechinia* Willd. sect. Parviflorae (Lamiaceae). Syst Bot 10:134–146.

Hartigan JA 1973 Minimum mutation fits to a given tree. Biometrics 29:53.

Heilbuth JC 2000 Lower species richness in dioecious clades. Am Nat 156:221–241.

Huelsenbeck JP, R Nielsen, JP Bollback 2003 Stochastic mapping of morphological characters. Syst Biol 52:131–158.

Huelsenbeck JP, F Ronquist 2001 MrBayes: Bayesian inference of phylogenetic trees. Bioinformatics 17:754–755.

Jones A, M Burd 2001 Vegetative and reproductive variation among unisexual and hermaphroditic individuals of *Wurmbea dioica* (Colchicaceae). Aust J Bot 49:603–609.

Jones RT, AP Martin 2006 Testing for differentiation of microbial communities using phylogenetic methods: accounting for uncertainty of phylogenetic inference and character state mapping. Microb Ecol 52:408–417.

Kohn JR, SW Graham, B Morton, JJ Doyle, SCH Barrett 1996 Reconstruction of the evolution of reproductive characters in Pontederiaceae using phylogenetic evidence from chloroplast DNA restriction-site variation. Evolution 50:1454–1469.

Krahenbuhl M, YM Yuan, P Kupfer 2002 Chromosome and breeding system evolution of the genus *Mercurialis* (Euphorbiaceae): implications of ITS molecular phylogeny. Plant Syst Evol 234:155–170.

Leschen RAB, TR Buckley 2007 Multistate characters and diet shifts: evolution of Erotylidae (Coleoptera). Syst Biol 56:97–112.

Levin RA, JS Miller 2005 Relationships within the tribe Lycieae (Solanaceae): paraphyly *of Lycium* and multiple origins of gender dimorphism. Am J Bot 92:2044–2053.

Lewis LA, PO Lewis 2005 Unearthing the molecular phylodiversity of desert soil green algae (Chlorophyta). Syst Biol 54:936–947.

Lewis PO 2001 A likelihood approach to estimating phylogeny from discrete morphological character data. Syst Biol 50:913–925.

Lloyd DG 1975 Breeding systems in *Cotula*. IV. Reversion from dioecy to monoecy. New Phytol 74:125–145.

Macfarlane TD 1980 A revision of *Wurmbea* (Liliaceae) in Australia. Brunonia 3:145–208.

——— 1986 Two new species of *Wurmbea* (Colchicaceae or Liliaceae s. lat.) from south western Australia. Nuytsia 5:407–413.

——— 1987 *Wurmbea* (Liliaceae). Flora Aust 45:387–405.

Macfarlane TD, SJ van Leeuwen 1996 *Wurmbea saccata* (Colchicaceae), a lepidopteran-pollinated new species from Western Australia. Nuytsia 10:429–435.

Maddison DR, WP Maddison 2001 MacClade, version 4.03. Sinauer, Sunderland, MA.

Maddison WP 1997 Gene trees in species trees. Syst Biol 46:523–536.

Maddison WP, LL Knowles 2006 Inferring phylogeny despite incomplete lineage sorting. Syst Biol 55:21–30.

Maddison WP, DR Maddison 2005 Mesquite: a modular system for evolutionary analysis, version 1.06. http://www.mesquiteproject.org.

Mannen J, FA Natalie, F Ehrendorfer 1994 Phylogeny of Rubiaceae-Rubieae inferred from the sequence of a cpDNA intergenic region. Plant Syst Evol 190:195–211.

Miadlikowska J, F Lutzoni 2004 Phylogenetic classification of peltigeralean fungi (Peltigerales, Ascomycota) based on ribosomal RNA small and large subunits. Am J Bot 91:449–464.

Miller JS 2002 Phylogenetic relationships and the evolution of gender dimorphism in *Lycium* (Solanaceae). Syst Bot 27:416–428.

Mooers AO, D Schluter 1999 Reconstructing ancestor states with maximum likelihood: support for one- and two-rate models. Syst Biol 48:623–633.

Mooers AO, SM Vamosi, D Schluter 1999 Using phylogenies to test macroevolutionary hypotheses of trait evolution in cranes (Gruinae). Am Nat 154:249–259.

Moore LB, E Edgar 1970 *Iphigenia* (Colchicaceae). Flora of New Zealand. II. Indigenous Tracheophyta. Government Printer, Wellington.

Nielson R 2002 Mapping mutations on phylogenies. Syst Biol 51:729–739.

Nordenstam B 1978 The genus *Wurmbea* in Africa except the Cape region. Notes R Bot Gard Edinb 36:211–233.

——— 1986 The genus *Wurmbea* (Colchicaceae) in the Cape region. Opera Bot 87:1–41.

——— 1998 Colchicaceae. Pages 175–185 *in* K Kubitzki, ed. The families and genera of vascular plants. Vol III. Flowering plants, monocotyledons: Lilianae (except Orchidaceae). Springer, New York.

Obbard DJ, SA Harris, RJA Buggs, JR Pannell 2006 Hybridization, polyploidy, and the evolution of sexual systems in *Mercurialis* (Euphorbiaceae). Evolution 60:1801–1815.

Olmstead RG 1995 Species concepts and plesiomorphic species. Syst Bot 20:623–630.

Olmstead RG, JA Sweere 1994 Combining data in phylogenetic systematics: an empirical approach using three molecular data sets in the Solanaceae. Syst Biol 43:467–481.

Pagel M 1994 Detecting correlated evolution on phylogenies: a general method for the comparative analysis of discrete characters. Proc R Soc B 255:37–45.

——— 1999 The maximum likelihood approach to reconstructing ancestral character states of discrete characters on phylogenies. Syst Biol 48:612.

Pagel M, A Meade, D Barker 2004 Bayesian estimation of ancestral characters states on phylogenies. Syst Biol 53:673–684.

Posada D, KA Crandall 1998 MODELTEST: testing the model of DNA substitution. Bioinformatics 14:817–818.

Rambaut A 1998 Se-Al (Sequence Alignment Editor), version 10α1, computer program and documentation. Department of Zoology, Oxford University, Oxford. http://tree.bio.ed.ac.uk/software/seal/.

Ramsey M, G Vaughton 2001 Sex expression and sexual dimorphism in subdioecious *Wurmbea dioica* (Colchicaceae). Int J Plant Sci 162:589–597.

——— 2002 Maintenance of gynodioecy in *Wurmbea biglandulosa* (Colchicaceae): gender differences in seed production and progeny success. Plant Syst Evol 232:189–200.

Ramsey M, G Vaughton, R Peakall 2006a Does inbreeding avoidance maintain gender dimorphism in *Wurmbea dioica* (Colchicaceae)? J Evol Biol 19:1497–1506.

——— 2006b Inbreeding avoidance and the evolution of gender dimorphism in *Wurmbea biglandulosa* (Colchicaceae). Evolution 60:529–537.

Renner SS, RE Ricklefs 1995 Dioecy and its correlates in the flowering plants. Am J Bot 82:596–606.

Renner SS, HS Won 2001 Repeated evolution of dioecy from monoecy in Siparunaceae (Laurales). Syst Biol 50:700–712.

Ronquist F 2004 Bayesian inference of character evolution. Trends Ecol Evol 19:475–481.

Rudall PJ, KL Stobart, W-P Hong, JG Conran, CA Furness, CG Kite, MW Chase 2000 Consider the lilies: systematics of Liliales. Pages 347–359 *in* KL Wilson, DA Morrison, eds. Monocots: systematics and evolution. CSIRO, Sydney.

Sakai AK, SG Weller 1999 Gender and sexual dimorphism in flowering plants: a review of terminology, biogeographic patterns, ecological correlates, and phylogenetic approaches. Pages 1–31 *in* MA Geber, TE Dawson, LF Delph, eds. Gender and sexual dimorphism in flowering plants. Springer, New York.

Sakai AK, SG Weller, WL Wagner, PE Soltis, DE Soltis 1997 Phylogenetic perspective on the evolution of dioecy: adaptive radiation in the endemic Hawaiian genera *Schiedea* and *Alsinidendron* (Caryophyllaceae: Alsinoideae). Pages 455–473 *in* TJ Givnish, KJ Sytsma, eds. Molecular evolution and adaptive radiation. Cambridge University Press, New York.

Sargent RD, JC Vamosi 2008 The influence of canopy position, pollinator syndrome, and region on evolutionary transitions in pollinator guild size. Int J Plant Sci 169:39–47.

Savolainen VJ, F Manen, E Douzery, R Spichiger 1994 Molecular phylogeny of families related to Celestrales based on *rbcL* 5′ flanking sequences. Mol Phylogenet Evol 3:27–37.

Shimodaira H, M Hasegawa 1999 Multiple comparisons of log-likelihoods with applications to phylogenetic inference. Mol Biol Evol 16:1114–1116.

Simmons MP, KM Pickett, M Miya 2004 How meaningful are Bayesian support values? Mol Biol Evol 21:188–199.

Soltis PE, DE Soltis, SG Weller, AK Sakai, WL Wagner 1996 Molecular phylogenetic analyses of the Hawaiian endemics *Schiedea* and *Alsinidendron* (Caryophyllaceae). Syst Bot 21:365–379.

Suzuki Y, GV Glazko, M Nei 2002 Overcredibility of molecular phylogenetics obtained by Bayesian phylogenetics. Proc Natl Acad Sci USA 99:16138–16143.

Swofford DL 2002 PAUP*: phylogenetic analysis using parsimony (*and other methods). Version 4.0b10. Sinauer, Sunderland, MA.

Sytsma KJ, JF Smith, PE Berry 1991 The use of chloroplast DNA to assess biogeography and evolution of morphology, breeding systems, and flavonoids in *Fuchsia* sect. *Skinnera* (Onagraceae). Syst Biol 16:257–269.

Taberlet P, L Gielly, G Pautou, J Bouvet 1991 Universal primers for amplification of three non-coding regions of chloroplast DNA. Plant Mol Biol 17:1105–1109.

Thompson JD, DG Higgins, TJ Gibson 1994 CLUSTAL W: improving the sensitivity of progressive multiple sequence alignment through sequence weighting, position-specific gap penalties and weight-matrix choice. Nucleic Acids Res 22:4673–4680.

Thomson JD, J Brunet 1990 Hypotheses for the evolution of dioecy in seed plants. Trends Ecol Evol 5:11–16.

Vamosi JC, SP Otto 2002 When looks can kill: the evolution of sexually dimorphic floral display and the extinction of dioecious plants. Proc R Soc B 269:1187–1194.

Vamosi JC, SP Otto, SCH Barrett 2003 Phylogenetic analysis of the ecological correlates of dioecy in angiosperms. J Evol Biol 16:1006–1018.

Vanderpoorten A, B Goffinet 2006 Mapping uncertainty and phylo-

genetic uncertainty in ancestral character state reconstruction: an example in the moss genus *Brachytheciastrum*. Syst Biol 55:957–971.

Vaughton G, M Ramsey 1998 Floral display, pollinator visitation and reproductive success in the dioecious perennial herb *Wurmbea dioica* (Liliaceae). Oecologia 115:93–101.

——— 2002 Evidence of gynodioecy and sex ratio variation in *Wurmbea biglandulosa* (Colchicaceae). Plant Syst Evol 232:167–179.

——— 2003 Self-compatibility and floral biology in subdioecious *Wurmbea dioica* (Colchicaceae). Aust J Bot 51:39–45.

——— 2004 Dry environments promote the establishment of females in monomorphic populations of *Wurmbea biglandulosa* (Colchicaceae). Evol Ecol 18:323–341.

Vinnersten A 2003 Tracing history: phylogenetic, taxonomic, and biogeographic research in the colchicum family. PhD diss. Uppsala University.

Vinnersten A, K Bremer 2001 Age and biogeography of major clades in Liliales. Am J Bot 88:1695–1703.

Vinnersten A, J Manning 2007 A new classification of Colchicaceae. Taxon 56:171–178.

Vinnersten A, G Reeves 2003 Phylogenetic relationships within Colchicaceae. Am J Bot 90:1455–1462.

Wagner WL, SG Weller, AK Sakai 1995 Phylogeny and biogeography in *Schiedea* and *Alsinidendron* (Caryophyllaceae). Pages 221–258 *in* WL Wagner, VA Funk, eds. Hawaiian biogeography: evolution on a hot spot archipelago. Smithsonian Institution, Washington, DC.

Webb CJ, D Kelly 1993 The reproductive biology of the New Zealand flora. Trends Ecol Evol 8:442–447.

Weiblen GD, RK Oyama, MJ Donoghue 2000 Phylogenetic analysis of dioecy in monocotyledons. Am Nat 155:46–58.

Weller SG, AK Sakai 1999 Using phylogenetic approaches for the analysis of plant breeding system evolution. Annu Rev Ecol Syst 30:167–199.

Weller SG, WL Wagner, AK Sakai 1995 A phylogenetic analysis of *Scheidea* and *Alsinidendron* (Caryophyllaceae: Alsinoideae): implications for the evolution of breeding systems. Syst Bot 20:315–337.

Zhang L-B, MP Simmons, A Kocyan, SS Renner 2006 Phylogeny of the Cucurbitales based on DNA sequences of nine loci from three genomes: implications for morphological and sexual system evolution. Mol Phylogenet Evol 39:305–322.

Int. J. Plant Sci. 169(1):157–168. 2008.
1058-5893/2008/16901-0013$15.00   DOI: 10.1086/523357

THE EVOLUTIONARY MAINTENANCE OF SEXUAL REPRODUCTION: EVIDENCE FROM THE ECOLOGICAL DISTRIBUTION OF ASEXUAL REPRODUCTION IN CLONAL PLANTS

Jonathan Silvertown

Department of Life Sciences, The Open University, Walton Hall, Milton Keynes MK7 7EH, United Kingdom

In theory, females that reproduce asexually should enjoy a twofold advantage in fitness over sexual females, yet sex remains the predominant mode of reproduction in virtually all eukaryotes. The evolutionary maintenance of sex is especially puzzling in clonal plants because the transition from sexual to exclusively asexual reproduction is an ever-present possibility in these species. In this article, I use published data on the genotypic diversity of populations of clonal plants to test five hypotheses about the ecological situations that limit or favor clonal reproduction in vascular plants. The data were drawn from 248 studies covering 69,000 individuals in >2000 populations of 218 species in 74 plant families. The tests showed the following: (1) the frequency of clonality increases with population age, indicating that clonal reproduction is limited by disturbance; (2) clonal reproduction is limited by dispersal; clones are more frequent in aquatic and apomictic species in which the dispersal of clonally produced propagules is less limiting; (3) clones are more frequent in populations of rare or endangered species; (4) populations of alien plants have higher frequencies of clonality; and (5) clones are more frequent at the edges of species' geographical ranges. Thus, it appears that the ultimately successful clonal plant would be a rare, aquatic, alien apomict living in an undisturbed, geographically marginal habitat. Since this combination of circumstances is so restrictive, it is perhaps better regarded as a sign of sexual failure than as a recipe for clonal success.

Keywords: apomixis, aquatic plants, clonal plants, dispersal, evolution of sex, genotypic diversity.

Introduction

The maintenance of sexual reproduction remains an evolutionary problem for which there are many contending solutions (Williams 1975; Kondrashov 1993; Otto and Lenormand 2002, Rice 2002). The puzzle is why females should continue sexual reproduction when this dilutes their genetic contribution to the next generation by half compared to asexual reproduction (Maynard Smith 1978). The problem is similar for outcrossing hermaphrodites, a category that includes the majority of seed plants. In fact, plants present the problem in its most acute form because sexual reproduction persists even though so many plants are clonal and capable of asexual reproduction (Lloyd 1980). Somatic embryogenesis is phylogenetically ancient in the plant kingdom (Mogie and Hutchings 1990), but only very rarely do clonal plants become entirely asexual (Eckert 2001; Pandit and Babu 2003). Why is the transition to complete asexuality so rare, even in organisms so well equipped to reproduce clonally?

The combination of sexual and asexual reproduction within plant life histories presents an opportunity as well as a challenge. The opportunity is to compare the relative success of recruitment through sexual and asexual routes across a broad range of clonal species in order to test under what conditions one mode of reproduction is favored over the other. It would be possible, through parentage analysis (Smouse et al. 1999; Burczyk et al. 2006), to measure the relative success of sexual

versus asexual reproduction directly, though at the time of writing, parentage analysis had not been used for this purpose. As an alternative, the genotypic diversity of a population can be used as an indirect measure of reproductive success via sexual and asexual routes (see "Methods").

Asexual reproduction in plants occurs in two fundamentally different forms: vegetative reproduction and agamospermy. In clonal plants, vegetative reproduction produces new ramets by budding from roots, rhizomes, stems, storage organs such as tubers or (more rarely) leaves, or inflorescences. The earlier literature on clonal plants tended to emphasize the ecological role of vegetatively produced organs in the growth of individual plants and populations and to neglect the evolutionary implications that arise from the fact that the same organs effect asexual reproduction (Eckert 1999).

The vegetative progeny, such as bulbils, of clonal plants tend to be better provisioned than seeds, or they start life as miniature versions of their mothers, complete with their own root systems. How long the vascular connection between mother and daughter endures varies within (Alpert 1999) and between species (Van Groenendael et al. 1997). Obligate vegetative reproduction, without even occasional sexual reproduction, may be selected against (and hence be rare) because of the mutational load that may accumulate (Kondrashov 1994; Caetano Anolles 1999; Paland and Lynch 2006), although mechanisms exist to purge mutations within individual plants (Thomas 2002), and these lower mutational load (Orive 2001).

Agamospermy is parthenogenetic seed production, also referred to as apomixis (in the narrow sense; Askers and Jerling 1992). Apomictic seeds are clones of the mother plant, but they are packaged and dispersed in the trappings of sexually produced progeny. Because apomictic seeds share the same dispersal mechanisms and require the same conditions for establishment as sexually produced seeds, they can be used to test the performance of clones without the normal dispersal handicap of vegetative reproduction (Mogie et al. 1990). Some apomictic (pseudogamous) species require pollination as a trigger for seed development, even though pollen makes no genetic contribution to the seed (Richards 2003). Stable coexistence of apomictic and sexual seed production within the same population appears to be difficult to achieve, and mixtures of the two reproductive modes are thought not to occur (Bengtsson and Ceplitis 2000; Nakayama et al. 2002; Whitton et al. 2008), although apomicts may very occasionally produce seeds by outcrossing (Richards 2003; Thompson et al. 2008). Apomixis is much rarer than vegetative reproduction among flowering plants (Whitton et al. 2008) and is often accompanied by polyploidization (Bierzychudek 1987; Richards 2003). One possible reason for this is that polyploidy protects apomicts from the expression of recessive mutations that accumulate in asexual genomes (Archetti 2004). A full discussion of apomixis is given by Whitton et al. (2008).

Successive reviews of genetic variation in clonal plants (Ellstrand and Roose 1987; Widén et al. 1994; Diggle et al. 1998; Khudamrongsawat et al. 2004) have all amply demonstrated that such populations are not genetically depauperate. Theoretical studies confirm that genetic diversity, as measured by alleles per locus or heterozygosity, is not expected to decrease and may even increase with increasing asexuality if clones are heterozygous (Balloux et al. 2003). However, both models and data show that the diversity of genotypes is reduced by asexual reproduction and that some populations may become monomorphic.

Models that examine the conditions under which sexual and asexual reproduction can coexist have found that sexual reproduction will persist if there is temporal variation in resource supply that clones cannot track (Weeks 1993). A life-history strategy combining sexual and asexual reproduction can be stable if the environment varies and the two modes of reproduction are successful in different circumstances (Bengtsson and Ceplitis 2000). Both these models emphasize the inflexibility of clonal reproduction when the environment varies. The view that clonally reproducing organisms are evolutionarily handicapped by a lack of genetic variation has been challenged on the grounds that clones often harbor hidden variation that enables them to adapt (Lushai et al. 2003). If this is true for clonal plants, then the persistence of sexual reproduction is all the more a puzzle.

The converse of the argument that environmental change favors sexual reproduction is that a lack of change should favor clonality. Two types of environmental change should be distinguished: episodic change, such as that caused by habitat disturbances such as fire or tree falls, and continuous change, such as climatic warming. Following Bender et al. (1984), I use the terms "pulse" and "press," respectively, for these two types of perturbation. Pulse perturbations occur in all kinds of plant communities and provide essential recruitment opportunities for many species that usually colonize from seed. As time since the last disturbance elapses, opportunities for further colonization from seed diminish, and the size of the most

successful clonal genotypes that have already established increases. Competition between clones is expected to lead to a fall in genotypic diversity over time (Gray 1987). The rate of clonal spread during this phase is an important contribution to the fitness of the genet (Pan and Price 2001). Press perturbation is no doubt as common as pulse perturbation, but it is much more difficult to test its effect on the relative success of sexual versus clonal reproduction in perennial plants because controls for gradual environmental change are rarely available over a sufficient period.

The long- and short-term advantages of sexual reproduction and, therefore, the corresponding disadvantages of asexual reproduction, have been reviewed elsewhere (Maynard Smith 1978; Bell 1981; Rice 2002; Otto and Gerstein 2006). I confine myself here to the possible limitations and advantages of clonal reproduction that lend themselves to testing by the use of field data.

Clones may be handicapped with respect to sexually produced offspring as follows:

1. Clones are prone to habitat disturbance. Test: sex should be favored over clonal reproduction in disturbed environments, and clones should increase in frequency with the time elapsed since the last disturbance.
2. Clones are poor dispersers. Test: (a) Vegetatively spread clones should be more successful in aquatic habitats because vegetative organs are more easily transported in water than on land (Barrett et al. 1993; Grace 1995); (b) apomicts produce genetically identical (=clonal) seeds that are not handicapped in dispersal with respect to sexually produced seeds. Apomictic populations, therefore, should have higher ratios of clonal/sexual individuals than clonal plants that reproduce vegetatively.

Conversely, clones may be at an advantage with respect to sexually produced offspring in the following situations:

3. In small populations, where sexual fecundity may be lowered by intraclone incompatibility (Nuortila et al. 2002; Honnay and Bossuyt 2005). Test: clones should be more frequent in rare/endangered populations than in common ones.
4. In alien environments, where sexual reproduction may require multiple colonization (Baker 1959). Test: clones should be more frequent in populations of aliens than in populations of natives.
5. At the edge of geographical ranges, where sexual reproduction is subject to physiological limitations. Test: clones should be more frequent in populations at the edge of geographical ranges.

In this article, I analyze the frequency of clonal reproduction in populations of more than 200 plant species to test these five predictions.

Methods

Study Selection

The aim was to include all studies that reported a value for the ratio genotypes per individual sampled (G/N) for one or more clonal plant populations that have been published since the first review of this subject was conducted 20 years ago (Ellstrand and Roose 1987). I used the ISI Citation Index to identify more than 400 articles that cited Ellstrand and Roose (1987) and then examined each of these studies to identify those that contained suitable data. A small number of relevant studies that did not cite (Ellstrand and Roose 1987) were found in the reference lists of the articles that did do so and among articles that cited them. This search was stopped when it ceased to yield any more unknown studies that had been published before January 1, 2006. I omitted the studies of 21 species used in Ellstrand and Roose's (1987) survey so that the results of my review would be independent of theirs.

Previous surveys of genotypic diversity (Ellstrand and Roose 1987; Widén et al. 1994; Diggle et al. 1998; Hangelbroek et al. 2002) were limited in the number of species covered or confined themselves to studies of populations "in which sexual recruitment is extremely limited" (Ellstrand and Roose 1987, p. 123). I did not apply this restriction because of the difficulty of knowing what it meant in any particular case and because we now know that clonal plants recruit episodically (Eriksson 1989) and that this can easily be missed.

The literature review identified 356 potentially informative studies of population genetic structure in clonal plants, 108 of which had to be rejected because they provided insufficient information. The most common deficiency was the absence of data on genotype frequencies (as distinct from gene frequencies). The 248 qualifying studies between them sampled more than 69,000 individuals in more than 2000 populations representing 218 species in 74 plant families. Some of the larger families, such as Poaceae (29 studies), Asteraceae (18 studies), Cyperaceae (17 studies), and Rosaceae (16 studies), were strongly represented, but relative to their size, so too were smaller families of aquatic plants, such as Alismataceae (six studies), Potamogetonaceae (six studies), Lemnaceae (five studies), Zosteraceae (four studies), and Posidoniaceae (four studies).

More than half of the studies used allozyme markers (58%; 146 studies), 54 studies (22%) used RAPDs, 23 (9%) used nuclear microsatellites, and 17 (7%) used amplified fragment length polymorphisms (AFLPs). Other types of markers used were intersimple sequence repeat (ISSR), RFLP, and chloroplast microsatellite. Nearly half (48%) of the studies were of terrestrial herbs, 17% were of aquatic herbs, 17% were of shrubs, and 12% were of trees. Regarding mode of asexual reproduction, 102 cases were rhizomatous, 30 suckered, 20 were stoloniferous, and 19 were apomictic. Other modes of vegetative reproduction included layering, bulbils, bulbs, tubers, turions, lignotubers, and fragmentation.

Indexes of Clonality and Independent Variables

Four indexes of clonality were used:

1. The ratio G/N of genotypes (G) identified in a population to the number of individuals sampled (N), which (Ellstrand and Roose 1987) called the "proportion distinguishable."
2. An index of genotypic identity (I_G), suggested by Weeks (1993) and closely related to Simpson's index of genotypic diversity.

3. The frequency of monoclonal populations belonging to a species (m).
4. A simple binary variable of presence/absence of clonal populations belonging to a species (M).

It is important to note what these indexes are and what they are not. They are population-level measurements of the net outcome of all processes that in the history of a population affected its genotypic diversity, including vegetative reproduction, sexual reproduction, and immigration by sexually and asexually produced propagules. The indexes are not direct estimates of the fitness of plants that reproduce sexually versus the fitness of those that reproduce vegetatively. However, I do assume that the indexes give an indirect estimate of the likely reproductive success of plants via sexual and asexual routes.

The indexes were correlated with one another (see "Results"), and none of them is unbiased, but all have the practical advantage that they are reported in or can be calculated from many studies. The index G/N reflects the ratio of recruitment into the population via sexual versus asexual reproduction if the population is at demographic equilibrium and if all identical genotypes are in fact clonemates. There are two reasons why the latter condition could be incorrect. First, if too few polymorphic loci have been used in clone discrimination, some apparent clonemates may have undetected differences at loci that were not sampled. The presence of this problem can be diagnosed by increasing the number of loci until the number of clones detected asymptotes. Second, some proportion of apparent clonemates may actually be the product of sexual reproduction. This problem is less easily solved, though there are several possible solutions based on allele frequencies and intersample distances (Parks and Werth 1993; Harada et al. 1997; Stenberg et al. 2003; Meirmans and Van Tienderen 2004; Halkett et al. 2005). Some recent studies of genotypic diversity in clonal plants have used these clone-discrimination techniques (e.g., Ivey and Richards 2001; Douhovnikoff and Dodd 2003), but the majority have not. In most cases, the data required to apply clone discrimination techniques retrospectively are not published. Incorrect clone discrimination is less likely to be a problem as the number of marker loci increases. A source of bias that could inflate G/N and other estimates of clonal diversity is somatic mutation (Klekowski 2003). This is particularly likely to inflate the number of genotypes scored using DNA markers such as microsatellites that are sensitive to sequence differences of a single base pair between samples. The influence of somatic mutation on genotype diversity has been considered, though rarely quantified, by a number of studies (Tuskan et al. 1996; Esselman et al. 1999; Torimaru et al. 2003; Van der Hulst et al. 2003; Nagamitsu et al. 2004; Kameyama and Ohara 2006; Kjolner et al. 2006). Somatic mutation is often undetectable (Kameyama and Ohara 2006) or occurs only at a very low rate (Cloutier et al. 2003).

The index of genotypic identity (I_G) is the probability that any two individuals drawn at random from a population have the same genotype (Bengtsson 2003). In entirely clonal populations, this will approach unity, while in entirely sexual populations, it will tend to zero, so I_G can be used as an index of clonality for a population. If a population has been at demographic equilibrium for sufficient time, I_G approaches a value

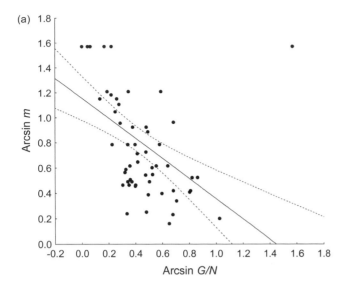

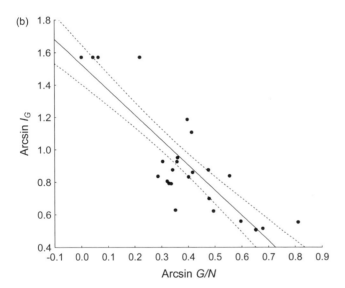

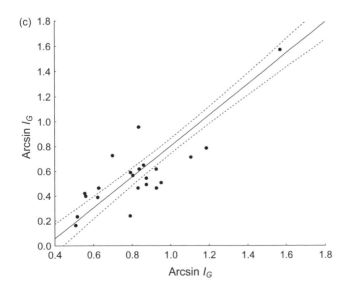

independent of starting conditions. When the proportion of individuals derived from sexual reproduction, σ, is small ($\sigma < 0.10$) and the population size is sufficiently large ($N > 10$), genotypic identity can be approximated (Bengtsson 2003) as

$$I_G \approx \frac{1}{2\sigma N + 1}.$$

Following the example of Ellstrand and Roose (1987) and the recommendation of Parker (1979), virtually all studies of genotypic diversity included in this survey calculated Simpson's index of diversity (D), corrected for finite sample size. Conveniently, like I_G, this also gives the probability that any two individuals drawn at random from a population have the same genotype:

$$D = \sum \frac{n_i(n_i - 1)}{N(N - 1)},$$

where n_i is the number of individuals belonging to the ith genotype and N is the total number of individuals. Simpson's index is invariably reported as the complement of D, so it increases from 0 to 1 as diversity increases. The complement of reported values of Simpson's index was therefore used to estimate I_G.

Statistical Analysis

Statistical models of each of the four indexes of clonality were run with a common set of independent variables. These variables were growth form (terrestrial herb, aquatic herb, shrub, tree, vine, succulent), rarity (scored "rare" if the source described the species as rare, endangered, or narrow endemic and scored "not rare" in all other cases), alien status (alien vs. native), and type of clonality (apomictic vs. vegetative). Interactions among independent variables were not examined.

Four covariates were also used to control for sampling bias. Since DNA markers tend to yield more polymorphic loci than allozymes, and anonymous DNA markers (RAPDs, ISSRs, AFLPs) are prone to artefactual variation caused by sample contamination and PCR error, use of the different markers could clearly influence the number of genotypes detected in a population. Possible bias from the use of different types of genetic markers was controlled for by coding whether studies used DNA markers or allozymes. Bias also could have arisen because some studies sampled populations using a spacing interval between samples designed to reduce the probability of sampling the same clone more than once. These studies could have been omitted altogether, but since the data showed that the same clones were in fact often resampled, I included the studies and flagged them with a binary covariate instead. Two other possible sources of bias were included as covariates in initial models: number of populations sampled for a species and an index of the spatial scale of sampling used.

Species that were the subject of more than one independent study were combined so that no species was represented more

Fig. 1 Correlations among the three continuously distributed indexes of clonality. Dashed lines are 95% confidence limits for linear regression lines. *a*, Arcsin $m = 1.1564 - .7977 \times$ arcsin genotype/sample ratio (G/N), $r = -.5074$; *b*, arcsin $I_G = 1.5264 - 1.548 \times$ arcsin (G/N), $r = -.8940$; *c*, arcsin $m = -.4384 + 1.2391 \times$ arcsin I_G, $r = 0.9494$.

than once in the data set. The sole exception where studies were not combined was the aquatic species *Butomus umbellatus*, because one study examined native European populations (Kirschner et al. 2004) and the other one studied alien North American populations (Eckert et al. 2003).

A combined value of G/N was calculated from individual study values of G_i and N_i as $\Sigma G_i / \Sigma N_i$. A species value of D was calculated as the arithmetic mean of D_i for individual studies. The percentage of polymorphic populations for a species was calculated as $(\Sigma p_i / \Sigma s_i) \times 100$, where p_i was the number of polymorphic populations and s_i was the total number of populations in an individual study.

All statistical tests were performed using STATISTICA, release 7 (StatSoft 2000). Separate general linear models of the arcsin-transformed values of G/N, I_G, and m were run. All independent variables and covariates were included in initial models, which were then repeated with nonsignificant variables dropped sequentially to obtain an optimum model. A generalized linear model with a binomial distribution and a logit link function for presence/absence of monoclonal populations (M) was run using the same sequential procedure. In tests of the disturbance and latitude hypotheses, the index G/N was compared between young and old populations and between southern and northern populations, respectively, across all relevant studies and using Wilcoxon matched-pairs tests (StatSoft 2000).

Results

Correlation among Clonality Indexes

As expected, the indexes of clonality were correlated with each other, but the relationships among the three continuously distributed variables (G/N, I_G, m) showed that they were not equivalent (fig. 1). Arcsin I_G was highly correlated with arcsin G/N ($r = -0.894$, $n = 29$, $P < 0.001$) and m ($r = 0.949$, $n = 29$, $P < 0.001$), but the sample size was small, so this variable is of limited use within the current data set. Arcsin G/N and arcsin m were not highly correlated with each other even though the relationship was significant ($r = -0.507$, $n = 61$, $P < 0.001$; fig. 1).

Change in G/N with Population Age

Twenty-two studies included populations with different times since last disturbance or populations of different successional ages. Two studies, both of coastal grasses (Bockelmann et al. 2003; Travis and Hester 2005), included sufficient populations of different age for regression to be used to test how G/N changed with time. In *Elymus athericus* (Bockelmann et al. 2003), regression of G/N on age for nine populations for which an age was given by the authors showed no significant change over time ($G/N = 0.0003$ age $+ 0.8777$, $r^2 = 0.067$). The results of Travis and Hester (2005) for *Spartina alterniflora* are mentioned in the "Discussion." The remaining 20 studies analyzed fewer than five population types (usually only two). A significant decline in G/N occurred with population age in the sample of 20 studies between the youngest ($\overline{X} = 0.604$, SE $= 0.058$) and oldest populations ($\overline{X} = 0.437$,

SE $= 0.068$; Wilcoxon matched-pairs test: $n = 20$, $Z = 2.24$, $P = 0.025$). Two of the three strong exceptions to this trend (fig. 2) were seagrasses.

Monoclonal Populations

In generalized linear models of M in 169 species, growth form ($\chi^2 = 11.567$, df $= 3$, $P = 0.009$), rarity ($\chi^2 = 10.55$, df $= 1$, $P = 0.001$), and the number of populations ($\chi^2 = 20.59$, df $= 1$, $P < 0.0001$) were all highly significant. In a comparison of growth forms, the majority of aquatic herbs had monoclonal populations, while the majority of species with other growth forms did not (fig. 3). A general linear model of arcsin m (the proportion of populations that were monoclonal) also showed that rarity had a highly significant effect ($F_{1, 163} = 11.700$, $P < 0.001$), in addition to growth form ($F_{3, 163} = 4.806$, $P = 0.003$) and number of populations ($F_{1, 163} = 7.876$, $P = 0.006$; fig. 4). The mean value of m computed from the model was 22.46% (SD $= 0.84$%) for the rare species in the sample, compared with mean m of only 5.57% (SD $= 0.14$%) in populations of species that were not rare. The sample of 169 species excluded eight species of succulents and vines because of the small sample size for these two growth forms, but analyses that included these species gave almost identical results.

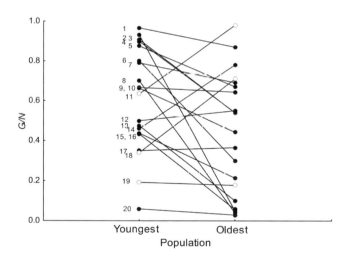

Fig. 2 Ratio of genotypes/sample (G/N) compared between young and old populations drawn from 20 studies of 19 clonal species. Seagrasses are indicated by open circles. Key to species: 1 = *Populus nigra* (Barsoum et al. 2004), 2 = *Calamagrostis epigejos* (Lehmann 1997), 3 = *Veratrum album* (Kleijn and Steinger 2002), 4 = *Oryza rufipogon* (Xie et al. 2001), 5 = *Empetrum hermaphroditum* (Szmidt et al. 2002), 6 = *Agrostis stolonifera* (Kik et al. 1990), 7 = *Cirsium arvense* (Sole et al. 2004), 8 = *Solidago altissima* (Maddox et al. 1989), 9 = *Psammochloa villosa* (Li and Ge 2001), 10 = *Andropogon gerardii* (Keeler et al. 2002), 11 = *Zostera marina* (Reusch et al. 2000), 12 = *Brachypodium pinnatum* (Schlapfer and Fischer 1998), 13 = *Rhododendron ferrugineum* (Pornon et al. 2000), 14 = *Circaea lutetiana* (Verburg et al. 2000), 15 = *Carex lasiocarpa* (McClintock and Waterway 1993), 16 = *Festuca rubra* (Rhebergen et al. 1988), 17 = *Carex pellita* (McClintock and Waterway 1993), 18 = *Posidonia oceanica* (Jover et al. 2003), 19 = *Zostera marina* (Rhode and Duffy 2004), and 20 = *Uvularia perfoliata* (Kudoh et al. 1999).

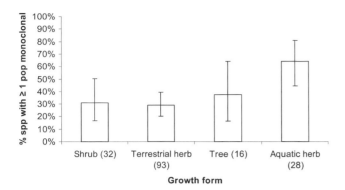

Fig. 3 Percentage of species in each of four growth forms that possessed at least one monoclonal population; 95% binomial confidence limits corrected for continuity are shown. Numerals in parentheses are sample sizes (*n*).

Ecological Correlates of G/N

A sample of 200 species was analyzed; vines and succulents were excluded because few of these plants were represented. In the final general linear model of the arcsin-transformed ratio G/N, growth form ($F_{3,193} = 5.066$, $P = 0.002$), type of clonality ($F_{1,193} = 7.424$, $P = 0.007$), alien status ($F_{1,193} = 10.318$, $P = 0.002$), and marker type ($F_{1,193} = 24.597$, $P < 0.0001$) were all significant (fig. 5).

Ecological Correlates of I_G

Values of I_G were available for 106 species in the data set. In the final general linear model of arcsin-transformed values of I_G, just two variables were significant: type of clonality ($F_{1,103} = 5.449$, $P = 0.0215$) and rarity ($F_{1,103} = 7.1332$, $P = 0.009$) (fig. 6).

Effect of Latitude on G/N

A barely significant decline in G/N occurred with increasing latitude. In the sample of 17 studies, southern populations had a higher mean value of G/N ($\overline{X} = 0.502$, SE = 0.055) than northern ones ($\overline{X} = 0.402$, SE = 0.069; Wilcoxon matched-pairs test: $n = 17$, $Z = 1.965$, $P = 0.0495$; fig. 7).

Discussion

Twenty years ago, Ellstrand and Roose (1987, p. 127) concluded, "The striking fact from the data we compiled is that the vast majority of species studied are multiclonal." A later review of a larger sample of 45 species (Widén et al. 1994) reached much the same conclusion. Others have questioned these conclusions on the grounds that the samples were subject to a variety of sources of bias (Eckert 2001; Honnay and Bossuyt 2005). In this study, I controlled for bias by using marker type (DNA markers vs. allozymes), scale of sampling, sampling design, and number of populations sampled as covariates and by analyzing a data set that was nearly five times larger than any used before. The results showed that Ellstrand and Roose's "striking fact" holds true as a broad generaliza-

tion, but it must be qualified by exceptions that form some clear ecological patterns.

In the current data set, clones were more prevalent in older, longer-established populations than in younger populations of the same species (fig. 2), indicating that the absence of disturbance favors asexual over sexual recruitment. The exceptions to this trend are informative. There were only three seagrasses in the sample, but in two of them, the ratio G/N increased rather than decreased with age. Vegetative parts of aquatic plants are much more easily dispersed than are clonal fragments of terrestrial plants, so the pattern that occurs on land, where populations tend to be established by sexual propagules and expand by asexual reproduction, may be reversed in water. Two studies of chronosequences of coastal grasses were not included in my sample. These plants are not aquatic, but their populations are regularly inundated, and fragments can be dispersed by water. In one study, *Spartina alterniflora* (Travis and Hester 2005) G/N declined very slowly with population age over a period of 1500 yr, while in the other, *Elymus athericus* (Bockelmann et al. 2003) G/N did not change significantly (see "Results"). In no case did the trend toward lower G/N with age terminate in a monoclonal population, indicating that very long timescales free from disturbance may be needed to allow clonal reproduction to prevail over sexual reproduction.

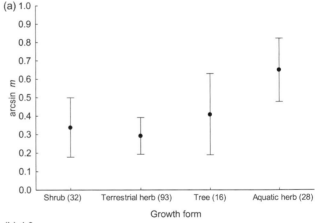

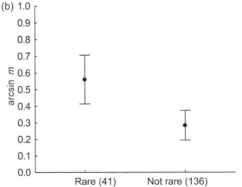

Fig. 4 Means and 95% confidence intervals of arcsin *m* for species with different growth forms (*a*) and rare versus other status (*b*). Numerals in parentheses are sample sizes (*n*).

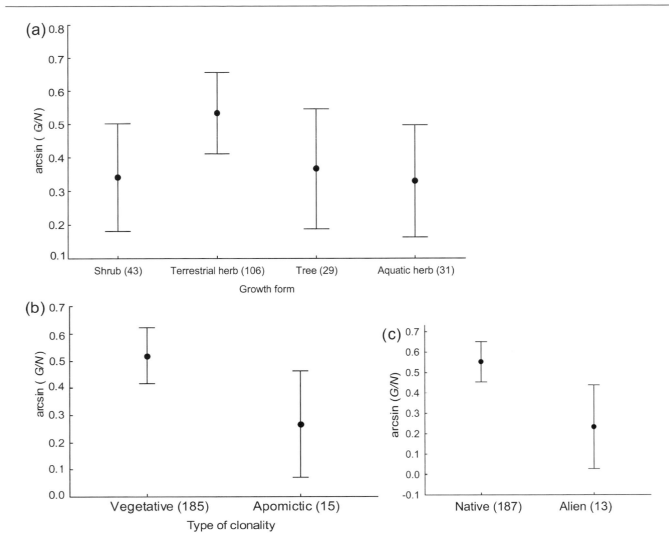

Fig. 5 Means and 95% confidence intervals of arcsin (genotype/sample ratio G/N) for species with different growth forms (*a*), vegetative versus apomictic asexual reproduction (*b*), and native versus alien status (*c*). Numerals in parentheses are sample sizes (*n*).

It follows from the greater dispersability of plant fragments in water than on land that aquatic plant populations should be more clonal, because new populations can more easily be founded by single genotypes. This prediction was strongly supported by results for the clonality indexes M and m. The majority of aquatic herbs in which more than one population was studied contained at least one population that was monoclonal (fig. 3). By contrast, in terrestrial herbs, shrubs, and trees, the majority of species contained no monoclonal populations. Aquatic herbs also had a higher proportion of monoclonal populations (m) than other growth forms (fig. 4). These trends were not apparent when the other two clonality indexes were used, G/N and I_G. The sample of aquatic herbs for which I_G was known was small ($n = 13$), so the test had low power and no conclusion can be drawn for this index, but this was not the case for the G/N sample ($n = 31$). In the study as a whole, the indexes m and G/N were not highly correlated (fig. 1*a*), suggesting that they measure different processes.

The results of the test of the disturbance hypothesis might explain why the monoclonality indexes (m, M) and G/N give different results for aquatic herbs. I suggested that seagrasses were exceptions to the decline in G/N with time because aquatic populations frequently establish from single vegetative fragments and because they acquire rather than lose clonal diversity with time. If the rate at which G/N rises over time in aquatic plants is faster than the rate at which G/N declines over time in terrestrial species, it is possible for aquatic and terrestrial species to have similar average values of G/N, even though they differ significantly in values of M and m. In short, aquatic populations are easily and frequently founded by single clones and move rapidly away from monoclonality, while terrestrial species move very slowly toward it from a sexual beginning.

Because apomicts can disperse their clonal progeny in seeds, thus avoiding the handicap that affects vegetative clonal offspring on land, they too should have higher frequencies of clonality. This was borne out by G/N (fig. 5*b*) and I_G (fig. 6*a*), but not by M or m. It would appear from this that apomicts behave like other seed-producing terrestrial plants as regards the frequency of monoclonal populations but that apomictic

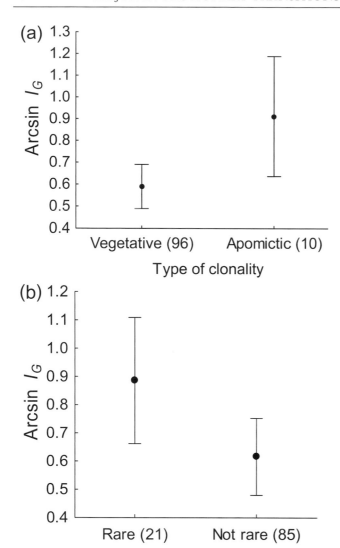

Fig. 6 Means and 95% confidence intervals of arcsin I_G for species with vegetative versus apomictic asexual reproduction (*a*) and rare versus other status (*b*). Numerals in parentheses are sample sizes (*n*).

latitude effect shown in figure 7 (no. 16). In nonclonal animals and plants, threatened species often have lower genetic diversity than related nonthreatened taxa (Eckert 2001; Klekowski 2003; Spielman et al. 2004).

Failure of sexual reproduction is common in small plant populations (Leimu et al. 2006). Is low genotypic diversity merely a passive by-product of sexual failure, or is there active selection in favor of clonal reproduction in such populations? The evidence on this point is slim, but one study found evidence for a trade-off between sexual function and vegetative growth in the clonal aquatic plant *Decodon verticillatus* (Dorken et al. 2004). Sexual reproduction has repeatedly been lost in marginal populations of *D. verticillatus*, presumably because this enhances clonal growth and increases fitness. I am not aware of any other studies that have demonstrated this trade-off in a plant where sexual reproduction has been lost in small or marginal populations. However, the trade-off has been found in other clonal plants (Bullock et al. 1995; Sutherland and Vickery 1988; van Kleunen et al. 2002) and may be general, in which case selection for loss of sexual reproduction would not be confined to the one example that is known. However, if a trade-off between

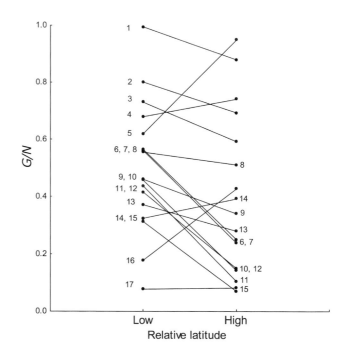

Fig. 7 Ratios of genotypes/sample size (*G/N*) compared between southern and northern populations. (Latitudes have been reversed for the Southern Hemisphere species 6 and 8). Species and sources of data are: 1 = *Corylus avellana* (Persson et al. 2004), 2 = *Zostera marina* (Reusch et al. 2000), 3 = *Cirsium heterophyllum* (Jump et al. 2003), 4 = *Titanotrichum oldhamii* (Wang et al. 2004), 5 = *Carex lugens* (Stenstrom et al. 2001), 6 = *Posidonia oceanica* (Jover et al. 2003), 7 = *Carex sifolia* (Stenstrom et al. 2001), 8 = *Eucalyptus curtisii* (Smith et al. 2003), 9 = *Cirsium arvense* (Jump et al. 2003), 10 = *Carex bigelowii* (Stenstrom et al. 2001), 11 = *Decodon verticillatus* (Dorken and Eckert 2001), 12 = *Carex stans* (Stenstrom et al. 2001), 13 = *Carex lasiocarpa* (McClintock and Waterway 1993), 14 = *Carex pelita* (McClintock and Waterway 1993), 15 = *Podostemum ceratophyllum* (Philbrick and Crow 1992), 16 = *Filipendula rubra* (Aspinwall and Christian 1992), and 17 = *Allium vineale* (Ceplitis 2001).

populations do have lower clonal diversity, as one would expect. The difference between apomicts and aquatic plants could be due to ecological differences between their propagules (Lloyd 1984) and a lower establishment rate of apomicts. Apomicts have propagules (seeds) that are small, compared to those of aquatics, and these disperse vegetative fragments, turions, and so forth.

The prediction that rare species would be more clonal was borne out by analyses of three of the four indexes: *M*, *m* (fig. 4), and I_G (fig. 6*b*). Not all rare species have small population size (Rabinowitz 1981), though the association between rarity and small populations did seem to recur frequently in my sample and reduced population size is the likeliest cause of low genotypic diversity. In *Filipendula rubra* (Aspinwall and Christian 1992), a species that was classified as rare in the data set, the mean value of *G/N* was 0.58 in large populations but only 0.13 in small ones, a highly significant difference ($F_{1,21} = 12.77$, $P = 0.002$) that was independent of the sex and clonal growth is general and selection for increased clonal growth is typical of small and marginal populations, the advantage may only be short-lived because small populations have a high probability of extinction (Henle et al. 2004; Matthies et al. 2004).

In passing, it is worth reflecting on what triggers the failure of sexual reproduction when vegetative (clonal) reproduction appears to be more robust. Of course, it is possible that this perception of failure is biased by the much greater attention that is given to sexual than to vegetative reproduction, but there is an intrinsic functional reason why sexual reproduction ought to be more vulnerable. Sexual reproduction can take place only in circumstances where vegetative growth is possible. However, this dependence is not symmetrical. Indeed, growth usually incurs a cost when sexual reproduction occurs (Silvertown and Dodd 1999). Hence, there are bound to be circumstances in which plants are physiologically capable of growth but not of reproduction, but there are no circumstances where the reverse is true. Plant vegetative growth and clonal reproduction are so closely linked (often involving the same organs) that it is almost inevitable that clonal reproduction will sometimes be physiologically possible even if sexual reproduction is not.

As predicted by Baker's law (Baker 1959), alien species had significantly lower values of G/N than natives (fig. 5c), though the other clonality indexes did not show a significant effect. Also as predicted, geographically marginal populations had lower values of G/N than populations in the hinterland of species' distributions (fig. 7), although this effect was only barely significant. No apomicts were present among the 17 species in my sample, but the association of apomixis with geographically marginal populations (geographic parthenogenesis) is already well established (Bierzychudek 1987; Peck et al. 1998; Horandl 2006) and occurs, for example, in the apomict *Antennaria rosea* (Bayer 1990). (This case was excluded from my sample because the source does not report values of G/N.)

Finally, I want to address possible sources of bias in this analysis. Major sources of bias that have concerned researchers in this field before, such as scale of sampling and type of genetic marker, were controlled within the analysis. The sample size used was large enough to include previously neglected factors such as life form and rarity and to compare types of clonality (apomixis vs. vegetative). However, the samples were still too small to analyze interactions among dependent variables. Phylogenetic effects (Silvertown and Dodd 1997) were eliminated in tests of two hypotheses, population age (1) and geographic range (5), by making comparisons within species, but phylogeny was not included in the other tests. A future phylogenetic reanalysis might produce results different from those reported here, though experience shows that phylogenetic and nonphylogenetic analyses often produce similar results (Ricklefs and Starck 1996).

The results of this analysis of more than 2,000 populations suggests that the ultimate clonal plant would be a rare, aquatic, alien apomict living in an undisturbed, geographically marginal habitat. This is such a restrictive set of ecological conditions that it is perhaps better regarded as a recipe for the failure of sexual reproduction than as clonal success. Revisiting the question that motivated this study, it is now clear that the ecological distribution of more extreme clonality tells us where sex fails, not why it persists. I suggest that clonal reproduction is not a substitute for sex but merely prolongs the time to extinction when sex is absent. In that case, genetic mechanisms, rather than any of the ecological factors I have examined, probably hold the answer to why sexual reproduction appears to be indispensable to long-term success.

Acknowledgments

I am grateful to Norm Ellstrand, Ove Eriksson, and an anonymous reviewer for comments on the manuscript and to Kevin McConway for statistical advice. In memoriam J.M.S.

Literature Cited

Alpert P 1999 Clonal integration in *Fragaria chiloensis* differs between populations: ramets from grassland are selfish. Oecologia 120:69–76.

Archetti M 2004 Recombination and loss of complementation: a more than two-fold cost for parthenogenesis. J Evol Biol 17:1084–1097.

Askers S, L Jerling 1992 Apomixis in plants. CRC, Boca Raton, FL.

Aspinwall N, T Christian 1992 Clonal structure, genotypic diversity, and seed production in populations of *Filipendula rubra* (Rosaceae) from the north central United States. Am J Bot 79:294–299.

Baker HG 1959 Reproductive methods as factors in speciation in flowering plants. Cold Spring Harbor Symp Quant Biol 24:177–191.

Balloux F, L Lehmann, T de Meeus 2003 The population genetics of clonal and partially clonal diploids. Genetics 164:1635–1644.

Barrett SCH, CG Eckert, BC Husband 1993 Evolutionary processes in aquatic plant populations. Aquat Bot 44:105–145.

Barsoum N, E Muller, L Skot 2004 Variations in levels of clonality among *Populus nigra* L. stands of different ages. Evol Ecol 18:601–624.

Bayer RJ 1990 Patterns of clonal diversity in the *Antennaria rosea* (Asteraceae) polyploid agamic complex. Am J Bot 77:1313–1319.

Bell G 1981 The masterpiece of nature. Chapman & Hall, London.

Bender EA, TJ Case, ME Gilpin 1984 Perturbation experiments in community ecology. Ecology 65:1–13.

Bengtsson BO 2003 Genetic variation in organisms with sexual and asexual reproduction. J Evol Biol 16:189–199.

Bengtsson BO, A Ceplitis 2000 The balance between sexual and asexual reproduction in plants living in variable environments. J Evol Biol 13:415–422.

Bierzychudek P 1987 Patterns in plant parthenogenesis. Pages 197–217 in SC Stearns, ed. The evolution of sex and its consequences. Birkhäuser, Basel.

Bockelmann AC, TBH Reusch, R Bijlsma, JP Bakker 2003 Habitat differentiation vs. isolation-by-distance: the genetic population structure of *Elymus athericus* in European salt marshes. Mol Ecol 12:505–515.

Bullock JM, B Clear Hill, J Silvertown, M Sutton 1995 Gap colonization as a source of grassland community change: effects of gap size and grazing on the rate and mode of colonization by different species. Oikos 72:273–282.

Burczyk J, WT Adams, DS Birkes, IJ Chybicki 2006 Using genetic markers to directly estimate gene flow and reproductive success parameters in plants on the basis of naturally regenerated seedlings. Genetics 173:363–372.

Caetano Anolles G 1999 High genome-wide mutation rates in vegetatively propagated bermudagrass. Mol Ecol 8:1211–1221.

Ceplitis A 2001 The importance of sexual and asexual reproduction in the recent evolution of *Allium vineale*. Evolution 55:1581–1591.

Cloutier D, D Rioux, J Beaulieu, DJ Schoen 2003 Somatic stability of microsatellite loci in eastern white pine, *Pinus strobus* L. Heredity 90:247–252.

Diggle PK, S Lower, TA Ranker 1998 Clonal diversity in alpine populations of *Polygonum viviparum* (Polygonaceae). Int J Plant Sci 159:606–615.

Dorken ME, CG Eckert 2001 Severely reduced sexual reproduction in northern populations of a clonal plant, *Decodon verticillatus* (Lythraceae). J Ecol 89:339–350.

Dorken ME, KJ Neville, CG Eckert 2004 Evolutionary vestigialization of sex in a clonal plant: selection versus neutral mutation in geographically peripheral populations. Proc R Soc B 271:2375–2380.

Douhovnikoff V, RS Dodd 2003 Intra-clonal variation and a similarity threshold for identification of clones: application to *Salix exigua* using AFLP molecular markers. Theor Appl Genet 106:1307–1315.

Eckert CG 1999 Clonal plant research: proliferation, integration, but not much evolution. Am J Bot 86:1649–1654.

——— 2001 The loss of sex in clonal plants. Evol Ecol 15:501–520.

Eckert CG, K Lui, K Bronson, P Corradini, A Bruneau 2003 Population genetic consequences of extreme variation in sexual and clonal reproduction in an aquatic plant. Mol Ecol 12:331–344.

Ellstrand NC, ML Roose 1987 Patterns of genotypic diversity in clonal plant species. Am J Bot 74:123–131.

Eriksson O 1989 Seedling dynamics and life histories in clonal plants. Oikos 55:231–238.

Esselman EJ, L Jianqiang, DJ Crawford, JL Windus, AD Wolfe 1999 Clonal diversity in the rare *Calamagrostis porteri* ssp. *insperata* (Poaceae): comparative results for allozymes and random amplified polymorphic DNA (RAPD) and intersimple sequence repeat (ISSR) markers. Mol Ecol 8:443–451.

Grace JB 1995 The adaptive significance of clonal reproduction in angiosperms: an aquatic perspective. Aquat Bot 44:159–180.

Gray AJ 1987 Genetic change during succession in plants. Pages 273–293 *in* AJ Gray, MJ Crawley, PJ Edwards, eds. Colonization, succession and stability. Blackwell, Oxford.

Halkett F, JC Simon, F Balloux 2005 Tackling the population genetics of clonal and partially clonal organisms. Trends Ecol Evol 20: 194–201.

Hangelbroek HH, NJ Ouborg, L Santamaria, K Schwenk 2002 Clonal diversity and structure within a population of the pondweed *Potamogeton pectinatus* foraged by Bewick's swans. Mol Ecol 11:2137–2150.

Harada Y, S Kawano, Y Iwasa 1997 Probability of clonal identity: inferring the relative success of sexual versus clonal reproduction from spatial genetic patterns. J Ecol 85:591–600.

Henle K, KF Davies, M Kleyer, C Margules, J Settele 2004 Predictors of species sensitivity to fragmentation. Biodivers Conserv 13:207–251.

Honnay O, B Bossuyt 2005 Prolonged clonal growth: escape route or route to extinction? Oikos 108:427–432.

Horandl E 2006 The complex causality of geographical parthenogenesis. New Phytol 171:525–538.

Ivey CT, JH Richards 2001 Genotypic diversity and clonal structure of everglades sawgrass, *Cladium jamaicense* (Cyperaceae). Int J Plant Sci 162:1327–1335.

Jover MA, L del Castillo-Agudo, M Garcia-Carrascosa, J Segura 2003 Random amplified polymorphic DNA assessment of diversity in western Mediterranean populations of the seagrass *Posidonia oceanica*. Am J Bot 90:364–369.

Jump AS, FI Woodward, T Burke 2003 *Cirsium* species show disparity

in patterns of genetic variation at their range-edge, despite similar patterns of reproduction and isolation. New Phytol 160:359–370.

Kameyama Y, M Ohara 2006 Predominance of clonal reproduction, but recombinant origins of new genotypes in the free-floating aquatic bladderwort *Utricularia australis* f. *tenuicaulis* (Lentibulariaceae). J Plant Res 119:357–362.

Keeler KH, CF Williams, LS Vescio 2002 Clone size of *Andropogon gerardii* Vitman (big bluestem) at Konza Prairie, Kansas. Am Midl Nat 147:295–304.

Khudamrongsawat J, R Tayyar, JS Holt 2004 Genetic diversity of giant reed (*Arundo donax*) in the Santa Ana River, California. Weed Sci 52:395–405.

Kik C, J Van Andel, W Van Delden, W Joenje, R Bijlsma 1990 Colonization and differentiation in the clonal perennial *Agrostis stolonifera*. J Ecol 78:949–961.

Kirschner J, I Bartish, Z Hroudova, L Kirschnerova, P Zakravsky 2004 Contrasting patterns of spatial genetic structure of diploid and triploid populations of the clonal aquatic species, *Butomus umbellatus* (Butomaceae), in central Europe. Folia Geobot 39:13–26.

Kjolner S, SM Sastad, C Brochmann 2006 Clonality and recombination in the arctic plant *Saxifraga cernua*. Bot J Linn Soc 152:209–217.

Kleijn D, T Steinger 2002 Contrasting effects of grazing and hay cutting on the spatial and genetic population structure of *Veratrum album*, an unpalatable, long-lived, clonal plant species. J Ecol 90: 360–370.

Klekowski EJ 2003 Plant clonality, mutation, diplontic selection and mutational meltdown. Biol J Linn Soc 79:61–67.

Kondrashov AS 1993 Classification of hypotheses on the advantage of amphimixis. J Hered 84:372–387.

——— 1994 Mutation load under vegetative reproduction and cytoplasmic inheritance. Genetics 137:311–318.

Kudoh H, H Shibaike, H Takasu, DF Whigham, S Kawano 1999 Genet structure and determinants of clonal structure in a temperate deciduous woodland herb, *Uvularia perfoliata*. J Ecol 87:244–257.

Lehmann C 1997 Clonal diversity of populations of *Calamagrostis epigejos* in relation to environmental stress and habitat heterogeneity. Ecography 20:483–490.

Leimu R, P Mutikainen, J Koricheva, M Fischer 2006 How general are positive relationships between plant population size, fitness and genetic variation? J Ecol 94:942–952.

Li A, S Ge 2001 Genetic variation and clonal diversity of *Psammochloa villosa* (Poaceae) detected by ISSR markers. Ann Bot 87:585–590.

Lloyd DG 1980 Benefits and handicaps of sexual reproduction. Evol Biol 13:69–111.

——— 1984 Variation strategies of plants in heterogeneous environments. Biol J Linn Soc 21:357–385.

Lushai G, HD Loxdale, JA Allen 2003 The dynamic clonal genome and its adaptive potential. Biol J Linn Soc 79:193–208.

Maddox GD, RE Cook, PH Wimberger, S Gardescu 1989 Clone structure in four *Solidago altissima* (Asteraceae) populations: rhizome connections within genotypes. Am J Bot 76:318–326.

Matthies D, I Brauer, W Maibom, T Tscharntke 2004 Population size and the risk of local extinction: empirical evidence from rare plants. Oikos 105:481–488.

Maynard Smith J 1978 The evolution of sex. Cambridge University Press, Cambridge.

McClintock KA, MJ Waterway 1993 Patterns of allozyme variation and clonal diversity in *Carex lasiocarpa* and *C. pellita* (Cyperaceae). Am J Bot 80:1251–1263.

Meirmans PG, PH Van Tienderen 2004 GENOTYPE and GENODIVE: two programs for the analysis of genetic diversity of asexual organisms. Mol Ecol Notes 4:792–794.

Mogie M, MJ Hutchings 1990 Phylogeny, ontogeny and clonal growth in vascular plants. Pages 3–22 *in* J van Groenendael, H de Kroon, eds. Clonal growth in plants. SPB, The Hague.

Mogie M, JR Latham, EA Warman 1990 Genotype-independent aspects of seed ecology in *Taraxacum*. Oikos 59:175–182.

Nagamitsu T, M Ogawa, K Ishida, H Tanouchi 2004 Clonal diversity, genetic structure, and mode of recruitment in a *Prunus ssiori* population established after volcanic eruptions. Plant Ecol 174:1–10.

Nakayama Y, H Seno, H Matsuda 2002 A population dynamic model for facultative agamosperms. J Theor Biol 215:253–262.

Nuortila C, J Tuomi, K Laine 2002 Inter-parent distance affects reproductive success in two clonal dwarf shrubs, *Vaccinium myrtillus* and *Vaccinium vitis-idaea* (Ericaceae). Can J Bot 80:875–884.

Orive ME 2001 Somatic mutations in organisms with complex life histories. Theor Popul Biol 59:235–249.

Otto SP, AC Gerstein 2006 Why have sex? the population genetics of sex and recombination. Biochem Soc Trans 34:519–522.

Otto SP, T Lenormand 2002 Resolving the paradox of sex and recombination. Nat Rev Genet 3:252–261.

Paland S, M Lynch 2006 Transitions to asexuality result in excess amino acid substitutions. Science 311:990–992.

Pan JJ, JS Price 2001 Fitness and evolution in clonal plants: the impact of clonal growth. Evol Ecol 15:583–600.

Pandit MK, CR Babu 2003 The effects of loss of sex in clonal populations of an endangered perennial *Coptis teeta* (Ranunculaceae). Bot J Linn Soc 143:47–54.

Parker ED 1979 Ecological implications of clonal diversity in parthenogenetic morphospecies. Am Zool 19:753–762.

Parks JC, CR Werth 1993 A study of spatial features of clones in a population of bracken fern, *Pteridium aquilinum* (Dennstaedtiaceae). Am J Bot 80:537–544.

Peck JR, JM Yearsley, D Waxman 1998 Explaining the geographic distributions of sexual and asexual populations. Nature 391:889–892.

Persson H, B Widen, S Andersson, L Svensson 2004 Allozyme diversity and genetic structure of marginal and central populations of *Corylus avellana* L. (Betulaceae) in Europe. Plant Syst Evol 244:157–179.

Philbrick CT, GE Crow 1992 Isozyme variation and population-structure in *Podostemum ceratophyllum* Michx (Podostemaceae): implications for colonization of glaciated North America. Aquat Bot 43:311–325.

Pornon A, N Escaravage, P Thomas, P Taberlet 2000 Dynamics of genotypic structure in clonal *Rhododendron ferrugineum* (Ericaceae) populations. Mol Ecol 9:1099–1111.

Rabinowitz D 1981 Seven forms of rarity. Pages 205–217 *in* H Synge, ed. The biological aspects of rare plant conservation. Wiley, Chichester.

Reusch TBH, WT Stam, JL Olsen 2000 A microsatellite-based estimation of clonal diversity and population subdivision in *Zostera marina*, a marine flowering plant. Mol Ecol 9:127–140.

Rhebergen LJ, J Theeuwen, JAC Verkleij 1988 The clonal structure of *Festuca rubra* in adjacent maritime habitats. Acta Bot Neerl 37:467–473.

Rhode JM, JE Duffy 2004 Relationships between bed age, bed size, and genetic structure in Chesapeake Bay (Virginia, USA) eelgrass (*Zostera marina* L.). Conserv Genet 5:661–671.

Rice WR 2002 Experimental tests of the adaptive significance of sexual recombination. Nat Rev Genet 3:241–251.

Richards AJ 2003 Apomixis in flowering plants: an overview. Philos Trans R Soc B 358:1085–1093.

Ricklefs RE, JM Starck 1996 Applications of phylogenetically independent contrasts: a mixed progress report. Oikos 77:167–172.

Schlapfer F, M Fischer 1998 An isozyme study of clone diversity and relative importance of sexual and vegetative recruitment in the grass *Brachypodium pinnatum*. Ecography 21:351–360.

Silvertown J, M Dodd 1997 Comparing plants and connecting traits. Pages 3–16 *in* J Silvertown, M Franco, JL Harper, eds. Plant life

histories: ecology, phylogeny and evolution. Cambridge University Press, Cambridge.

——— 1999 The demographic cost of reproduction and its consequences in balsam fir (*Abies balsamea*). Am Nat 154:321–332.

Smith S, J Hughes, G Wardell-Johnson 2003 High population differentiation and extensive clonality in a rare mallee eucalypt: *Eucalyptus curtisii*–conservation genetics of a rare mallee eucalypt. Conserv Genet 4:289–300.

Smouse PE, TR Meagher, CJ Kobak 1999 Parentage analysis in *Chamaelirium luteum* (L.) Gray (Liliaceae): why do some males have higher reproductive contributions? J Evol Biol 12:1069–1077.

Sole M, W Durka, S Eber, R Brandl 2004 Genotypic and genetic diversity of the common weed *Cirsium arvense* (Asteraceae). Int J Plant Sci 165:437–444.

Spielman D, BW Brook, R Frankham 2004 Most species are not driven to extinction before genetic factors impact them. Proc Natl Acad Sci USA 101:15261–15264.

StatSoft 2000 STATISTICA for Windows (computer program manual). StatSoft, Tulsa, OK.

Stenberg P, M Lundmark, A Saura 2003 MLGsim: a program for detecting clones using a simulation approach. Mol Ecol Notes 3:329–331.

Stenstrom A, BO Jonsson, IS Jonsdottir, T Fagerstrom, M Augner 2001 Genetic variation and clonal diversity in four clonal sedges (*Carex*) along the Arctic coast of Eurasia. Mol Ecol 10:497–513.

Sutherland S, RK Vickery Jr 1988 Trade-offs between sexual and asexual reproduction in the genus *Mimulus*. Oecologia 76:330–335.

Szmidt AE, MC Nilsson, E Briceno, O Zackrisson, XR Wang 2002 Establishment and genetic structure of *Empetrum hermaphroditum* populations in northern Sweden. J Veg Sci 13:627–634.

Thomas H 2002 Ageing in plants. Mech Ageing Dev 123:747–753.

Thompson SL, G Choe, K Ritland, J Whitton 2008 Cryptic sex within male-sterile polyploid populations of the Easter daisy, *Townsendia hookeri*. Int J Plant Sci 169:183–193.

Torimaru T, N Tomaru, N Nishimura, S Yamamoto 2003 Clonal diversity and genetic differentiation in *Ilex leucoclada* M. patches in an old-growth beech forest. Mol Ecol 12:809–818.

Travis SE, MW Hester 2005 A space-for-time substitution reveals the long-term decline in genotypic diversity of a widespread salt marsh plant, *Spartina alterniflora*, over a span of 1500 years. J Ecol 93:417–430.

Tuskan GA, KE Francis, SL Russ, RH Romme, MG Turner 1996 RAPD markers reveal diversity within and among clonal and seedling stands of aspen in Yellowstone National Park, USA. Can J For Res 26:2088–2098.

Van der Hulst RGM, THM Mes, M Falque, P Stam, JCM Den Nijs, K Bachmann 2003 Genetic structure of a population sample of apomictic dandelions. Heredity 90:326–335.

Van Groenendael JM, L Klimes, J Klimesova, RJJ Hendriks 1997 Comparative ecology of clonal plants. Pages 191–209 *in* J Silvertown, M Franco, JL Harper, eds. Plant life histories: ecology, phylogeny and evolution. Cambridge University Press, Cambridge.

van Kleunen M, M Fischer, B Schmid 2002 Experimental life-history evolution: selection on the allocation to sexual reproduction and its plasticity in a clonal plant. Evolution 56:2168–2177.

Verburg R, J Maas, HJ During 2000 Clonal diversity in differently-aged populations of the pseudo-annual clonal plant *Circaea lutetiana* L. Plant Biol 2:646–652.

Wang CN, M Moller, QCB Cronk 2004 Population genetic structure of *Titanotrichum oldhamii* (Gesneriaceae), a subtropical bulbiliferous plant with mixed sexual and asexual reproduction. Ann Bot 93:201–209.

Weeks SC 1993 The effects of recurrent clonal formation on clonal invasion patterns and sexual persistence: a Monte Carlo simulation of the frozen niche-variation model. Am Nat 141:409–427.

Whitton J, CJ Sears, EJ Baack, SP Otto 2008 The dynamic nature of apomixis in the angiosperms. Int J Plant Sci 169:169–182.

Widén B, N Cronberg, M Widén 1994 Genotypic diversity, molecular markers and spatial distribution of genets in clonal plants: a literature survey. Folia Geobot Phytotaxon 29:245–263.

Williams GC 1975 Sex and evolution. Princeton University Press, Princeton, NJ.

Xie ZW, YQ Lu, S Ge, DY Hong, FZ Li 2001 Clonality in wild rice (*Oryza rufipogon*, Poaceae) and its implications for conservation management. Am J Bot 88:1058–1064.

Int. J. Plant Sci. 169(1):169–182. 2008.
1058-5893/2008/16901-0014$15.00   DOI: 10.1086/523369

THE DYNAMIC NATURE OF APOMIXIS IN THE ANGIOSPERMS

Jeannette Whitton,* Christopher J. Sears,* Eric J. Baack,* and Sarah P. Otto[†]

*Department of Botany, University of British Columbia, Vancouver, British Columbia V6T 1Z4, Canada; and
†Department of Zoology, University of British Columbia, Vancouver, British Columbia V6T 1Z4, Canada

Apomixis, the asexual production of seed, is a trait estimated to occur in fewer than 1% of flowering plant species, with an uneven distribution among lineages. In the past decade, targeted research efforts have aimed at clarifying the genetic basis of apomixis, with the goal of engineering or breeding apomictic crops. Recent work suggests a simple genetic basis for apomixis, but it also indicates that natural populations of apomicts are much more complex than is often assumed. For example, in nature, nearly all apomicts that go through a megagametophyte stage (gametophytic apomicts) are polyploid, while their sexual relatives are typically diploid. Although populations have been characterized as obligately sexual or apomictic, it is increasingly clear that many plant populations exhibit some variation in reproductive mode. Many apomicts retain residual sexual function as pollen donors and thus have the potential to spread apomixis via male gametes, thereby increasing the genetic diversity observed within apomictic populations. Here, we summarize our current understanding of the genetic basis and transmission of apomixis. We use insights from previous case studies and models for the spread of asexuality to explore the potential for establishment and spread of apomixis in nature.

Keywords: apomixis, asexual reproduction, cost of sex, Crepis, polyploidy.

Introduction

Sexual reproduction is a nearly universal characteristic of angiosperms. Despite the overwhelming importance and broad occurrence of sexuality, most plant species (other than annuals) are capable of some form of asexual propagation, though in the majority of cases, this remains an augmentation to sexual reproduction rather than the dominant reproductive mode (Richards 2003; Silvertown 2008). In a number of groups, however, asexuality has become predominant through evolutionary transitions from sexuality. Whereas some plants have evolved asexuality through the degeneration of sexual structures in favor of vegetative means of propagation (Eckert 2002; Silvertown 2008), the more common route to asexuality is through the evolution of apomixis, the production of clonal seed in the absence of fertilization (Richards 1986). First described by Smith (1841), apomixis is now reported in more than 300 species in more than 40 angiosperm families (Asker and Jerling 1992). Thus, evolutionary transitions to asexual reproduction have occurred repeatedly in flowering plants, with only rare shifts in the reverse direction (Chapman et al. 2003).

Asexual reproduction in plants has received steady attention from ecologists and evolutionary biologists during the past several decades, including summaries in classic works by Stebbins (1950), Harper (1977), Grant (1981), and Richards (1986), among many others. In recent years, significant attention has been given to determining the genetic basis of apomixis (reviews in Grimanelli et al. 2001; Ozias-Akins 2006), with an eye to the development of apomictic crops. Another area of active research focuses on understanding natural variation in apomictic complexes, notably in Taraxacum (van der Hulst et al. 2000; Verduijn et al. 2004; Meirmans et al. 2006), Hieracium (Gadella 1987; Bicknell et al. 2003; Houliston and Chapman 2004), and Crataegus (Muniyama and Phipps 1979a, 1979b, 1984a, 1984b; Dickinson and Phipps 1986; Talent and Dickinson 2005, 2007). Here, we focus on the genetic, population genetic, and ecological factors that affect the dynamics of establishment and spread of asexual lineages. We are especially interested in the contributions of occasional sexual reproduction to the establishment and spread of asexuality and specifically in the role that pollen plays in this context. Maynard Smith (1978, p. 66) noted the potential importance of pollen in the spread of apomixis, but this area remained largely unexplored before work by Mogie (1992). Since this time, substantial insights into the spread of apomixis have been gleaned from theoretical and empirical work, and we summarize these findings and identify areas for future work.

A Brief Description of Apomictic Phenomena

Throughout this review, "apomixis" is used synonymously with "agamospermy" (Stebbins 1950), referring to the asexual production of seed. Other terms describing apomixis are defined in table 1. Apomixis is unknown in gymnosperms, although cleavage polyembryony, the cloning or twinning of embryos from the same sexual zygote, is well documented (Mogie 1992). The pathways by which apomictic seeds are produced are divided into three broad categories (table 1; Stebbins 1950): adventitious embryony, diplospory, and apospory. The latter two categories are referred to collectively as gametophytic apomixis (Stebbins 1950; Grant 1981; Asker and Jerling 1992). In each case, apomictic embryos are derived from maternal genetic

Table 1

Description of Apomictic Phenomena in Flowering Plants

Mode of apomixis	Origin of embryo	Endosperm development	Ploidy	Pollen	Genetic basis	Frequency of apomictic seeds	Taxa
Adventitious embryony	Somatic tissue surrounding ovule; sexual ovule must be fertilized	Requires development of sexually fertilized seed	Usually diploid	Necessary for sexual reproduction	Little studied; one locus dominant?	??	Widely distributed, especially Rutaceae, Celestraceae, and Orchidaceae
Gametophytic apomixis:							
Diplospory	Unreduced megaspore mother cell gives rise to unreduced megagametophyte	Usually autonomous	Rarely diploid	May transmit apomixis	Two unlinked loci	High	Asteraceae
Apospory	Nucellus gives rise to unreduced megagametophyte	Usually pseudogamous	Rarely diploid	Necessary for endosperm development	Single linkage block; 1+ dominant loci	High	Poaceae and Rosaceae

material, but each is associated with different probabilities of producing sexual progeny, different selection pressures to maintain male fertility, and, consequently, different expected levels of genetic diversity within populations.

In adventitious embryony, embryos develop from somatic cells (either from the nucellus or from the integument of the ovule) rather than from the megagametophyte. In gametophytic apomixis, unreduced megagametophytes are produced that subsequently develop into embryos. In this case, the cell that gives rise to the megagametophyte can have one of two origins. In diplospory, the unreduced megagametophyte is produced by the modification or circumvention of meiosis in the megaspore mother cell—the same cell that would give rise to sexual megagametophytes. In apospory, the unreduced megagametophyte arises through mitotic divisions of a cell of the nucellus, usually in conjunction with or following degeneration of the sexual megagametophyte. Diplospory and apospory describe the processes generating the mature megagametophyte; apomixis results when the unfertilized egg undergoes parthenogenetic development to produce the embryo.

Some apomicts require pollen for proper seed maturation (the alternative is called autonomous apomixis; Nygren 1967). In most of these cases, pollen is necessary for the proper development of the endosperm, with at least one of the pollen nuclei fusing with at least one of the polar nuclei of the megagametophyte (Richards 1986). This phenomenon is known as pseudogamy. Adventitious embryony is usually pseudogamous. Among gametophytic apomicts, pseudogamy is prevalent among aposporous apomicts, whereas autonomous endosperm formation is more common with diplospory (Richards 1986). The requirement for fertilization of the endosperm selects for the maintenance of at least some viable pollen (Noirot et al. 1997). Apomicts with autonomous endosperm formation tend to produce less viable pollen and, in some cases, are male sterile (Meirmans et al. 2006; Thompson and Whitton 2006; Thompson et al. 2008). In rare cases, apomicts have been shown to require pollination to stimulate seed development, even though neither the embryo nor the endosperm is fertil-

ized (Bicknell et al. 2003). In the following, we describe the major classes of apomicts in more detail and discuss the implications of each for pollen maintenance and for the production of mixtures of sexual and asexual seed.

Adventitious Embryony

Unlike gametophytic apomixis, adventitious embryony has the potential to occur in parallel with sexual megagametophyte development (Nygren 1967; Asker and Jerling 1992). In most cases that have been examined, the sexual embryo sac appears to develop normally, and pollination followed by double fertilization initiates both sexual embryo and endosperm development. Once the sexual embryo is initiated, additional somatic embryos develop from cells of the nucellus or the integuments (Naumova 1993). A large proportion of taxa with adventitious embryony can produce multiple embryos per seed (polyembryony), one of which may be sexual (Richards 2003). In some cases, these embryos have been shown to compete for the resources of the endosperm (Grant 1981; Naumova 1993), with the percentage of asexual seed varying according to the outcome of the competition. The relative success of sexual versus asexual embryos therefore varies among species and conditions. For example, the percentage of asexual seed ranges from 33% in Eureka lemon to 100% in Dancy mandarin among *Citrus* cultivars (Reuther et al. 1968, in Grant 1981).

Numerous embryological observations of at least occasional adventitious embryony led some authors to conclude that this is the most taxonomically widespread form of apomixis (Naumova 1993; Carman 1997), despite being the least well known in terms of both its importance in nature and the genetic basis of apomixis. With the occurrence of sexual and asexual processes alongside each other, parallels are often drawn between adventitious embryony and vegetative reproduction because asexuality is often facultative in both cases (Grant 1981; Nogler 1984). It is unclear whether this is a fair representation, however, because so little is known about the frequency of asexual seed production in taxa with adventitious embryony. Furthermore, the phenomena of adventitious apomixis and vegetative reproduction have very different implications for dispersal and for the production of seeds capable of surviving inhospitable seasons (Richards 2003).

Another interesting feature of adventitious apomicts is that they are frequently diploid (or of the same ploidy as their sexual relatives), a situation that is exceedingly rare among gametophytic apomicts (Asker and Jerling 1992; Koltunow 1993). That so little is known about the population biology of adventitious embryony (Naumova 1993) limits our ability to draw inferences about this phenomenon.

Gametophytic Apomixis

The two forms of gametophytic apomixis, apospory and diplospory, have been studied in far greater depth than adventitious embryony and are better understood (Richards 1986; Mogie 1992). Among the most striking features of gametophytic apomixis is its strong correlation with polyploidy: of examples in more than 126 genera (Carman 1997), less than a handful are reported to include diploids (Nogler 1984; Mogie 1992; Schranz et al. 2006), and some of these cases are not supported by recent studies (e.g., *Potentilla argentea*; Holm et al. 1997). Whereas odd ploidy levels (e.g., triploidy, pentaploidy) are often reliable predictors of the presence of apomixis, tetraploidy is the most common ploidy level among apomicts (Asker and Jerling 1992). Explanations for the association between gametophytic apomixis and polyploidy are numerous and varied; they include mechanisms that focus on the ecology of polyploids and apomicts (Stebbins 1950), the genetic consequences of polyploidy and apomixis (Lokki 1976; Manning and Dickson 1986), the genetic basis of apomixis (Mogie 1992), and the shared role of unreduced gamete formation (Harlan and deWet 1975). We return to this association in "What Is the Genetic Basis of Apomixis?"

Diplospory

Diplosporous embryos are the product of either a complete omission of meiosis or an abnormal meiosis that yields unreduced products, one of which develops mitotically into the unreduced megagametophyte. Because diplosporous megagametophytes are derived from the megaspore mother cell, diplospory interferes directly with sexual reproduction, and thus individuals exhibiting diplospory are more likely to be obligately asexual, assuming that whatever is interfering with normal meiosis is expressed uniformly in all ovules. Diplospory can lead to genetically variable offspring if crossing over or automixis (Thompson and Ritland 2006) occurs before the cessation of meiosis.

Apospory

In aposporous apomicts, the unreduced megagametophyte is derived from a somatic cell of the ovule (rather than the megaspore mother cell), which undergoes mitotic divisions to form a megagametophyte. Meiosis in the megaspore mother cell usually appears normal, but products may begin to degenerate soon after meiosis is completed (Albertini et al. 2001). Aposporous apomicts have been shown to produce some sexual offspring through fertilization of reduced and unreduced megagameto-

phytes (Bicknell et al. 2003). In some cases, however, even if mature sexual embryo sacs are produced and the aposporous initials fail to develop, sexual embryos may not be viable (Nogler 1984). Whether this represents anomalies in sexual development that contributed to the selective advantage of apomixis or reflects a breakdown in sexual reproduction that occurred after apomixis arose is unknown.

The distinction between apospory and diplospory is not as sharp as the previous descriptions would suggest. This is best illustrated in the Rosaceae, where apospory and diplospory have been described in the same individuals (Muniyamma and Phipps 1984*a*). This intriguing observation may arise from the difficulty of distinguishing somatic and generative cells in the multicellular female archesporium (the part of the nucellus that gives rise to the megaspore mother cell) of many Rosaceae (Asker 1980). However, there are several species in which both diplospory and apospory occur, leading some workers to speculate that these two modes of megagametophyte formation share a common genetic basis (Mogie 1992).

Taxonomic Occurrence of Apomixis

Apomixis occurs sporadically among the ca. 457 angiosperm families (APG 2003). Carman (1997) lists more than 330 genera with apomixis, with more than two-thirds of these being taxa with adventitious embryony. The frequency of asexual reproduction in taxa reported to produce adventitious embryos is poorly documented; thus, this number provides less information about the importance of asexuality in nature than do the numbers for gametophytic apomicts. Of ca. 126 genera known to include gametophytic apomicts (Carman 1997), roughly three-fourths of cases occur in just three families—the Rosaceae, Poaceae, and Asteraceae, which together comprise ca. 15% of angiosperm species (Richards 1986; Asker and Jerling 1992; APG 2003). Estimating the total number of apomictic species is inherently problematic because of the wide differences of opinion on the taxonomic treatment of apomicts (Richards 2003), but it is unlikely that more than 1% of flowering plant species are substantially apomictic. Predominant mechanisms of apomixis also differ among taxonomic groups: members of the Rosaceae and Poaceae are most often aposporous, whereas apomictic Asteraceae are commonly diplosporous (table 1). Reports of adventitious embryony are rare in these three families (Asker and Jerling 1992; Naumova 1993). Adventitious embryony is most frequent in the Rutaceae, Celastraceae, and Orchidaceae (Naumova 1993). Although adventitious embryony is most common in tropical or subtropical trees and shrubs, gametophytic apomixis is described as most common in temperate perennial herbs (Asker and Jerling 1992). However, no surveys of apomixis to date correct for possible correlations between taxonomy and growth forms or geographic distribution.

Regardless of the precise numbers, the broad taxonomic distribution indicates that apomixis, in all of its various forms, has arisen multiple times. It also suggests that the different forms of apomixis either are more likely to arise or are more likely to establish in some groups than in others. What might account for such patterns? Different explanations have been offered for the distribution of different forms of apomixis. For example, it has recently been observed that pollen limitation is more severe in tropical taxa, especially among self-incompatible species and trees (Vamosi et al. 2006), which might selectively favor adventitious embryony as a strategy to augment reproductive output. However, the taxonomic distribution of apomixis may be determined by appropriate genetic opportunities rather than being favored by ecological circumstances. Members of the Rosaceae, Poaceae, and Asteraceae have been shown to produce high frequencies of unreduced gametes, especially in hybrids (Ramsey and Schemske 1998, their supplementary table 1), which might provide repeated opportunities for the evolution of gametophytic apomixis. However, the same is true of other taxa (e.g., Lilliaceae, Convolvulaceae) in which apomixis is not particularly common. Further study is necessary to confirm—or refute—these hypotheses.

What Is the Genetic Basis of Apomixis?

The promise and prospect of harnessing apomixis as a tool for use in agriculture and plant breeding have resulted in the proliferation of studies aimed at identifying genes that contribute to the control of apomixis, with apomictic grasses thus far receiving the most attention (Ozias-Akins 2006). Most genetic studies rely on the production of meiotically reduced viable pollen in apomicts to study segregation of the trait in F_1 progeny from crosses with sexually reproducing female parents. A number of excellent reviews provide detailed summaries of the genetics of apomixis (Grimanelli et al. 2001; Grossniklaus et al. 2001; Ozias-Akins 2006). As in these works, we focus here on more recent studies using molecular genetic approaches.

The genetic basis of adventitious embryony has apparently been studied only in *Citrus* (Garcia et al. 1999). In this case, the 3 : 1 ratio of apomictic to sexual offspring suggests the action of a single locus, with the apomixis allele being dominant. However, quantitative trait locus (QTL) analysis identifies at least three independently segregating markers that appear to be associated with apomixis in this system, with additional loci involved in the control of polyembryony. The QTL analysis relied on relatively few markers (69) and progeny (50), and so these conclusions are somewhat preliminary.

Gametophytic apomixis is typically dissected into two traits, which potentially have different genetic bases: unreduced megagametophyte formation (through apospory or diplospory) and parthenogenetic development of the embryo. In the case of autonomous apomixis, proper endosperm development without fertilization represents a third genetic change required for apomixis. Studies of the genetic basis of apospory in seven species (Poaceae: *Pennisetum squamulatum*, *Pennisetum ciliare*, *Panicum maximum*, *Brachiaria* sp., *Paspalum notatum*; Ranunculaceae: *Ranunculus* sp.; Asteraceae: *Hieracium* sp.) have found that aposporous megagametophyte development and parthenogenesis cosegregate as a single dominant locus (see references in Ozias-Akins 2006). Some of these studies have shown that the region controlling aposporous apomixis has reduced recombination, with multiple markers and potentially multiple genes falling within the linkage group (Ozias-Akins 2006). The only clear example of segregation of apospory from parthenogenesis is in *Poa pratensis* (Albertini et al. 2001; Matzk et al. 2004). In this case, whereas apospory and parthenogenesis

cosegregated in the majority of progeny, two individuals showed signs of apospory without parthenogenesis, supporting the idea that at least two loci inherited as a single linkage block control aposporous apomixis.

The genetic control of diplosporous apomixis is well studied in three systems: *Erigeron* and *Taraxacum* in the Asteraceae and *Tripsacum* in the Poaceae (Ozias-Akins 2006). In two of these cases, control of apomixis involves two unlinked loci separately controlling the production of diplosporous megagametophytes and their parthenogenetic development (van Dijk et al. 1999; Noyes and Rieseberg 2000; Noyes et al. 2007). In *Tripsacum*, although the two components of apomixis appear to co-occur on a single linkage group, recombination between them has been detected (Grimanelli et al. 1988). However, only in recent work in *Erigeron* (Noyes et al. 2007) does recombination appear to occur freely, with nearly equal proportions of the four expected phenotypic combinations from a cross between a sexual seed parent and an apomict-sexual hybrid.

Interesting trends emerge from the genetic data obtained thus far. First, the evidence consistently suggests that most of the alleles controlling apomixis are dominant. This may not be surprising, given the association between apomixis and polyploidy, which, in a tetraploid, would require four copies of recessive alleles to be expressed. Mechanistically, dominance suggests that apomixis involves an active trigger of adventitious embryony, unreduced gamete formation, and parthenogenesis rather than inactivating mutations that knock out normal functions, which tend to be recessive. This is contrary to the views of Grimanelli et al. (2001), who suggest that apomixis arises as a deregulation of sexual function.

A puzzle about the genetics of apomixis raised by Asker (1980) and others is the requirement for nearly simultaneous transitions in the formation of megagametophytes and in embryo development. The chance that two mutations causing these two shifts would occur soon after each other within a small nascent apomictic population seems prohibitive; yet, without the simultaneous emergence of the two traits, apomixis seems unlikely. For example, the development of functional diplosporous megagametophytes without parthenogenesis would lead to embryos with increasing ploidy in each generation, whereas the converse would produce haploids. This puzzle led Asker and Jerling (1992) to argue in favor of a single-gene mechanism, with one mutation causing both unreduced egg formation and parthenogenesis. To date, however, several studies (cited previously) implicate multiple loci, whereas no study has yet to demonstrate that only a single gene is involved (although this may turn out to be the case in some of the examples involving a single linkage group). One way around this puzzle is suggested by the observation that parthenogenesis may be triggered by the presence of unreduced megagametophytes, at least to some extent. Among the segregants of *Erigeron* examined by Noyes and Rieseberg (2000), offspring inheriting the parthenogenesis alleles did not express the trait unless diplospory was also present, a possibility also suggested by the work of Albertini et al. (2001) in *P. pratensis*. However, the most recent work on *Erigeron* (Noyes et al. 2007) has found autonomous development of embryos and endosperm in *Erigeron* without diplospory; aneuploidy may be responsible for the incomplete penetrance observed in the earlier work. Even so, parthenogenesis was more extensive when diplospory was also present. These studies suggest that it is possible that a sexual diploid could harbor alleles for parthenogenesis with little to no penetrance and that the expression of such alleles could increase once unreduced ovules are produced.

Another possible route to the coestablishment of unreduced embryo production and parthenogenesis is through hybridization, a long-recognized correlate of apomixis (Stebbins 1950; Asker and Jerling 1992). In a number of examples, hybrids produce a much higher frequency of unreduced gametes than do their parents (Ramsey and Schemske 1998, their supplementary table 1), with frequencies reaching 86%. This suggests that meiosis is deregulated to some extent in newly formed hybrids, leading to an automatic increase in unreduced gametes. Simultaneously, parthenogenesis might be favored in hybrids to avoid the production of lower-fitness F_2 offspring resulting from genetic incompatibilities. If unreduced gamete formation in hybrids leads to triploid formation, then later-generation hybrids could also have decreased fitness due to segregation problems, further favoring parthenogenesis. In addition, recurrent hybridization also provides a continual supply of individuals of different genetic constitutions, increasing the chance that a lineage capable of both unreduced gamete formation and parthenogenesis arises.

Another observation of interest is that the alleles or linkage blocks contributing to apomixis are, at least in some cases, lethal in haploid pollen (Nogler 1984; Roche et al. 2001). Because gametophytic apomixis is typically associated with regular meiosis in pollen (Mogie 1992; see recent examples in Bicknell et al. 2003; Noyes et al. 2007; Talent and Dickinson 2007), diploid carriers would be unable to pass such alleles to their offspring through pollen. This suggests that the mutations leading to apomixis likely established in lineages either that had previously become polyploid (so that the mutant allele gained fitness through pollen) or that were predisposed to parthenogenetic development (so that the mutant allele gained fitness through apomictic seeds). This assumes that haploid lethality is associated with apomixis from the beginning. Alternatively, haploid pollen of apomicts might be viable initially and then might gradually deteriorate due to the relaxation of selection pressure on pollen.

The Expression of Apomixis in Polyploids

These findings also shed some light on the association between gametophytic apomixis and polyploidy. Because the formation of unreduced megagametophytes is a characteristic feature of gametophytic apomicts, as well as a key element in the production of polyploids in sexuals (Harlan and deWet 1975; Ramsey and Schemske 1998), it is tempting to infer that unreduced egg formation is the causal link between apomixis and polyploidy. However, whether this is the causal link and exactly how it might work remain unclear.

One possibility is that a mutation causing a high frequency of unreduced gametes first becomes established within a sexual population, generating a high frequency of polyploids but also selecting for parthenogenetic development to avoid low-fitness offspring of even higher ploidy levels. However, one problem with this model is that it is difficult to explain how a mutation generating a high frequency of unreduced gam-

etes would survive for long within a sexual population. In a sexual diploid, not only would such a mutation generate a high frequency of triploid offspring, but also pollen fitness would be severely reduced if alleles for the production of unreduced megagametophytes are lethal in haploid gametes (Nogler 1984; Roche et al. 2001).

Another possible causal link between unreduced gametes and polyploidy assumes that apomixis first becomes established in diploids by a simultaneous increase in the frequency of unreduced gametes and parthenogenesis (as discussed previously). In such a nascent apomictic population, occasional fertilization of unreduced eggs may occur, generating higher ploidy levels until such time that stricter mechanisms to avoid fertilization have evolved, which would produce the association between apomixis and polyploidy as well as patterns of multiple ploidy levels observed in many apomictic groups. Although this model does not preclude the persistence of diploid apomicts, one would predict that ploidy levels would rise by a ratchet mechanism (Meyers and Levin 2006) if increases in ploidy are much more common than reductions. Indeed, the modeling results of Meyers and Levin (2006) suggest that because of the irreversibility of increases in ploidy level, polyploids can achieve high frequencies within lineages over time even if they are at a selective disadvantage. In a system with apomixis, diploid apomicts would be subject to stochastic loss, with little possibility of replenishment through ploidy reduction. In addition, any factor that limited the persistence or spread of apomixis genes in diploids would accentuate this pattern.

It remains possible, however, that the link between polyploidy and apomixis is not causally related to unreduced gamete formation. Rather, apomixis and polyploidy might be induced by yet another factor. For example, both apomixis and polyploidization could be selectively favored in hybrids between two species as mechanisms to avoid hybrid breakdown and to ensure balanced segregation of parental genomes. Such polyploids, originating from crosses between different species, are known as allopolyploids, whereas those polyploids originating within one species are termed autopolyploids. Autopolyploidy has traditionally been thought to contribute rarely to long-lasting polyploid lineages, but this has recently been called into question. Soltis et al. (2007) point out several cases in which autopolyploid origins are more parsimonious. Determining whether there is a bias toward allopolyploidy among apomicts will help to shed light on the causal mechanisms linking apomixis and polyploidy. If most apomicts are allopolyploid, we would infer that hybridization is the key causal link between apomixis and polyploidy. If allopolyploidy is not overrepresented among apomicts, it suggests that either polyploidy predated apomixis or apomixis predated the complete exclusion of pollen (see above).

Phylogenetic Patterns in the Origin and Spread of Apomixis

It is difficult to determine the rate of appearance of apomictic lineages via mutation because apomixis can also spread via hybridization. Apomixis often appears repeatedly in closely related species assemblages known as agamic complexes (Grant 1981). This pattern could indicate that apomixis arises more readily in some groups or that it spreads more easily in these groups. Our work in two North American agamic complexes in the Asteraceae is used here to illustrate the challenges in interpreting the origins and spread of apomixis in systems with complex phylogenetic histories.

The North American *Crepis* agamic complex includes seven taxonomic species and more than 100 apomictic forms. This group was studied extensively in the 1920s and 1930s by Babcock and Stebbins, who conducted morphological and cytological surveys to characterize patterns of variation. Seven of the species include both sexual diploids and polyploids that can be sexual but are more often apomictic. Although most polyploid apomicts are assigned to one of these seven taxonomic species along with diploid sexuals, polyploids often combine characteristics of two or more diploids. The remaining two species, *Crepis barbigera* and *Crepis intermedia*, are termed "agamospecies," known only as polyploid apomicts with distinct morphological characteristics. All taxonomic species in the complex are connected through an intricate web of hybrid apomictic polyploids, and no part of the complex is completely genetically isolated from the rest (Babcock and Stebbins 1938; Whitton 1994). We are using phylogenetic analyses of populations within the agamic complex to infer evolutionary relationships among members of the complex, to clarify taxonomic treatments, and to shed light on the origins and spread of apomixis. Here, we describe results of a study examining phyloge-

netic affinities of the agamospecies *C. barbigera* using variation in chloroplast DNA.

Crepis barbigera is most common in the dry regions of central and eastern Washington and occurs less frequently in adjacent Idaho and Oregon. Across its range, it overlaps with a number of other species in the agamic complex. The most common taxa occurring within the range of *C. barbigera* are *Crepis acuminata*, *Crepis atribarba*, and *Crepis modocensis* ssp. *rostrata*. Babcock and Stebbins hypothesized that *C. barbigera* is a complex derivative of these three taxa. Because *C. intermedia*, *Crepis occidentalis*, and *C. modocensis* spp. *modocensis* are also sympatric with these taxa and overlap morphologically, we also included these species in our study.

Three strongly supported nodes emerge from our analysis (fig. 1). We find that populations identified as *C. barbigera* (following the keys of Babcock and Stebbins 1938) occur in at least two distinct clusters, one northern (Washington) and one southern (Oregon), with populations of the other taxa, including *C. atribarba* and *C. intermedia*, intermingled within these two clades. Populations of *C. modocensis* ssp. *modocensis* and *C. occidentalis* occupy a somewhat distinct position on these trees. Although there is taxonomic uncertainty about whether northern and southern populations of *C. barbigera* should be treated as conspecific (C. Sears, personal communication), the general lack of correspondence between the chloroplast phylogeny and the morphological species identity suggest that

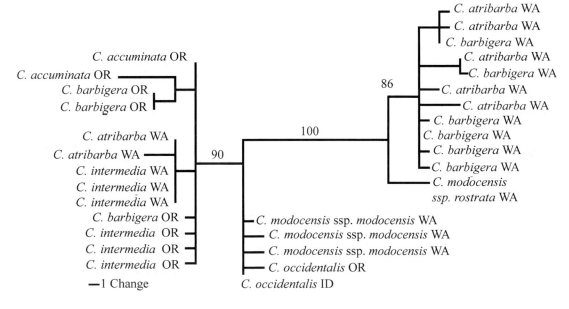

Fig. 1 Unrooted strict consensus of tree from 120 maximum parsimony analysis of four cpDNA regions for samples of the North American *Crepis* complex from Washington and Oregon. Bootstrap support values of more than 50% are indicated above the nodes. Sampled individuals represent seven populations of *Crepis barbigera*, two populations of *Crepis acuminata*, four populations of *Crepis atribarba*, two populations of *Crepis modocensis*, four populations of *Crepis intermedia*, and two populations of *Crepis occidentalis*. Genomic DNA was isolated from field-collected leaf samples dried in silica gel using a modified CTAB protocol (Michiels et al. 2003). Four chloroplast DNA regions (*trn*G, *Rpl*16, *rps*16, and *trn*M-*trn*s; Shaw et al. 2005), covering ca. 3507 bp, were sequenced in both directions using Big Dye chemistry and were run on an Applied Biosystems 3730S 48-capillary sequencer at the University of British Columbia Nucleic Acid Protein Service unit. Raw sequences were proofread and edited in Sequencher, version 4.2.2, and were manually aligned in Se-Al, version 2.0a (Rambaut 1996). Nucleotide substitutions and unambiguous indels were used as characters in phylogenetic analysis. Thirty-five phylogenetically informative characters were present in the combined data set. The data were analyzed using PAUP, version 4.0b (Swofford 2003), under the maximum parsimony criterion with heuristic search, using 1000 replicates and tree-bisection-branch swapping on the best trees only.

polyploid apomicts in *Crepis* have a complex history of hybridization, in agreement with previous studies (Babcock and Stebbins 1938; Holsinger et al. 1999). The existence of gene flow within the complex suggests that apomixis might have spread via hybridization. It is also possible that diploid hybrids have increased frequencies of unreduced gamete formation and that this contributes to both increases in ploidy level and enhanced expression of apomixis. Although seemingly less parsimonious, it is also possible that apomixis has arisen de novo multiple times. Our understanding will be greatly enhanced if we can follow the evolutionary history of genes underlying apomixis. In such an interconnected reticulating complex, we would predict that apomixis alleles have spread extensively.

In contrast to the situation in *Crepis*, apomixis in *Townsendia* occurs in the absence of evidence of extensive hybridization, but recent work suggests that apomixis can also become repeatedly established in this group. In *Townsendia*, apomixis occurs in approximately half of the ca. 30 taxa, where it is restricted to tetraploids. All polyploids are described as autopolyploids using morphological criteria (Beaman 1957). Autopolyploid origins of apomicts are indicated in *Townsendia hookeri* (Thompson and Whitton 2006) because there is no evidence of morphological hybridity in diploid or polyploid populations. All sampled *T. hookeri* polyploids have chloroplast haplotypes that either are identical to or recently derived from those of diploid conspecifics. Phylogenetic analysis of *T. hookeri* chloroplast DNA suggests a minimum of four evolutionary transitions from sexual diploidy to apomictic polyploidy (Thompson and Whitton 2006), but it seems more plausible that the alleles for apomixis spread into these four lineages rather than invoking four distinct origins of apomixis alleles. Alternatively, apomixis alleles may represent a shared ancestral polymorphism that gains expression in polyploid populations of *T. hookeri* (and possibly other apomictic townsendias). It is difficult to explain how apomixis alleles could persist in diploids, unless the expression of apomixis is somehow ploidy dependent. Once again, we find that a clear understanding of trait evolution will be greatly advanced by the ability to track the history of apomixis alleles.

Pathways for the Spread of Apomixis

Given that an apomict arises within a sexual population, there are two primary ways for asexuality to spread over space: direct dispersal via apomictic seed and indirect transmission via pollen. In addition, apomixis genes can spread via any sexual seeds produced by an apomictic parent and through vegetative propagation. Vegetative spread tends to be geographically circumscribed and is less likely to differ between apomicts and related sexuals (but see O'Connell and Eckert 1999).

Direct Dispersal

The direct dispersal of apomictic progeny is the most obvious and straightforward means by which the trait could spread over space. Indeed, it is commonly assumed that asexual plants should spread at an even higher rate than sexual relatives because they avoid the costs of sex (Maynard Smith 1978). However, this assumption may not be valid as surveys conducted in a number of apomicts (e.g., *Antennaria* [Bayer

1989] and *Townsendia* [Thompson et al. 2008]) find that single clones tend to be geographically restricted to one or a small number of populations (summarized originally in Ellstrand and Roose 1987). Further studies are needed to confirm these findings, which suggest that apomictic lineages rarely become widespread through seed dispersal alone.

That asexuals do not appear to have a major advantage through proliferation suggests the need to reexamine assumptions regarding the costs of sex. If an asexual mutant makes no investment in male function and is able to reallocate these resources to seed production, then the asexual lineage is predicted to have a direct twofold transmission advantage. This assumes that competing sexual species allocate reproductive resources equally to male and female functions, as is predicted to occur among outcrossing species. One factor that could reduce the cost of sex paid by related sexuals is selfing, as allocation to male function is predicted and observed to be lower among selfers (Ritland and Ritland 1989; Parachnowitsch and Elle 2004). However, apomixis tends to arise from self-incompatible progenitors (Asker and Jerling 1992; see Roy 1995 for a notable exception in *Boechera* [formerly *Arabis*] *holboellii*), which are likely to suffer from the full cost of sex. Of course, apomicts will gain benefits from avoiding the cost of sex only if the reallocation to female function accompanies the appearance of apomixis (Maynard Smith 1978; Mogie 1992). Because mutations that produce meiotic abnormalities in ovules often do not affect pollen development, the transition to apomixis need not reduce pollen investment. This suggests that the initial spread of apomixis may be little aided by reallocating resources away from male function. Empirical studies examining the immediate consequences for male fertility are needed in novel apomicts to quantify the extent to which they pay less of a cost of sex.

Over the longer term, as investment in pollen declines, apomictic lineages should begin to reap the full benefits of avoiding the costs of sex. Two studies demonstrate that established apomicts may gain a nearly twofold reproductive advantage. Meirmans et al. (2006) compared allocation patterns of naturally occurring male sterile and male fertile apomictic dandelions. They found significantly greater (but not twofold) reallocation in the form of fruit number in male sterile plants. O'Connell and Eckert (2001) provide evidence of a twofold advantage in apomictic females of the dioecious *Antennaria parlinii*. We thus expect a transition to occur as pollen investment declines, with young apomictic lineages having little fertility advantage over related sexual populations, whereas older apomicts gain from a reduced cost of sex.

Indirect Transmission via Pollen

The previous discussion assumes that pollen production only slows the spread of apomixis, but this need not be true. Ironically, the genes for maternal clonality can be transmitted via male gametes, and this mode of transmission may well be important in the establishment and spread of apomixis. That the genes for apomixis can be transferred to sexuals via the pollen of apomicts has long been known. Experimental crosses have made use of the pollen of apomicts beginning in the 1940s (Asker and Jerling 1992; e.g., Tas and Van Dijk 1999; Brock 2004), and this has become the prevalent means of studying the inheritance of apomixis (Noyes and Rieseberg 2000).

Although individual crossing studies provide proof of principle for the inheritance of apomixis via the pollen of apomicts, studies of wild-collected individuals are just beginning to uncover the importance of this mechanism in nature. Surveys of apomictic populations have documented patterns of isozyme and cpDNA variation consistent with multiple origins of apomixis or transmission of apomixis via pollen (Roy and Rieseberg 1989; Sharbel and Mitchell-Olds 2001; Thompson and Whitton 2006). However, so far these data do not allow assessment of the frequency of transmission of apomixis via pollen.

The transmission of apomixis genes to sexuals via pollen may be of long-term importance for the spread of apomixis, even if these events are relatively infrequent. Thus, although it is straightforward to examine the reproductive mode of open-pollinated seed of sexuals for evidence of transmission of apomixis, tools that facilitate screening of large numbers of progeny may be necessary to allow detection of rare events and document this pathway in nature. Two recently developed approaches could prove essential in understanding the role of pollen in the transmission of apomixis. Flow cytometric seed screening (FCSS; Matzk et al. 2000) can be used to assess the relative ploidy of endosperm and embryo and thus to determine the genetic origin of seeds. For example, a tetraploid autonomous apomict is expected to have a 4 : 8 embryo : endosperm ploidy ratio (assuming that two unreduced polar nuclei fuse to form the endosperm), while a sexual tetraploid would have a 4 : 6 ratio (assuming that two reduced polar nuclei fuse with a reduced pollen nucleus). To date, the use of this approach has focused mainly on estimating the rate of sexual and asexual reproduction in apomicts (Matzk et al. 2000), but these data can also indicate that pollen from apomicts sires seed (Talent and Dickinson 2007). FCSS can be applied to natural or to experimental populations of mixed sexuals and asexuals. Mártonfiová (2006) applied FCSS to experimentally produced progenies, involving crosses with sexual diploids and apomictic triploids and tetraploids in *Taraxacum* sect. *Ruderalia*. If triploids were used as pollen parents, all 11 diploid mothers examined produced some triploid progeny. However, none of the 35 open-pollinated diploid mothers produced triploid seed. This suggests that detection of gene flow between sexual diploids and apomictic polyploids in nature may require intensive sampling. One unfortunate limitation of this approach is that seeds are sampled destructively, precluding later evaluation of the reproductive mode of the offspring. In an alternative approach, Bicknell et al. (2003) made use of transgenic *Hieracium pilosella* individuals carrying antibiotic resistance to track the reproductive dynamics of two facultatively apomictic individuals. Although this approach is limited to systems in which transformation protocols have been worked out, it allows screening of large numbers of seeds and efficient recovery of rare sexual offspring under experimental conditions.

To the extent that pollen from apomicts fertilizes ovules in sexual individuals and produces apomictic offspring, apomictic lineages will gain a transmission advantage even if they do not reallocate resources from male to female function (Maynard Smith 1978; Holsinger 2000; Britton and Mogie 2001). Apomicts are 100% related to the seeds that they produce directly

(compared with 50% in obligate outcrossers) as well as being 50% related to the seeds that they sire via pollen. If pollen success is similar among apomicts and sexuals, this would generate a 3 : 2 transmission advantage that parallels the transmission advantage enjoyed by selfers arising within outcrossing populations (Maynard Smith 1978; Uyenoyama 1984; Holsinger 2000). The magnitude of the transmission advantage will depend on levels of pollen viability in new apomicts, about which little is known. In established apomictic populations of *T. hookeri*, pollen viability averaged 17%, with the highest viability being 42% (Thompson and Whitton 2006). However, Noyes et al. (2007) produced a novel apomict with 98% viable pollen.

Maynard Smith (1978, p. 41) noted that as the frequency of apomixis rises, transmission via pollen becomes less efficient, and thus, one would expect pollen fertility to decline (through drift or selection; Maynard Smith 1978; Eckert 2002). Because apomixis would not enjoy a transmission advantage within a highly selfing population (in which outcross pollen is rarely incorporated), the observation that apomixis typically arises from self-incompatible species (Asker and Jerling 1992) suggests that the initial transmission advantage arising from producing pollen as well as apomictic seeds might be important to the spread of apomixis. Interestingly, apomicts also tend to maintain pollinator attraction features and are not associated with reduced flower size, again suggesting that pollen transmission continues to be important in these taxa.

When pollen from apomicts fertilizes sexual ovules, this allows the genes for apomixis to move into multiple genetic backgrounds; this is perhaps even more important than the transmission advantage gained. As pointed out by Mogie (1992), this movement might be key to the evolutionary success of apomicts. Gene flow via pollen improves the chance that apomixis will be associated with a genotype of high fitness, increasing the probability of successful establishment. The spread of apomixis via pollen also increases the genetic variability of asexual populations, improving their ability to respond to selection and to adapt to new habitats. Indeed, patterns and levels of diversity within and among apomictic populations reveal substantial variation (Ellstrand and Roose 1987; Silvertown 2008). Thus, the occasional capture of different sexual genetic backgrounds via pollen flow from apomicts may provide many of the evolutionary benefits of sex and recombination to predominantly asexual plant populations (further discussed in "Apomixis and the Evolution of Sex").

Apomixis genes may also be transmitted via sexually fertilized seeds produced by otherwise apomictic plants (Asker and Jerling 1992; Matzk et al. 2000; Bicknell et al. 2003; Richards 2003). If this process involves meiotically reduced eggs, the apomict would lose its transmission advantage with respect to these seeds (which bear only half of the genes of the maternal apomictic parent). If unreduced eggs were involved, the apomict would continue to transmit the same number of genes to the next generation via these seeds, but the seeds would have an increased ploidy level and might well be less fit. Thus, we would expect a lower transmission advantage in apomicts that often act as maternal parents of sexually produced seed. However, to the extent that this pathway allows apomixis to be expressed in novel genetic backgrounds, it too may increase the likelihood of persistence.

Geographical Parthenogenesis

That sexual and asexual relatives tend to have nonoverlapping ranges in both plants and animals has long been noted, resulting in a pattern termed "geographical parthenogenesis" by Vandel (1928). The distributions of a number of taxa reveal that asexuals are likely to occupy latitudes and elevations higher than those occupied by their sexual counterparts. Data relating to geographical parthenogenesis in plants are summarized by Bierzychudek (1985) and Hörandl (2006).

Explanations for the existence of geographical parthenogenesis generally fall into either ecological or demographic categories. Ecological explanations are based on differential ecological tolerances or competitive abilities of sexuals and asexuals. For example, it has been proposed that asexuals may comprise general-purpose genotypes (Lynch 1984) that can occupy a wide range of ecological conditions but are relatively poor at tracking extreme or specialized habitats. Support for the existence of general-purpose genotypes in plants is, however, sparse. Although it might seem unlikely that multiple genetically distinct lineages of apomicts would have hit on similar general-purpose genotypes, independent lineages of apomicts might exhibit similar ecological features as a result of parallel polyploidization events; polyploidy often induces specific morphological changes, altered ecological tolerances, and shifts in gene expression (Otto and Whitton 2000; Adams and Wendel 2005). We are aware of no studies to date that examine the ecological response of multiple lineages of a single apomict.

Demographic explanations attribute the distribution of sexuals and asexuals to the relative dispersal and establishment abilities of apomicts. In addition to the ability to found populations with a single propagule, apomicts may be favored during recolonization of previously glaciated habitats because of the scarcity of potential pollen sources and pollinators. We would thus predict that apomicts that colonize deglaciated areas would more often be autonomous or be able to use self pollen to initiate endosperm. Once established in these habitats, apomictic plants could exclude sexual relatives. owing to frequency-dependent fitness effects described below.

Because in plants geographical parthenogenesis tends to involve gametophytic apomixis and because such apomicts tend to be polyploid (unlike with adventitious embryony), Bierzychudek (1985) suggested that polyploidy rather than asexuality may be the key contributor to the distributional patterns. Stebbins (1950) also observed that sexual polyploids tend to have more northerly and higher-elevation distributions than related diploid sexuals. Brochmann et al. (2004) confirmed that polyploids are more prevalent in the arctic but noted that this trend applied mainly to higher ploidy levels (greater than tetraploid) and to floras in previously glaciated regions. More frequently, authors suggest that asexuality rather than polyploidy is key in establishing the pattern (Hörandl 2006; Thompson and Whitton 2006). The relative contributions of apomixis and polyploidy to geographic patterns will be best explored in systems in which both sexual and apomictic polyploids are known, such as in the *Antennaria parlinii* complex (Bayer and Stebbins 1983).

Polyploidy and apomixis may also interact in a manner that contributes to geographical parthenogenesis, through a phenomenon known as destabilizing hybridization (Lynch 1984). Interploidy crosses in sexually reproducing groups can generate strong reproductive incompatibilities through the formation of inviable or sterile offspring (e.g., triploid block; Ramsey and Schemske 1998; Burton and Husband 2000). Where sexual diploids and asexual polyploids co-occur, hybridization asymmetrically reduces the fitness of sexual diploids because sexuals are more likely than apomicts to incorporate pollen from other ploidy levels, leading to offspring of odd ploidy levels. Thus, all else being equal, the presence of asexuals can destabilize sexual populations. Sexuals might therefore persist only in portions of the range where they have a selective advantage over asexuals. In this regard, it is noteworthy that in both *Crepis* and *Townsendia*, diploids tend to occupy habitats that could be described as more specialized. For example, in *T. hookeri*, diploids occur on Niobrara and Pierre shales, with soils that are significantly higher in selenium, aluminum, and other elements known to restrict the distribution of plants in other groups (E. M. Stacey and J. Whitton, unpublished data). The restriction of diploids to these unusual substrates may indicate that they hold a selective advantage under these conditions.

Apomixis and the Evolution of Sex

One of the longest-standing questions in evolutionary biology is why sex is so prevalent. That apomixis arises repeatedly in plants but fails to take substantial hold suggests that plants may shed light on the benefits of sex and the costs of asexuality. Here, we discuss how the emerging data on apomixis informs the literature on the evolution of sex and vice versa.

To structure this discussion, we focus on three interrelated questions: Why are there no ancient asexual angiosperms? Why do few plant populations exhibit mixtures of sexual and asexual individuals? And why don't most plants reproduce apomictically while engaging in occasional sex?

Why Are There No Ancient Asexual Angiosperms?

Although apomixis may appear fixed or nearly so within particular populations or portions of a species range, the only known instance of a phylogenetically isolated apomictic lineage in seed plants is *Houttuynia*, a small (perhaps monotypic) genus in Saururaceae, first reported as apomictic by Okabe (1930). However, it is not clear that the entire lineage is apomictic. Even if it is, whether the loss of sex occurred early or late in its evolutionary history is not known. All remaining angiosperm apomicts are at the tips of the tree of life, with no other higher-level taxa (families, genera) that are asexual. A number of well-studied temperate apomicts have geographic distributions consistent with an origin following Pleistocene glaciation, suggesting that these apomicts may be of very recent origin (Hörandl 2006; Thompson and Whitton 2006). Despite this phylogenetic distribution, there are indications that the tendency to produce apomictic lineages may have arisen early in the diversification of some deeper lineages. For example, apomixis in the Rosaceae occurs within species of numerous genera. Recent phylogenetic results (Potter et al. 2007) reveal that *Rubus*, which contains a number of apomicts, diverged early in the evolutionary history of subfamily Rosoideae. This suggests that the tendency to generate apomicts may have appeared early within this lineage. Although a predisposition toward apomixis might persist for long periods of evolutionary time, apomictic lineages, once produced, have only a shallow phylogenetic history in seed plants, with no truly ancient asexual taxa.

Why apomictic lineages of angiosperms appear to be restricted to the very tips of evolutionary branches remains unclear. Ancient asexual lineages appear in other groups, most notably bdelloid rotifers and darwinulid ostracods but also among ferns, where vegetatively reproducing *Vittaria* sp. and *Trichomanes* sp. are estimated to be ca. 10 million years old (Farrar 1990).

Recent theoretical work on the evolution of sex has highlighted the importance of maintaining sex and recombination in populations that are limited in size because genetic variation can be rapidly depleted by the combined action of drift and selection (see reviews in Otto and Gerstein 2006; de Visser and Elena 2007). Genetic variation in fitness is particularly limited in asexuals because of their inability to bring together high-fitness alleles that reside in different individuals. Consequently, any advantage that apomicts might initially have is thought to decrease over time because of (*a*) the accumulation of deleterious mutations (Muller 1932; Keightley and Otto 2006), (*b*) a reduction in the rate of adaptation due to beneficial mutations that arise in different individuals and cannot be brought together in apomicts (the Fisher-Muller hypothesis; Morgan 1913; Fisher 1930; Muller 1932; Barton and Otto 2005), and (*c*) a greater loss of beneficial mutations because they happen to arise in genomes carrying deleterious mutations (the ruby-in-the-rubbish hypothesis; Fisher 1930; Peck 1994). In each case, the fortune of apomicts is predicted to decline over evolutionary timescales because of the difficulty of combining fit alleles.

In this context, there are two potential ways for obligate asexual species to avoid extinction. First, an asexual lineage could avoid the accumulation of deleterious mutations if mutation rates happen to be particularly low. Such a lineage would be unable to adapt rapidly, but if its environment were sufficiently stable, it might avoid extinction. Some data suggest that darwinulid ostracods might have hit on this long-term solution because ostracods have unusually low mutation rates (Butlin and Menozzi 2000; Schön et al. 2003). Second, if an asexual species has a sufficiently large population size and a high migration rate among populations, genetic variation within local populations could remain high enough to allow asexuals to evolve at a rate akin to that of sexual competitors (Ladle et al. 1993; Judson 1995; Martin et al. 2006; Salathé et al. 2006). This is perhaps the route taken by bdelloid rotifers because analyses of sequence data suggest that dispersal rates are high, genetic diversity levels are similar to those of sexual populations, and the ability to respond to selection is similar to that of sexual monogonont rotifers (Birky et al. 2005).

So why aren't there ancient asexual plants? Perhaps the special circumstances enabling asexuals to persist over long periods of evolutionary time do not co-occur in plants. Estimates obtained to date of the genome-wide deleterious mutation rate are high in angiosperms (typically above 0.1 per generation; Drake et al. 1998; Schoen 2005). Furthermore, as a result of limited seed dispersal and low rates of gene flow via pollen, strong spatial structure with isolation by distance is likely to exist in many apomictic species (Meirmans et al. 2003), suggesting that genetic variation within local populations is not replenished at high rates by migration (Thompson and Whitton 2006; Thompson et al. 2008).

Why Do Few Plant Populations Exhibit Mixtures of Sexual and Asexual Individuals?

Characterization of reproductive mode in apomictic complexes suggests that it is uncommon for local populations to be composed of substantial numbers of both primarily sexual and primarily apomictic individuals (Beaman 1957; Gadella 1987; Asker and Jerling 1992; O'Connell and Eckert 1999). Where co-occurrence has been documented, it often results from secondary contact between sexuals and apomicts, producing complex sexual-asexual dynamics (Verduijn et al. 2004; Noyes and Allison 2005; Thompson et al. 2008; K. M. Dlugosch, C. J. Sears, and J. Whitton, unpublished manuscript). A notable exception occurs in *Taraxacum*, in which ploidy cycles, with accompanying switches between sexuality and apomixis, seem to arise on shorter timescales (Meirmans et al. 2003; Verduijn et al. 2004), based on studies of standing variation.

That populations tend to be predominantly apomictic or predominantly sexual is consistent with models exploring the outcome of competition between asexuals and sexuals. In a number of ecological models, it is assumed that whatever advantage(s) sex has in the long run can be modeled as a fixed fitness advantage (Joshi and Moody 1995, 1998; Bengtsson and Ceplitis 2000; Britton and Mogie 2001; Carrillo et al. 2002). Under these assumptions, whether apomicts spread to fixation depends on the assumed fitness advantage of sexuals, the relative pollen output of the apomict, the ability of apomicts to reallocate male function, and the relative success of male gametes from apomicts (Joshi and Moody 1995). In both spatial (Britton and Mogie 2001; Carrillo et al. 2002) and nonspatial (Joshi and Moody 1995; Bengtsson and Ceplitis 2000) versions of these models, it is difficult to maintain sexuals and asexuals together indefinitely. Thus, these models predict that either sexuals or asexuals will win out within a population, depending on the exact balance of selective forces.

These models assume that the benefits of sex are fixed, which is unlikely to be true. For example, newly formed apomicts may suffer little or no loss of viability compared with sexuals because there has been little time for mutation accumulation. In contrast to these ecological models, evolutionary models track the dynamics of selected loci in sexual and asexual populations and allow the benefits of sex to emerge from differences in the response to selection. Nevertheless, evolutionary models also typically observe that either sexuality or asexuality comes to dominate a population. For example, spatially explicit models of individuals subject to recurrent deleterious mutation find that asexuals beat out sexuals when dispersal rates are high or mutation is low, while the reverse is true when dispersal is local and mutation is common (Peck 1994; Salathé et al. 2006). Similarly, in classical host-parasite models, Hamilton (1980; Hamilton et al. 1990) found that sexuals took over the population under some conditions (many parasites per host; strong selection) but that asexuals otherwise dominated.

Although local populations tend to be apomictic or sexual, both sexual and apomictic populations typically coexist on a regional or specieswide scale. This observation is consistent with the previously discussed models if the conditions favoring sex over apomixis vary over space. Furthermore, even if one reproductive mode ultimately outcompetes the other, this process may take a long time. For example, adjacent clumps of sexuals and asexuals coexisted for thousands of years in the simulations of Britton and Mogie (2001), especially when apomictic pollen viability is low and therefore apomicts only gradually invade sexual populations from the edges.

More theoretical work is needed, however, because these models fail to capture all of the key natural history features of plant apomixis. For example, models comparing the evolutionary dynamics of mutation accumulation in sexuals versus asexuals have yet to consider the occasional transmission of apomixis via pollen. The recurrent capture of sexual genotypes via pollen might allow a stable coexistence between sexual and asexual populations, which is seen when asexuality arises by recurrent mutation (Salathé et al. 2006). Further models are also needed that explore the sorts of complex spatial patterns of selection experienced by many species with apomixis. For example, strong selection in harsh sites might favor particular genotypes, and apomicts, by avoiding recombination, might do well under these conditions. Yet, in other sites, the ability of sexual populations to respond more rapidly to selection might allow sexuals to outcompete asexuals (e.g., because of the advantages of bringing together fit alleles carried by different individuals). Thus, although most previous theory suggests that sexual and asexual individuals should not coexist over the long term, it may well be that more realistic models incorporating features commonly encountered in plant apomicts might lead to models that are more favorable to coexistence.

Why Don't Most Plants Reproduce Apomictically While Engaging in Occasional Sex?

Apomixis is typically associated with the abandonment or great diminution of recombination, but the handful of studies aimed at detecting and quantifying recombination in apomicts have found evidence for at least occasional sex in apomicts (van der Hulst et al. 2000, 2003; Chapman et al. 2004; Thompson et al. 2008). In models that allow variable levels of sex, it is commonly observed that a little bit of sex goes a long way, so it would seem that the advantages of sex and recombination could be achieved in predominantly asexual lineages that undergo occasional recombination (Green and Noakes 1995; Hurst and Peck 1996; Peck and Waxman 2000). If this were true, then why aren't all species predominantly apomicts with a small degree of sexual recombination?

Theoretical analyses have identified some conditions under which high levels of sex are favored over lower levels. For example, when hosts and parasites coevolve and are subject to strong selection, lineages with low levels of sex are invaded by lineages with high levels of sex (Peters and Lively 2007). When selection varies over space, migration can generate combinations of alleles (linkage disequilibria) that are not locally favored, which can select for high levels of sex and recombination (Lenormand and Otto 2000). When heterozygote advantage and inbreeding are present, variants with higher rates of sex can be favored over those with lower rates of sex (Peck and Waxman 2000; Dolgin and Otto 2003). Finally, even just recurrent selection against deleterious mutations can select for

high levels of sex and recombination because a little recombination is not sufficient to release genetic variation hidden by linkage disequilibrium among loci in finite populations (Keightley and Otto 2006). Thus, there are a reasonable number of models that can explain why high levels of sex have evolved and been maintained in most plants, even though reproductive systems exist that would allow apomixis with low levels of sex.

As theoretical studies accumulate that can explain the evolutionary maintenance of high levels of sex, the question turns on its head, and we must ask what special circumstances allow apomixis to gain a foothold. The pattern of geographical parthenogenesis suggests that the most common of these special circumstances might occur when populations extend into harsh environments where certain combinations of alleles must be kept together. Apomixis would then allow the possibility of a "frozen genotype" that can survive and spread beyond the geographical limits faced by sexual populations, which are subject to ongoing gene flow that disrupts the necessary genotype(s) (Peck et al. 1998). Empirical studies demonstrating that sexuals fail to adapt at the edge of their species range because of gene flow from the core of the population would provide support for this hypothesis.

Conclusions

The emerging view of apomixis in flowering plants reveals this mode of reproduction to be highly dynamic. Apomictic populations harbor substantial genetic variation, and in a growing number of cases, including *Crepis* and *Townsendia*, evidence indicates that multiple genetic lineages have been captured from related sexual populations, either through fertilization by pollen from apomicts or through multiple origins of apomixis. This dynamic nature of apomixis might be critical to its establishment and spread. In particular, gene flow via pollen produced by apomicts has the potential to affect both the spread of the trait and the ability of apomicts to appear within novel genetic backgrounds. Still, our understanding of the role of pollen in natural apomicts is limited. Future empirical studies focusing on regions of co-occurrence of apomicts and sexuals, as well as studies of apomicts that are known to reproduce sexually at moderate frequencies, will provide important insights into the role of male fertility in sexual-asexual dynamics. Further studies are also needed to examine the properties of recently derived apomicts. For example, do new apomicts reallocate resources from male to female function and so pay less of a cost of sex? Do new apomicts have different ecological tolerances? And is this process idiosyncratic, depending on the exact genotype involved, or are there some features that apomicts have in common and that differentiate them from sexual relatives? Such studies will help to clarify the relative importance of demographic effects, ecological differences, and their interactions in the establishment and spread of apomixis, providing a broader and more complete understanding of the conditions favoring evolutionary transitions to asexuality in plants.

Acknowledgments

We thank Spencer Barrett for unwavering support during the preparation of this manuscript and Minako Kaneda for

Japanese translation of the article by Okabe (1930). Tim Dickinson, Spencer Barrett, and two anonymous reviewers provided comments that helped improve the manuscript. This work was funded in part by Natural Sciences and Engineering Research Council of Canada Discovery Grants to J. Whitton and S. P. Otto and by a fellowship from the National Evolutionary Synthesis Center (National Science Foundation grant EF-0423641) to S. P. Otto.

Literature Cited

Adams KL, JF Wendel 2005 Polyploidy and genome evolution in plants. Curr Opin Plant Biol 8:135–141.

Albertini E, A Porceddu, F Ferranti, L Reale, G Barcaccia, B Romano, M Falcinelli 2001 Apospory and parthenogenesis may be uncoupled in *Poa pratensis*: a cytological investigation. Sex Plant Reprod 14: 213–217.

APG (Angiosperm Phylogeny Group) 2003 An update of the Angiosperm Phylogeny Group classification for the orders and families of flowering plants: APG II. Bot J Linn Soc 141:399–436.

Asker S 1980 Gametophytic apomixis: elements and genetic regulation. Herditas 93:277–293.

Asker SE, L Jerling 1992 Apomixis in plants. CRC, Boca Raton, FL. 298 pp.

Babcock EB, GL Stebbins 1938 The American species of *Crepis*: their relationships and distribution as affected by polyploidy and apomixis. Carnegie Inst Wash Publ 504:1–199.

Barton NH, SP Otto 2005 Evolution of recombination due to random drift. Genetics 169:2353–2370.

Bayer RJ 1989 A taxonomic revision of the *Antennaria rosea* (Asteraceae: Inuleae: Gnaphaliinae) polyploid complex. Brittonia 41: 53–60.

Bayer RJ, GL Stebbins 1983 Distribution of sexual and apomictic populations of *Antennaria parlinii*. Evolution 37:555–561.

Beaman JH 1957 The systematics and evolution of *Townsendia*. Contrib Gray Herb Harv Univ 183:1–151.

Bengtsson BO, A Ceplitis 2000 The balance between sexual and asexual reproduction in plants living in variable environments. J Evol Biol 13:415–422.

Bicknell RA, SC Lambie, RC Butler 2003 Quantification of progeny classes in two facultatively apomictic accessions of *Hieracium*. Hereditas 38:11–20.

Bierzychudek P 1985 Patterns in plant parthenogenesis. Experientia 41:1255–1264.

Birky CWJ, C Wolf, H Maughan, L Herbertson, E Henry 2005 Speciation and selection without sex. Hydrobiologia 546:29–45.

Britton NF, M Mogie 2001 Poor male function favours the coexistence of sexual and asexual relatives. Ecol Lett 4:116–121.

Brochmann C, AK Brysting, IG Alsos, L Borgen, HH Grundt, AC Scheen, R Elven 2004 Polyploidy in arctic plants. Biol J Linn Soc 82:521–536.

Brock M 2004 The potential for genetic assimilation of a native dandelion species, *Taraxacum ceratophorum* (Asteraceae), by the exotic congener *T. officinale*. Am J Bot 91:656–663.

Burton TL, BC Husband 2000 Fitness differences among diploids, tetraploids, and their triploid progeny in *Chamerion angustifolium*: mechanisms of inviability and implications for polyploid evolution. Evolution 54:1182–1191.

Butlin RK, P Menozzi 2000 Open questions in evolutionary ecology: do ostracods have the answers? Hydrobiologia 419:1–14.

Carman JG 1997 Asynchronous expression of duplicate genes in angiosperms may cause apomixis, bispory, tetraspory, and polyembryony. Biol J Linn Soc 61:51–94.

Carrillo C, NF Britton, M Mogie 2002 Coexistence of sexual and asexual conspecifics: a cellular automaton model. J Theor Biol 217: 275–285.

Chapman H, GJ Houliston, B Robson, I Ilne 2003 A case of reversal: the evolution and maintenance of sexuals from parthenogenetic clones in *Hieracium pilosella*. Int J Plant Sci 164:719–728.

Chapman H, B Robson, ML Pearson 2004 Population genetic structure of a colonising, triploid weed, *Hieracium lepidulum*. Heredity 92:182–188.

de Visser JAGM, SF Elena 2007 The evolution of sex: empirical insights into the roles of epistasis and drift. Nat Rev Genet 8:139–149.

Dickinson TA, JB Phipps 1986 Studies in *Crataegus* (Rosaceae: Maloideae). XIV. The breeding system of *Crataegus crus-galli* sensu lato in Ontario (Canada). Am J Bot 73:116–130.

Dolgin ES, SP Otto 2003 Segregation and the evolution of sex under overdominant selection. Genetics 164:1119–1128.

Drake JW, B Charlesworth, D Charlesworth, JF Crow 1998 Rates of spontaneous mutation. Genetics 148:1667–1686.

Eckert CG 2002 The loss of sex in clonal plants. Evol Ecol 15:501–520.

Ellstrand NC, ML Roose 1987 Patterns of genotypic diversity in clonal plant species. Am J Bot 74:123–131.

Farrar DR 1990 Species and evolution in asexually reproducing independent fern gametophytes. Syst Bot 15:98–111.

Fisher RA 1930 The genetical theory of natural selection. Oxford University Press, Oxford.

Gadella TWJ 1987 Sexual tetraploid and apomictic pentaploid populations of *Hieracium pilosella* (Compositae). Plant Syst Evol 157: 219–245.

Garcia R, MJ Asins, J Forner, EA Carbonell 1999 Genetic analysis in *Citrus* and *Poncirus* by genetic markers. Theor Appl Genet 99: 511–518.

Grant V 1981 Plant speciation. 2nd ed. Columbia University Press, New York. 563 pp.

Green RF, DLG Noakes 1995 Is a little bit of sex as good as a lot? J Theor Biol 174:87–96.

Grimanelli D, O Leblanc, E Espinosa, E Perotti, DG De Leon, Y Savidan 1998 Mapping diplosporous apomixis in tetraploid *Tripsacum*: one gene or several genes? Heredity 80:33–39.

Grimanelli D, O Leblanc, E Perotti, U Grossniklaus 2001 Developmental genetics of gametophytic apomixis. Trends Genet 17:597–604.

Grossniklaus U, GA Nogler, PJ van Dijk 2001 How to avoid sex: the genetic control of gametophytic apomixis. Plant Cell 13:1491–1497.

Hamilton WD 1980 Sex vs. non-sex vs. parasite. Oikos 35:282–290.

Hamilton WD, R Axelrod, R Tanese 1990 Sexual reproduction as an adaptation to resist parasites: a review. Proc Natl Acad Sci USA 87: 3566–3573.

Harlan JR, JMJ deWet 1975 On Ö. Winge and a prayer: the origins of polyploidy. Bot Rev 4:361–390.

Harper JL 1977 Population biology of plants. Academic Press, London. 892 pp.

Holm S, L Ghatnekar, BO Bengtsson 1997 Selfing and outcrossing but no apomixis in two natural populations of diploid *Potentilla argentea*. J Evol Biol 10:343–352.

Holsinger KE 2000 Reproductive systems evolution in vascular plants. Proc Natl Acad Sci USA 97:7037–7042.

Holsinger KE, RJ Mason-Gamer, J Whitton 1999 Genes, demes, and plant conservation. Pages 23–46 *in* LF Landwebber, AP Dobson, eds. Genetics and the extinction of species. Princeton University Press, Princeton, NJ.

Hörandl E 2006 The complex causality of geographical parthenogenesis. New Phytol 171:525–538.

Houliston GJ, HM Chapman 2004 Reproductive strategy and population variability in the facultative apomict *Hieracium pilosella* (Asteraceae). Am J Bot 91:37–44.

Hurst LD, JR Peck 1996 Recent advances in understanding the evolution and maintenance of sex. Trends Ecol Evol 11:46–52.

Joshi A, ME Moody 1995 Male gamete output of asexuals and the dynamics of populations polymorphic for reproductive mode. J Theor Biol 174:189–197.

——— 1998 The cost of sex revisited: effects of male gamete output of hermaphrodites that are asexual in their female capacity. J Theor Biol 195:533–542.

Judson OP 1995 Preserving genes: a model of the maintenance of genetic variation in a metapopulation under frequency-dependent selection. Genet Res 65:175–191.

Keightley PD, SP Otto 2006 Interference among deleterious mutations favours sex and recombination in finite populations. Nature 443:89–92.

Koltunow AM 1993 Apomixis: embryo sacs and embryos formed without meiosis or fertilization in ovules. Plant Cell 5:1425–1437.

Ladle RJ, RA Johnstone, OP Judson 1993 Coevolutionary dynamics of sex in a metapopulation: escaping the Red Queen. Proc R Soc B 253:155–160.

Lenormand T, SP Otto 2000 The evolution of recombination in a heterogeneous environment. Genetics 156:423–438.

Lokki J 1976 Genetic polymorphism and evolution in parthenogenetic animals. VII. The amount of heterozygosity in diploid populations. Hereditas 83:57–64.

Lynch M 1984 Destabilizing hybridization, general purpose genotypes and geographical parthenogenesis. Q Rev Biol 59:257–290.

Manning JT, DPE Dickson 1986 Asexual reproduction, polyploidy, and optimal mutation rates. J Theor Biol 118:485–489.

Martin G, SP Otto, T Lenormand 2006 Selection for recombination in structured populations. Genetics 172:593–609.

Mártonfiová L 2006 Possible pathways of the gene flow in *Taraxacum* sect. *Ruderalia*. Folia Geobot 41:183–201.

Matzk F, A Meister, I Schubert 2000 An efficient screen for reproductive pathways using mature seeds of monocots and dicots. Plant J 21:97–108.

Matzk F, S Prodanovic, H Baumlein, I Schubert 2004 The inheritance of apomixis in *Poa pratensis* confirms a five locus model with differences in gene expressivity and penetrance. Plant Cell 17:13–24.

Maynard Smith J 1978 The evolution of sex. Cambridge University Press, Cambridge. 222 pp.

Meirmans PG, HJCM Den Nijs, PH Van Tienderen 2006 Male sterility in triploid dandelions: asexual females vs. asexual hermaphrodites. Heredity 96:45–52.

Meirmans PG, EC Vlot, JC Den Nijs, SB Menken 2003 Spatial ecological and genetic structure of a mixed population of sexual diploid and apomictic triploid dandelions. J Evol Biol 16:343–352.

Meyers LA, DA Levin 2006 On the abundance of polyploids in flowering plants. Evolution 60:1198–1206.

Michiels A, W van den Ende, M Tucker, L Van Riet, A Van Laere 2003 Extraction of high-quality genomic DNA from latex-containing plants. Anal Biochem 315:85–89.

Mogie M 1992 The evolution of asexual reproduction. Chapman & Hall, London. 276 pp.

Morgan TH 1913 Heredity and sex. Columbia University Press, New York. 282 pp.

Muller HJ 1932 Some genetic aspects of sex. Am Nat 66:118–138.

Muniyamma M, JB Phipps 1979a Cytological proof of apomixis in *Crataegus* (Rosaceae). Am J Bot 66:149–155.

——— 1979b Meiosis and polyploidy in Ontario species of *Crataegus* in relation to their systematics. Can J Genet Cytol 21:231–241.

——— 1984a Studies in *Crataegus*. 10. A note on the occurrence of diplospory in *Crataegus dissona* Sarg. (Maloideae, Rosaceae). Can J Genet Cytol 26:249–252.

——— 1984b Studies in *Crataegus*. 11. Further cytological evidence for the occurrence of apomixis in North American hawthorns. Can J Bot 62:2316–2324.

Naumova T 1993 Apomixis in angiosperms: nucellar and integumentary embryony. CRC, Boca Raton, FL. 144 pp.

Nogler GA 1984 Gametophytic apomixis. Pages 475–518 *in* BM Johri, ed. Embryology of angiosperms. Springer, Berlin.

Noirot M, D Couvet, S Hamon 1997 Main role of self-pollination rate on reproductive allocations in pseudogamous apomicts. Theor Appl Genet 95:479–483.

Noyes RD, JR Allison 2005 Cytology, ovule development, and pollen quality in sexual *Erigeron strigosus* (Asteraceae). Int J Plant Sci 166:49–59.

Noyes RD, R Baker, B Mai 2007 Mendelian segregation for two-factor apomixis in *Erigeron anuus* (Asteraceae). Heredity 98:92–98.

Noyes RD, LH Rieseberg 2000 Two independent loci control agamospermy (apomixis) in the triploid flowering plant *Erigeron annuus*. Genetics 155:379–390.

Nygren A 1967 Apomixis in the angiosperms. Handb Pflanzenphysiol 18:551–596.

O'Connell LM, CG Eckert 1999 Differentiation in sexuality among populations of *Antennaria parlinii* (Asteraceae). Int J Plant Sci 160:567–575.

——— 2001 Differentiation in reproductive strategy between sexual and asexual populations of *Antennaria parlinii* (Asteraceae). Evol Ecol Res 3:311–330.

Okabe S 1930 Über parthenogenesis bei *Houttuynia cordata*. Jpn J Genet 6:14–19.

Otto SP, AC Gerstein 2006 Why have sex? the population genetics of sex and recombination. Biochem Soc Trans 34:519–522.

Otto SP, J Whitton 2000 Polyploidy incidence and evolution. Annu Rev Genet 34:401–437.

Ozias-Akins P 2006 Apomixis: developmental characteristics and genetics. Crit Rev Plant Sci 25:199–214.

Parachnowitsch AL, E Elle 2004 Variation in sex allocation and male-female trade-offs in six populations of *Collinsia parviflora* (Scrophulariaceae s.l.). Am J Bot 91:1200–1207.

Peck JR 1994 A ruby in the rubbish: beneficial mutations, deleterious mutations and the evolution of sex. Genetics 137:597–606.

Peck JR, D Waxman 2000 What's wrong with a little sex? J Evol Biol 13:63–69.

Peck JR, JM Yearsley, D Waxman 1998 Explaining the geographic distributions of sexual and asexual populations. Nature 391:889–892.

Peters AD, CM Lively 2007 Short- and long-term benefits and detriments to recombination under antagonistic coevolution. J Evol Biol 20:1206–1217.

Potter D, T Eriksson, RC Evans, S Oh, JEE Smedmark, DR Morgan, M Kerr, et al 2007 Phylogeny and classification of Rosaceae. Plant Syst Evol 266:5–43, doi:10.1007/s00606-007-0539-9.

Rambaut A 1996 Se-Al: sequence alignment editor. http://evolve.zoo.ox.ac.uk.

Ramsey J, DW Schemske 1998 Pathways, mechanisms, and rates of polyploid formation in flowering plants. Annu Rev Ecol Syst 29:467–501.

Reuther W, LD Batchelor, HJ Webber, eds 1968 The citrus industry. Vol 2. Anatomy, physiology, genetics, and reproduction. 2nd ed. University of California Press, Berkeley.

Richards AJ 1986 Plant breeding systems. Unwin Hyman, London. 529 pp.

——— 2003 Apomixis in flowering plants: an overview. Philos Trans R Soc B 358:1085–1093.

Ritland C, K Ritland 1989 Variation of sex allocation among eight taxa of the *Mimulus guttatus* species complex (Scrophulariaceae). Am J Bot 76:1731–1739.

Roche D, WW Hanna, P Ozias-Akins 2001 Is supernumerary chromatin involved in gametophytic apomixis in polyploidy plants? Sex Plant Reprod 13:343–349.

Roy BA 1995 The breeding system of six species of *Arabis* (Brassicaceae). Am J Bot 82:869–877.

Roy BA, LH Rieseberg 1989 Apomixis in *Arabis holboellii* Hornemann. Heredity 80:506–509.

Salathé M, R Salathé, P Schmid-Hempel, S Bonhoeffer 2006 Mutation accumulation in space and the maintenance of sexual reproduction. Ecol Lett 9:941–946.

Schoen DJ 2005 Deleterious mutation in related species of the plant genus *Amsinckia* with contrasting mating systems. Evolution 59:2370–2377.

Schön I, K Martens, KV Doninck, RK Butlin 2003 Intraclonal genetic variation: ecological and evolutionary aspects. Biol J Linn Soc 79:93–100.

Schranz EM, L Kantama, H de Jong, T Mitchell-Olds 2006 Asexual reproduction in a close relative of *Arabidopsis*: a genetic investigation of apomixis in *Boechera* (Brassicaceae). New Phytol 171:425–438.

Sharbel TF, T Mitchell-Olds 2001 Recurrent polyploid origins and chloroplast phylogeny in the *Arabis holboellii* complex (Brassicaceae). Heredity 87:59–68.

Shaw J, EB Lickey, JT Beck, SB Farmer, W Liu, J Miller, KC Siripun, et al 2005 The tortoise and the hare. II. Relative utility of 21 noncoding chloroplast DNA sequences for phylogenetic analysis. Am J Bot 92:142–166.

Silvertown J 2008 The evolutionary maintenance of sexual reproduction: evidence from the ecological distribution of asexual reproduction in clonal plants. Int J Plant Sci 169:157–168.

Smith J 1841 Notice of a plant which produces perfect seeds without any apparent action of pollen. Trans Linn Soc Lond 18:509–512.

Soltis DE, PM Soltis, DW Schemske, JF Hancock, JN Thompson, BC Husband, WS Judd 2007 Autopolyploidy in angiosperms: have we grossly underestimated the number of species? Taxon 56:13–30.

Stebbins GL 1950 Variation and evolution in plants. Columbia University Press, New York. 643 pp.

Swofford DL 2003 PAUP*: phylogenetic analysis using parsimony (*and other methods). Version 4.0b 10. Sinauer, Sunderland, MA.

Talent N, TA Dickinson 2005 Polyploidy in *Crataegus* and *Mespilus* (Rosaceae, Maloideae): evolutionary inferences from flow cytometry of nuclear DNA amounts. Can J Bot 83:1268–1304.

——— 2007 Endosperm formation in aposporous *Crataegus* (Rosaceae, Spiraedoideae, tribe Pyreae): parallels to Ranunculaceae and Poaceae. New Phytol 173:231–249.

Tas ICQ, PJ Van Dijk 1999 Crosses between sexual and apomictic dandelions (*Taraxacum*). I. The inheritance of apomixis. Heredity 83:707–714.

Thompson SL, G Choe, K Ritland, J Whitton 2008 Cryptic sex within male-sterile polyploid populations of the Easter daisy, *Townsendia hookeri*. Int J Plant Sci 169:183–193.

Thompson SL, K Ritland 2006 A novel mating system analysis for modes of self-oriented mating applied to diploid and polyploid arctic Easter daisies (*Townsendia hookeri*). Heredity 97:119–126.

Thompson SL, J Whitton 2006 Patterns of recurrent evolution and geographic parthenogenesis within apomictic polyploid Easter daisies (*Townsendia hookeri*). Mol Ecol 15:3389–3400.

Uyenoyama MK 1984 On the evolution of parthenogenesis: a genetic representation of the "cost of meiosis." Evolution 38:87–102.

Vamosi JC, TM Knight, JA Streets, SJ Mazer, M Burd, TL Ashman 2006 Pollination decays in biodiversity hotspots. Proc Natl Acad Sci USA 103:956–1061.

Vandel A 1928 La parthénogenese géographique: contribution a l'étude biologique et cytologique de la parthénogenese naturelle. Bull Biol Fr Belg 62:164–182.

Van der Hulst RGM, THM Mes, JCM den Nijs, K Bachmann 2000 Amplified fragment length polymorphism reveal that population structure of triploid dandelions (*Taraxacum officinale*) exhibits both clonality and recombination. Mol Ecol 9:1–8.

Van der Hulst RGM, THM Mes, M Falque, P Stam, JCM Den Nijs, K Bachmann 2003 Genetic structure of a population sample of apomictic dandelions. Heredity 90:326–335.

van Dijk PJ, ICQ Tas, M Falque, T Barx-Schotman 1999 Crosses between sexual and apomictic dandelions (*Taraxacum*). II. The breakdown of apomixis. Heredity 83:715–721.

Verduijn MH, P van Dijk, JMM van Damme 2004 Distribution, demography and phenology of sympatric sexual and asexual dandelions (*Taraxacum officinale* L.): geographic parthenogenesis on a small scale. Biol J Linn Soc 82:205–218.

Whitton J 1994 Systematic and evolutionary investigation of the North American *Crepis* agamic complex. PhD diss. University of Connecticut, Storrs.

Int. J. Plant Sci. 169(1):183–193. 2008.
1058-5893/2008/16901-0015$15.00   DOI: 10.1086/523363

CRYPTIC SEX WITHIN MALE-STERILE POLYPLOID POPULATIONS OF THE EASTER DAISY, *TOWNSENDIA HOOKERI*

Stacey Lee Thompson,* Gina Choe,[†] Kermit Ritland,[‡] and Jeannette Whitton[†]

*Institut de Recherche en Biologie Végétale, Université de Montréal, 4101 Sherbrooke Est, Montréal, Québec H1X 2B2, Canada; †Department of Botany, University of British Columbia, Vancouver, British Columbia V6T 1Z4, Canada; and ‡Department of Forest Sciences, University of British Columbia, Vancouver, British Columbia V6T 1Z4, Canada

After a transition from sexuality to asexuality, the evolutionary dynamics in apomictic lineages will largely depend on the frequency of recombination. We evaluated the presence and extent of asexuality and recombination within populations of the Easter daisy, *Townsendia hookeri*, from the Yukon Territory, Canada. Amplified fragment-length polymorphism (AFLP) fingerprints were used to genotype 78 individuals from four populations. Multilocus AFLP genotypes from each population were subjected to four tests for deviations from free recombination among loci, and the long-term frequency of sexuality was estimated for each population with a novel procedure. In addition, a sample of individuals was surveyed for genome size using flow cytometry, and pollen was assayed for male fertility. One male-fertile, diploid population showed evidence of rampant recombination. Two male-sterile populations (i.e., with aborted anthers) were tetraploid and asexual. The remaining population was male-sterile and included both triploids and tetraploids. Evidence of both sexuality and asexuality was uncovered in this mixed-ploidy population, at an equilibrium rate of approximately three sexual events every two generations. The presence and extent of sexuality differed with ploidy, while cryptic sex was uncovered within a morphologically asexual population, thus reinforcing the power of genome surveys to assess reproductive dynamics at the limit of a plant's geographical range.

Keywords: agamospermy, apomixis, asexual reproduction, polyploidy, recombination, sexuality rate, *Townsendia hookeri*.

Introduction

Stolons, runners, rhizomes, tubers, bulbils, corms, layering, fragmentation, and apomixis are among the myriad alternative modes of asexual reproduction exhibited among angiosperms (Richards 1997). While many perennial plants adopt a mixture of sexual and asexual reproduction, in some cases asexuality predominates, with recombination and segregation occurring infrequently. In extreme cases, particularly at the margins of a range, some populations appear to abandon sexual reproduction for good (reviewed by Eckert [2002]). Observations on the potential fixation of sterility mutations in *Decodon verticillatus* prompted Eckert (2002) to invoke the "use-it-or-lose-it" hypothesis for the loss of sex in clonal plants. This vestigialization of sexual processes is hypothesized to occur once sex no longer confers adaptive benefits to an otherwise asexual organism and may be selectively neutral (e.g., Eckert et al. 1999) or advantageous (Dorken et al. 2004). Eckert (2002) hypothesized that near the limits of species' ranges, factors that severely limit sexual recruitment (e.g., lack of pollinators, short growing season) may tip the balance that would otherwise favor retention of sexuality, facilitating vestigialization of sexual traits in asexual plants

growing in these environments. Eckert's hypothesis was developed with vegetative asexuality (henceforth referred to as "clonality") in mind, but vestigialization of male function may also follow a transition to apomixis (more precisely, gametophytic agamospermy, the production of genetically identical clones through seed; see Whitton et al. 2008), if male function no longer contributes to fitness.

While the coexistence and interplay of sexual and clonal propagation has been part of the theoretical framework surrounding vegetative asexuality for some time (Harper 1977; Silander 1985), organisms that reproduce through apomixis have until recently been treated as exhibiting one fixed mode of reproduction (Asker and Jerling 1992; but see van der Hulst et al. 2003; Chapman et al. 2004). Most empirical studies of asexual organisms demonstrate that levels of genotypic diversity in standing populations are on par with those found in sexually reproducing relatives (Parker 1979; Ellstrand and Roose 1987; Hamrick and Godt 1989; Widén et al. 1994; Loxdale and Lushai 2003). It may no longer be surprising to find high levels of genotypic variation in predominantly asexual populations; however, the identification of sources of variation presents a more significant challenge (see Silvertown 2008). Genotypic diversity in asexual populations can have many origins, including somatic mutation, multiple origins of clones from sexuals, sporadic episodes of meiotic and mitotic recombination, and migration among genetically differentiated populations (van der Hulst et al. 2000). Several approaches have been developed to distinguish evidence for recombination, sexuality's hallmark, from patterns that are consistent with mutation accumulation in asexual lines (note that neither multiple origins nor migration can be distinguished with these methods). As the presence of even limited recombination is expected to affect the amount of genotypic diversity and the pace of adaptation within asexual lineages (Goddard et al. 2005), an understanding of the sources of variation in asexual lineages is critical to understanding the long- and short-term effects of evolutionary transitions to a primarily asexual lifestyle.

Among autonomous apomicts (those that no longer require pollen for endosperm or embryo initiation), many are known to undergo abnormalities in microsporogenesis leading to reduced male fertility (e.g., Beaman 1957; Smith 1963; Rozenblum et al. 1988; Meirmans et al. 2006), and thus it appears that in the absence of selection to maintain male fertility, plants that no longer "use it" can indeed "lose it." In more obvious cases, the presence of obligate asexuality can be inferred directly, for example, by determining that populations lack males in dioecious species or are missing various reproductive structures. However, males may be produced at excessively low frequencies ("spanandric" males; e.g., Schaefer et al. 2006), or sexual reproductive features may be produced only under specific ecological conditions (e.g., Reddy et al. 2000; Liu et al. 2001; Xing et al. 2003), rendering searches for evidence of sex unfruitful. It is indeed difficult to conclusively demonstrate that an organism has given up its former sexual strategy in favor of obligate asexuality, as the diagnosis of a lack of sex typically relies on negative evidence (but see Mark Welch and Meselson 2000; Mark Welch et al. 2004). However, even when sex is cryptic or infrequent, the signature of a rare round of recombination may linger for generations in the genomes of a largely asexual population (Burt et al. 1996; Gandolfi et al. 2003), making analyses of standing genotypic variation reliable and powerful for demonstrating that sex has indeed been lost completely.

Previous surveys of the Easter daisy, *Townsendia hookeri*, suggest that male traits, specifically pollen fertility, degenerate in polyploid populations (Thompson and Ritland 2006; Thompson and Whitton 2006). Embryological studies of a number of *Townsendia* species, including the Easter daisy, found all *Townsendia* polyploids to be obligately apomictic with autonomous endosperm formation (i.e., pollen is not required; Beaman 1954, 1957). More recent genetic analyses confirm that polyploids reproduce apomictically (at a mean population rate of 69%) and indicate that diploids do likewise (68%); however, their degree of outcrossing clearly varies (Thompson and Ritland 2006; also see Husband et al. 2008). The rare Easter daisy populations in the Yukon Territory, Canada (Douglas et al. 1981), represent the extreme northern limit of the species range (from Colorado northward throughout the montane regions of North America) and are highly disjunct from more southerly populations in Jasper, Alberta, more than 1200 km to the south. Rangewide surveys of pollen size and stainability (which has been shown to correlate strongly with viability and hence male fertility; Mayer 1991) indicate that diploid populations make small, viable pollen grains (rangewide mean viability = 73%, mean size = 20 μm; Thompson and Whitton 2006). Two of the six Yukon populations that have been examined also make small, viable pollen

(mean viability = 81%, mean size = 20.4 μm) and were therefore inferred to be diploid, while the remaining Yukon populations were morphologically male-sterile, aborting their anthers early in development (Thompson and Ritland 2006; Thompson and Whitton 2006). These male-sterile populations were hypothesized to be apomictic polyploids. Polyploid Easter daisies exhibit a cline in male fertility across the southern part of the range (range = 4.7%–42.3%, mean 17%; Thompson and Whitton 2006). The lack of functional male sex organs suggests that these populations are most likely asexual and have indeed lost their capacity for sexual reproduction, consistent with the prediction of the use-it-or-lose-it hypothesis.

In this study, we examine three male-sterile populations (described by Thompson and Ritland [2006]) and a fourth, male-fertile population of the Easter daisy from the Yukon Territory for genome size variation. We test for sexual and asexual reproduction in these populations and also estimate the long-term frequency of sexual events, using a newly developed procedure (Thompson 2007). Our expectation is that male-fertile populations are diploid and frequently undergo recombination, while male-sterile populations are tetraploid and asexual. In summarizing our findings, we evaluate alternative sources of genotypic variation and integrate our results with the broader issue of reproductive dynamics at the species' margin.

Material and Methods

Collections

Plant material was collected from four Yukon populations of *Townsendia hookeri*: populations 66 (Mile Thirteen Alaska Highway: 60°59′N, 135°10′W), 67 (Conglomerate Mountain: 61°37′N, 135°52′W), 68 (Tantalus Butte: 62°07′N, 136°15′W), and 71 (Tachäl Dhäl Kluane: 61°00′N, 138°32′W). As plants occur exclusively on disjunct substrate-exposed hilltops in the Yukon Territory, the geographic limits to Easter daisy populations are well defined. On each hilltop, individuals were sampled as evenly as possible to roughly maximize the distance among sampled individuals (typically several meters apart). Young leaves were frozen immediately in liquid nitrogen, then stored at −80°C for subsequent DNA extraction. Seeds were collected from dried flower heads to provide fresh seedling material for flow cytometry. Additional plants were dried for pollen examination and for voucher specimens, deposited at the University of British Columbia Herbarium (UBC).

Flow Cytometry

Seeds were germinated on moist filter paper in petri dishes, then transferred to seedling trays and allowed to grow for 6–8 wk. Ca. 40 mg of young leaf material was prepared for flow cytometry, according to the following protocol, optimized for *Chamerion angustifolium* (P. Kron and B. Husband, personal communication). Leaf samples from progeny of six mother plants per population were chopped in petri dishes over ice with a razor blade in 1 mL of the chopping buffer of Bino et al. (1992) with the modifications of Dart et al. (2004). Another 0.5 mL of buffer was added to the slurry, and the sample was mixed by pipetting and then filtered through a 30-μm CellTrics filter (Partec, Münster, Germany). The effluent was centrifuged at 13,000 rpm for 10 s and the supernatant removed. The pellet was resuspended in 0.3 mL of 1.12-mg/mL propidium iodide staining buffer (Arumuganathan and Earle 1991) and stained in the dark for ca. 45 min. *Pisum sativum* cv. Minerva Maple was used as an internal standard for calibration based on its established DNA content of 9.56 pg/2C nucleus (Johnson et al. 1999). Flow cytometry was performed on a BD FACS Scan benchtop analyzer (BD Biosciences, Mississauga, Canada) at the UBC Multi-User Flow Cytometry Facility (Biomedical Research Centre, University of British Columbia) according to their internal protocols. Peak means and variances were calculated from 5000 fluorescence events using FloJo Software (TreeStar, Corvallis, OR). Mean genome size values from individuals were subjected to K-means clustering according to the second-order Calinski-Harabasz criterion (Legendre and Legendre 1998). Means and quartiles for DNA content were also determined for populations, with significant differences among populations evaluated by the post hoc Tukey-Kramer HSD test in JMP IN 4.0.4 (SAS Institute, Cary, NC).

DNA Extraction

Genomic DNA was isolated from 20 plants per population by the CTAB (hexadecyltrimethylammonium bromide) protocol (Doyle and Doyle 1987) from ca. 0.1 g of leaf tissue, with the following modifications: volumes were reduced for extraction in 1.5-mL microfuge tubes, and sodium metabisulfite (1% w/v) was added to the isolation buffer. Isolated DNA was quantified by a Hoefer DyNAQuant 200 fluorometer (San Francisco) according to the manufacturer's directions, diluted to 10 ng/μL, and then used directly in PCR amplifications.

Amplified Fragment-Length Polymorphisms (AFLPs)

All individuals were AFLP fingerprinted with the method of Vos et al. (1995), with modifications developed by R. D. Noyes (Noyes and Rieseberg 2000). Denatured reaction products were run on an Applied Biosystems 377 DNA sequencer (Foster City, CA) at the University of British Columbia's Nucleic Acid and Protein Sequencing Unit, according to their internal protocols. Eight primer combinations (all permutations of Eco+[ACC/ACG] and Mse+[AGC/ACG/ACC/AAC]) were screened twice for three individuals from each of the four populations; then, two primer combinations were selected according to the optimality criteria of consistency across all individuals, repeatability for each individual, and level of polymorphism. All individuals were fingerprinted using EcoACC/MseAGC and EcoACG/MseAGC primers and then scored independently by the first and second authors for band presence/band absence using Genographer 1.6.0 (Benham 2001). Loci were considered reliable and subsequently analyzed only when scoring results for both researchers were consistent.

Tests for Genotypic Overrepresentation

The excess of a particular multilocus genotype among individuals within a population provides robust evidence of asexual reproduction (Tibayrenc et al. 1991). To test for this excess, we examined departures from a binomial distribution of genotypes, as expected under sexual reproduction. The probability of obtaining at least the observed number of individuals with a given identical multilocus genotype, P_{sex}, was calculated as

$$P_{sex} = \sum_{i=m}^{n} \frac{n!}{m!(n-m)!} g^m (1-g)^{n-m},$$

where g is the probability of the multilocus genotype under free recombination, m is the observed number of individuals with the identical multilocus genotype, and n is the total number of sampled individuals. Only one sample of each repeated multilocus genotype was included in the determination of band frequencies for this calculation. The significance of this overrepresentation was determined through comparison with a randomized distribution of P_{sex} values generated through simulation of 10,000 replicate populations under free recombination among loci with MLGsim (Stenberg et al. 2003). A significant p value implies that a repeated multilocus genotype is statistically improbable to have been derived via sexual means and is therefore consistent with arising through asexual reproduction.

Phylogenetic Tests for Asexuality

Tree-building methods tested for the absence of recombination among loci. As recombination generates observed homoplasy (Maynard Smith and Smith 1998), trees will be poorly resolved under sexual reproduction, while strict asexuality will result in a fully resolved tree that is significantly shorter than a randomized distribution of trees (Burt et al. 1996). Phylogenetic analyses were performed on unordered band genotypes from each population with PAUP*, version 4.0b10 (Swofford 2000). Depending on the number of AFLP genotypes detected per population, exhaustive ($\leq$10) or heuristic (>10) searches generated maximum parsimony trees. For heuristic searches, starting trees were obtained by simple stepwise addition of taxa, and TBR branch swapping was used on best trees. Tree lengths, homoplasy indexes (HIs), and the number of equally parsimonious trees were noted. Strict consensus trees were constructed when multiple parsimony trees were returned. Length scores of maximum parsimony trees are not intended to be directly compared among populations; instead, these scores are compared with those obtained under a randomized data distribution for the given population (i.e., where recombination is permitted among loci) in order to test for a significant lack of recombination. This was achieved through permutation tail probability tests (Faith and Cranston 1991), based on 1000 replicates, performed for each population in PAUP*.

Compatibility Analyses

For dominant data, compatibility analyses assume that the presence of all four combinations of band presence/absence within a population, for a given pair of loci (deemed an "incompatibility") indicates recombination (Burt et al. 1996). The number of incompatibilities is then tallied for all exhaustive locus pairings. Because this observed count is dependent on the number of variable loci, counts are not directly comparable among populations. Instead, differences between the observed incompatibility counts and those obtained from replicate pseudopopulations generated through resampling (drawing with replacement from pools of observed band frequencies within the population, as expected under free recombination) ascertain whether the lack of incompatibility within each population is significant. Compatibility analyses using PICA 4.0 (Wilkinson 2001) were performed for each population, with monomorphic loci excluded and only one representative of each AFLP genotype included. Matrix incompatibility and permutation tail probabilities (PTPs) based on 10,000 random data permutations were determined for each population with the MATRIX program, following the approach adopted by van der Hulst et al. (2000) and Eckert et al. (2003), among others.

Tests for Linkage Disequilibrium

The association among band states across loci was determined for the populations, assuming that significant associations result from a lack of sex. The following measure of association, adopted for use with dominant markers, was used:

$$r^2 = \frac{\sum_{k=1}^{N} \sum_{i=1}^{l} \sum_{j=i+1}^{l} (b_{ik} - f_i)^2 (b_{jk} - f_j)^2}{\sum_{k=1}^{N} \sum_{i=1}^{l} \sum_{j=i+1}^{l} f_i(1 - f_i) f_j(1 - f_j)},$$

where f_i is the frequency of banded genotypes at locus i, f_j is the frequency of banded genotypes at locus j, b_{ik} is an indicator variable for the presence ($b_{ik} = 1$) or absence ($b_{ik} = 0$) of bands at locus i in individual k, and likewise for b_{jk}. Loci with extreme genotype frequencies (outside 0.01–0.99) were excluded. This formula is similar to Agapow and Burt's (2001) index of multilocus linkage association, $\bar{r}_d$, which they show to be independent of the number of loci compared. This measure is also analogous to the measure of normalized linkage disequilibrium of Hill and Robertson (1968), except that presence/absence of bands replaces observed presence/absence of alleles. Significance was determined by comparing the value of observed association with a distribution of 99 permutations of genotypes among individuals, sampling with replacement at each locus. The variance of the permuted values was used to estimate the standard error of the original estimate.

Estimation of the Frequency of Sex

When the number of sexual recombination events per generation is low, mutation can generate significant genotypic variation in asexual populations. Building on the approach of Burt et al. (1996), Thompson (2007) showed that, assuming equilibrium conditions, the single-locus and multilocus measures of clonal identity can be used to estimate the long-term frequencies of sexuality within populations. The estimate of the per-generation rate of sexuality s in a population of size N is computed as

$$Ns = \frac{1 - f - fn[(1 - g)/(2g - 1)]}{2f(1 - f)},$$

where f is the mean pairwise multilocus clonality identity, g is the pairwise clonal identity at single loci, and n is the number of markers evaluated (Thompson 2007). All loci were included in calculations, and standard errors of estimates were found by jackknifing across individuals.

Results

Flow Cytometry

Mean genome sizes of individuals fell into three distinct clusters of values. One cluster comprised samples from population 68, and the second comprised values obtained from populations 66, 67, and 71, while the third cluster of values was exclusively found in three individuals from population 71. Means and quartiles for DNA content by population, in addition to clustering results, are depicted in figure 1, indicating distinct differences in DNA content per 2C nucleus among Yukon populations (fig. 1; Tukey-Kramer HSD test: $\alpha = 0.05$). These clusters likely correspond to diploid, tetraploid, and triploid values, respectively, although a lack of strict linearity may be due to our calibration with only one external standard. Population 68 had a mean DNA content of 13.22 ± 0.37 (SE) pg/2C nucleus, and this mean differed significantly from the means of all other populations.

Populations 66 and 67 had mean genome sizes of 25.93 ± 0.51 and 25.68 ± 0.45 pg DNA/2C nucleus, respectively. These population means did not differ significantly from one another and represent an approximate twofold increase in DNA content relative to population 68, suggesting that these populations consist of tetraploid plants. However, in population 71 (mean 23.15 ± 1.01 pg DNA/2C nucleus), there are two discrete clusters of 2C values (fig. 1). If the two discrete clusters of values are separated into two categories, one category

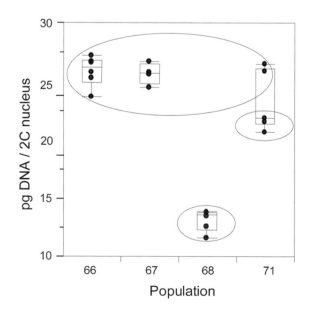

Fig. 1 DNA content per 2C nucleus as determined through flow cytometry for six individuals from each of four populations of the Easter daisy, *Townsendia hookeri*, from the Yukon Territory, Canada. Ovals indicate membership within the three ploidy clusters as indicated by K-means clustering. Box plots give the mean and quartiles for each population.

corresponds to approximately the tetraploid level (mean 25.45 ± 0.86 pg DNA/2C nucleus), and one corresponds to an approximately triploid value (mean 21.56 ± 0.29 pg DNA/2C nucleus). The mean of the tetraploid grouping does not differ significantly from the means from populations 66 and 67, while the triploid grouping mean differs significantly from those of the tetraploid grouping as well as all other population means. These are the first indications of triploidy in *T. hookeri*.

AFLP Fingerprinting

Out of hundreds of AFLP loci surveyed, only 58 loci were relatively consistent across individuals, repeatable for each individual, and variable among the four populations. Nevertheless, this number of loci is more than adequate to estimate mating system, as 18 is the minimum number of dominant-marker loci necessary to achieve robust estimates of outcrossing (Gaiotto et al. 1997). Thirty-two variable loci were scored using the primer combination EcoACC/MseAGC, and 26 variable loci were scored using the primer combination EcoACG/MseAGC. Two samples (one from population 67 and one from population 68) did not produce consistent amplifications and hence were excluded from further analyses. Within the 78 remaining samples, 39 multilocus genotypes were detected. Within population 66, five loci were polymorphic and six genotypes were detected; in population 67, 20 polymorphic loci defined eight genotypes; population 68 had 16 genotypes based on 18 loci, while population 71 had 13 genotypes and 27 polymorphic loci (table 1).

Overrepresentation of Genotypes

Overrepresentation statistics (table 2) indicate that all three of the repeated multilocus genotypes found in population 66, as well as the most common multilocus genotypes in populations 67 and 71, were significantly overrepresented. In addition, the most common multilocus genotypes from populations 66 and 67 were identical, further evidence of asexual reproduction. None of the repeated multilocus genotypes within population 68 was significantly overrepresented.

Parsimony Analyses

Maximum parsimony trees were constructed for each population (fig. 2). Analyses of populations 66 and 67 each resulted in one fully resolved tree (homoplasy index [HI] = 0),

of five and 20 steps, respectively. This is consistent with a lack of recombination, where genotypic variation can be attributed to mutation alone. However, PTP tests indicate that the tree length for population 66 does not differ significantly from that of a randomized distribution of trees; this lack of statistical support is likely due to the low level of genotypic variation detected within the population. Population 67, however, had a tree length that was significantly shorter than that of a randomized distribution of trees (PTP = 0.002), as expected under strict asexuality. In contrast, phylogenetic analysis of population 68 yielded 556 maximally parsimonious trees, with a length of 24 steps and an HI of 0.25. Population 71 was likewise rife with homoplasy (HI = 0.27), with 130 maximally parsimonious trees uncovered, with a tree length of 37 steps. For both populations 68 and 71, the length of optimal trees did not differ from that of a randomized distribution, as expected if sexual reproduction occurs within populations, and the comblike topology of the strict consensus of each set of poorly resolved trees can be observed (fig. 2).

Compatibility Analyses

Compatibility analyses (fig. 3) demonstrate that population 66 had a matrix incompatibility count of 0, although this result is nonsignificant, likely because of the paucity of polymorphic loci within the population. Population 67 likewise had a matrix incompatibility count of 0, and this was highly significant, indicating a lack of recombination within this population. Populations 68 and 71 had nonzero matrix incompatibility counts, with all four possible combinations of band presence and absence being detected for pairs of loci within populations. Permutation results of significance indicate that the genotypic structure of these populations is consistent with the occurrence of recombination and hence sex.

Associations among Loci

Associations of bands among loci, as measured by r^2 (table 1), indicate that population 67 had the strongest association among loci, with a value of 0.217, followed by population 71, which showed ca. 75% of the association of population 67. Population 66 showed the next-lowest association (0.097), but this value did not significantly differ from 0, probably because of the low number of polymorphic loci in this population, as stated above. Population 68 had the lowest estimate of association, at 0.073. Permutation tests

Table 1

Number of Multilocus Amplified Fragment-Length Polymorphism (AFLP) Genotypes Detected, the Number of Polymorphic AFLP Loci, Association Statistics, Measures of Clonal Identity, and Estimates of Effective Rates of Long-Term Sexuality (Ns) within Four Populations of the Easter Daisy (*Townsendia hookeri*) from the Yukon Territory, Canada

Population	Estimated ploidy	Sample size	Number of multilocus genotypes	Number of polymorphic loci, k	Association among loci, r^2 (SE)	Multilocus identity, f (SE)	Single-locus identity, g (SE)	Ns (SE)
66	4×	20	6	5	.097 (.029)	.368 (.028)	.015 (.001)	.645 (.096)
67	4×	19	8	20	.217 (.020)*	.327 (.031)	.046 (.004)	−.629 (.199)
68	2×	19	16	18	.073 (.009)*	.023 (.005)	.056 (.002)	20.950 (6.736)
71	3×, 4×	20	13	27	.157 (.012)*	.116 (.016)	.076 (.005)	1.478 (.637)

* Significant at $\alpha = 0.01$.

Table 2

Overrepresentation (P_{sex}) Statistics and Their
Significance for Four Populations of the
Easter Daisy (*Townsendia hookeri*) from
the Yukon Territory, Canada

Population, N	P_{sex}	p value
66:		
12	2.85×10^{-14}	.00004
3	1.77×10^{-3}	.047
2	1.83×10^{-6}	.00034
67:		
11	5.22×10^{-4}	.014
2	.292	.54
68:		
3	.330	.71
2	.162	.53
71:		
7	1.67×10^{-4}	.031
2	4.05×10^{-2}	.50

Note. N indicates the number of individuals with the identical multilocus genotype (for N > 1) observed within each population. Total sample sizes for populations are given in table 1.

indicate that estimates of association among loci were significant in populations 67, 68, and 71.

Estimation of the Frequency of Long-Term Recombination

Single-locus identities varied among populations by a factor of up to five (table 1), while multilocus identity was approximately an order of magnitude lower for the diploid population (68) than for the polyploid populations (66, 67, and 71). The resulting estimates of Ns (table 1) indicate that population 66 has a low rate of long-term sexuality, on the order of one sexual event every two generations in the entire population, while population 71 exhibits a higher degree of recombination across all loci, with estimates of three outcrossing events every two generations. For population 67, the estimated rate of long-term recombination was negative, indicating that equilibrium had not yet been reached. Population 68 had a significantly higher amount of sexuality, with more than 20 recombination events inferred in the population each generation, although it must be cautioned that the method used here produces the most reliable estimates when the rate of sexuality within the population is similar in order to the AFLP band mutation rate (Thompson 2007), a condition that was not met for this population (results not shown).

Discussion

Easter daisy populations in the Yukon Territory exhibit sexual reproduction, asexual reproduction, and a combination of the two. The extent of asexuality and recombination varies with population, male fertility, and ploidy level. Population 67 consisted of male-sterile tetraploid plants that are strictly asexual, despite much genotypic variability within this population (e.g., eight of 19, or 42% of sampled individuals, had unique genotypes). Population 66 showed evidence of asexuality (although this evidence was accompanied by

limited statistical power because of a paucity of variable loci) and a stable, low frequency of sexuality, on the order of one sexual mating every other generation in the entire population. The high-frequency genotypes observed within these two populations strongly suggest that entire genomes are being inherited as a unit (a consequence of asexual reproduction), as opposed to the independent allelic transmission expected when individuals reshuffle their genomes through sex. The case for an asexual origin of overrepresented genotypes becomes even stronger when these genotypes are shared among distinct localities, as found in populations 66 and 67. This is because the excessive genotypic frequencies were observed despite the differences in allele frequencies among replicate populations (Tibayrenc et al. 1991).

In strong contrast, population 68 consisted of diploid plants that made viable pollen at a rate of 67.67% (Thompson and Whitton 2006). Independent chromosome counts characterize population 68 as 2n=18 (Löve 1968; T. Mosquin, personal communication), the diploid number for *Townsendia* (Beaman 1954, 1957). Our tests confirm that sexual reproduction occurs in this population and estimate the equilibrium population rate of sexuality to be ca. 21 events per generation. Although the tests that we use in this study indicate that the signal of recombination among sampled genotypes is strong, they do not indicate that asexual reproduction never occurs in this population, as even very low levels of recombination lead to a detection of sexual reproduction. In addition to an outcrossing rate of 23%, progeny analyses from diploid population 68 indicated significant levels of apomictic reproduction (68%) and moderate but statistically insignificant levels of selfing and automixis (Thompson and Ritland 2006). Simulation studies indicate that an outcrossing rate of 23% is enough to cause a diploid population of size 100 with 50 dominant markers at a two-allele mutation rate of 0.0003 to report sexuality for the phylogenetic, compatibility, and overrepresentation tests used here (S. L. Thompson, unpublished data). The low levels of significant association among loci within population 68 are likely indicative of asexual reproduction; however, this linkage disequilibrium could also persist in the population because of the effects of selection, random drift within a small population, preequilibrium conditions, or biparental inbreeding.

A more complex situation exists within population 71, which exhibits evidence of asexuality, recombination, and ploidy-level variation. Ca. 100 plants examined over three field seasons were male-sterile, and flow cytometry, despite very low sample sizes, detected both tetraploid and triploid plants. Parsimony analyses yield a homoplastic tree, and compatibility analysis likewise suggests sexuality. By jackknifing genotypes and recalculating the matrix incompatibility (after removing the individuals that contributed to the incompatibility, as per van der Hulst et al. 2000), we identified four genotypes as recombinant within population 71 (data not shown). When these four individuals were removed from the parsimony analysis, a fully resolved tree resulted (data not shown), demonstrating that the remaining individuals within the population were consistent with the occurrence of asexual reproduction. This concurs with the overrepresentation and association results that implicate asexual reproduction. Estimates of long-term recombination rates indicate two

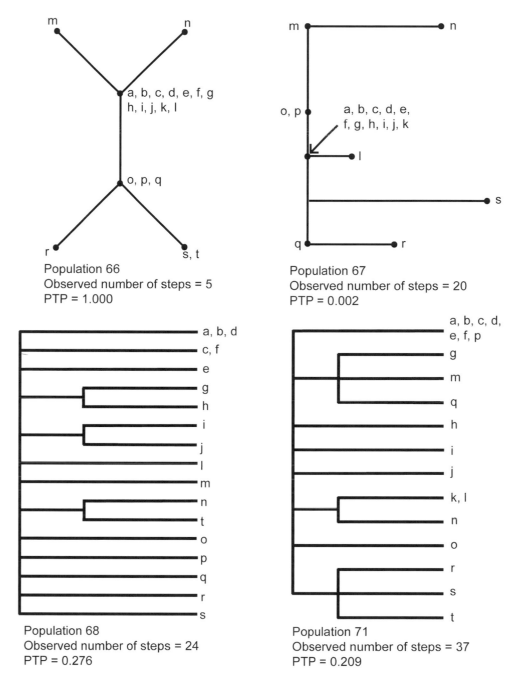

Fig. 2 Most parsimonious trees (populations 66 and 67) and strict consensus trees (populations 68 and 71) from Yukon populations of the Easter daisy, *Townsendia hookeri*, based on unordered amplified fragment length polymorphism genotypes. The observed number of steps in the most parsimonious tree(s) and permutation tail probabilities of the number of steps for a randomized distribution of trees are given.

sexual events every three generations. In addition, progeny arrays indicate moderate but significant levels of either selfing or automixis in this population (31%; Thompson and Ritland 2006). Taken together, these results portray a population with ploidy-level variation that undergoes both sexual and asexual reproduction. Residual sexuality appears to be below the threshold of direct observation, as pollen was not detected in our sampling. This emphasizes the power of surveying the genome with molecular markers for evidence of sexuality; marker

analyses indicate recombination, likely through automixis (Thompson and Ritland 2006), whereas morphological evidence alone would indicate asexuality.

Our analyses consider mutation and occasional bouts of sexual outcrossing as the determinants of genotypic variation within populations: several of these tests are likely to be biased in favor of detecting recombination (S. L. Thompson, unpublished data), and they represent a simplistic situation. Alternative sources of genotypic variation include multiple origins of clones from sexuals, migration, self-fertilization, automixis, and biparental inbreeding. Multiple origins are unlikely here, as a phylogenetic analysis of chloroplast DNA (cpDNA) sequences from across the entire range of Easter daisies indicates that the three polyploid populations examined in this study share a single haplotype (haplotype j; Thompson and Whitton 2006) and represent a single origin of polyploid asexuality from a haplotype in diploid population 68 (haplotype g; Thompson and Whitton 2006). Although one origin is a conservative estimate, there is no evidence suggesting that multiple origins of asexuality have occurred in the Yukon.

Migration among populations would affect our interpretation in at least two ways. First, if many genotypes have migrated among populations, the importance of mutation might be overstated. This is of minor importance in our data. We detected only one genotype that was common and shared between populations 66 and 67. Overall, patterns suggest that migration between populations is limited and thus does not contribute to overestimating the importance of mutation. Second, migration of sexual genotypes into asexual populations could also contribute to genotypic variation. A survey of chloroplast haplotype variation across the Yukon Territory showed that one of the putative recombinant genotypes in this population (as mentioned above) had a haplotype identical to that of the 20 individuals examined here from diploid population 68 (Thompson 2006), suggesting the possibility of seed flow between populations that vary in ploidy and mating system. We can speculate that the observed triploidy, previously unknown from this species, may have become established through hybridization between diploid and tetraploid cytotypes. If this is true, we might expect the occurrence of diploids within this population. As our sample sizes for flow cytometry were quite small for this population ($N = 6$), larger sampling of genome size values within this population would have to be performed in order to corroborate this. An alternate explanation for haplotype variation is incomplete lineage sorting of an ancestral cpDNA polymorphism. Although Easter daisy populations have been reported as being fixed for cpDNA haplotype (Thompson and Whitton 2006), deeper DNA sequencing reveals that variation indeed exists (six of 12 populations are polymorphic, based on a small sample size of $N = 3$ within each population; S. L. Thompson, unpublished data). This high degree of cpDNA polymorphism would have to be resolved before migration from a diploid population into population 71 can be inferred with any amount of confidence.

Other factors that could contribute to patterns of genotypic variation in standing populations are selfing, automixis, and biparental inbreeding. Joint estimates of the rates of outcrossing, selfing, automixis, and apomixis through genetic analyses of progeny suggest that automixis (specifically, recombination within the restitution nucleus during the abnormal first meiotic division of the megasporocyte found in *Ixeris*-type apomicts) may account for the low levels of recombination detected in populations 71 and 66 (Thompson and Ritland 2006). It should be noted that one limitation of the progeny results is that true self-fertilization cannot be distinguished with any confidence from biparental inbreeding (i.e., mating among identical clones) under the joint estimation procedure. The estimates of long-term sexuality rate given in this study allow for rare matings between clone mates (equivalent to biparental inbreeding, as opposed to strict self-fertilization), and our results suggest that sex between clone mates may occur infrequently within polyploid populations.

In addition to these biological sources of uncertainty, methodological limitations can also affect inferences of reproductive mode. As discussed by Eckert et al. (2003), dominance of marker polymorphisms can obscure minor genetic variation, as rare recessive alleles tend to be present only in heterozygotes and hence are not detected. Polyploidy further compounds this problem, as recessive alleles are detected only when completely homozygous, which can be quite rare for tetraploids (i.e., at frequency q^4 in a sexual tetraploid). Hence, for the same actual levels of genetic variation, levels of apparent sex should be generally less in tetraploids than in diploids when dominant markers are used. This is not a significant problem with our interpretations, as the presence of dominance would reduce the frequency at which all loci are recessive for a tetraploid, and thus our observation of sexual reproduction in population 71, consisting of triploids and tetraploids, is conservative and particularly noteworthy.

Our findings can be compared with the handful of other botanical studies that use an approach similar to our own on *Taraxacum* (Mes 1998; van der Hulst et al. 2000, 2003), *Allium vineale* (Ceplitis 2001), *Butomus umbellatus* (Eckert et al. 2003), *Hieracium* (Chapman et al. 2004), and *Saxifraga cernua* (Kjølner et al. 2006), all of which use complementary tests, as we do here, to increase confidence in conclusions. A common thread among these studies is that sexual reproduction was detected in polyploid populations that were initially believed to be strictly asexual. As noted by Thompson and Ritland (2006) and Thompson and Whitton (2006), when a novel polyploid is established from within a diploid population, a transition to asexuality would be favorable in the short term, as this would reduce the potential for matings with the diploid cytotype, thus decreasing the effects of the minority cytotype disadvantage (Levin 1975; Ramsey and Schemske 1998). When sex occurs at low frequencies, a polyploid organism has the benefits of asexuality from increased reproductive isolation plus the occasional benefits of sex, which include the production of new genotypic combinations, some of which may prove advantageous, as well as the benefit of purging their genomes of deleterious mutations by exposing them to selection over the long term (Muller 1964). The optimal relative frequencies of asexual and sexual reproduction that would promote neopolyploid establishment have yet to be determined.

Male fertility declines with increased latitude in polyploid Easter daisies throughout the range (Thompson 2006; Thompson and Whitton 2006), both with and without the inclusion of Yukon populations, in which anthers abort. It is unknown whether anther abortion in Yukon polyploid Easter daisies represents the extension of the clinal degradation observed in the rest of the range or an independent loss of male sexual function. In addition, it is unknown whether there is a genetic basis to the patterns in male fertility or whether it is solely environmentally determined, as observations in the Easter daisy are based on field conditions only. Tests of male fertility under a range of common-garden conditions are required in order to verify this.

Vestigialization of sexual function in asexual lineages is predicted to occur when sexual recruitment no longer

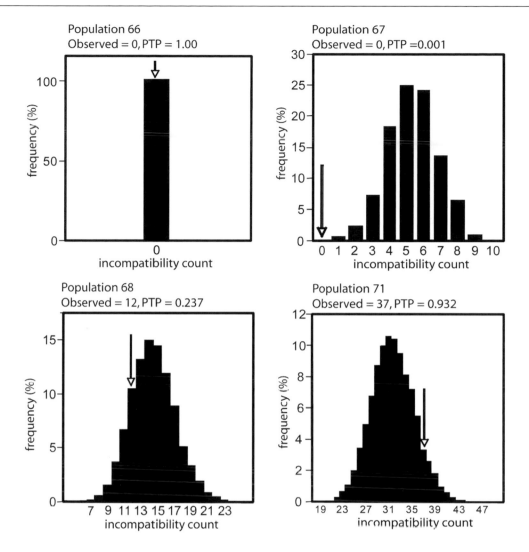

Fig. 3 Randomized incompatibility distributions of amplified fragment length polymorphism genotypes for four Yukon populations of the Easter daisy, *Townsendia hookeri*. Observed incompatibility counts are indicated by arrows, while permutation tail probabilities (PTPs) give levels of significance from 10,000 randomizations.

contributes to fitness (Eckert 2002). The lack of functional male sex organs in the Yukon could result from neutral processes or from antagonistic pleiotropic effects. Superficially, it might seem that male fitness should be neutral in autonomous apomicts like *Townsendia*, which do not require pollen to initiate embryo or endosperm development. However, vestigialization of pollen may be selectively favored if resources no longer expended on pollen are reallocated to other fitness components, such as apomictic seed or vegetative growth or performance. This could contribute to higher fitness in asexuals that could be favored at low densities (Peck et al. 1998) or to differential overwintering survival farther north, as has been suggested in sexually sterile clones of *Decodon verticillatus* (Dorken et al. 2004).

It should be emphasized that while surveys of standing variation such as ours may point to strictly asexual population structure, the above scenario suggests that this may result from more complex conditions than the straightforward fixation of asexual genotypes, as has been suggested for other forms of asexuality (e.g., Dorken et al. 2004). Lack of sex may, for example, reflect limited availability of pollen or limited pollinator activity rather than the complete loss of potential for recombination. For example, if a lack of pollinators in the north contributes to reduced sexual recruitment, male fertility may degenerate, further decreasing the potential for occasional sexual reproduction in apomicts and perhaps favoring increased expression of apomixis. Thus, populations that appear to be strictly asexual may achieve this state through the combined effects of decreased selection for male fertility and increased asexuality via female function.

We have demonstrated that genome surveys can elucidate recent contributions of sexual and asexual reproduction to patterns of standing variation in diploid and polyploid populations. We have found that while the accumulation of mutations within asexual lineages can account for most of the observed genotypic variation in apomictic populations, the signal of infrequent recombination can also be uncovered and its rate estimated. The detection of recombination in plant populations deemed asexual on morphological grounds suggests that predominantly asexual plants that appear to "lose it" can still manage to "use it."

Acknowledgments

Thanks to Mary Berbee, Anne Bruneau, Chris Eckert, Sally Otto, and anonymous reviewers for thoughtful comments on previous versions of the manuscript. Mariannick Archambault helped with graphics, while Bruce Bennett and Jesse Thompson assisted in the field. A collecting permit for Kluane National Park and a Scientists' and Explorers' License from the territorial government of the Yukon are both acknowledged. This research was funded by Natural Science and Engineering Research Council (NSERC) of Canada grants to all authors, as well as Northern Scientific Training Program grants to S. L. Thompson. S. L. Thompson received postdoctoral funding through a team Fonds Québécois de la Recherche sur la Nature et les Technologies (FQRNT) grant to Bernard Angers, Anne Bruneau, Luc Brouillet, Francois-Joseph Lapointe, and Pierre Legendre at the Université de Montréal.

Literature Cited

Agapow P-M, A Burt 2001 Indices of multilocus linkage disequilibrium. Mol Ecol Notes 1:101–102.

Arumuganathan K, ED Earle 1991 Estimation of DNA content of plants by flow cytometry. Plant Mol Biol Rep 9:229–233.

Asker SE, L Jerling 1992 Apomixis in plants. CRC, Boca Raton, FL. 298 pp.

Beaman JH 1954 Chromosome numbers, apomixis, and interspecific hybridization in the genus *Townsendia*. Madroño 12:169–180.

——— 1957 The systematics and evolution of *Townsendia* (Compositae). Contrib Gray Herb Harv Univ 183:1–151.

Benham JJ 2001 Genographer. http://hordeum.oscs.montana.edu/genographer.

Bino RJ, JN De Vries, HL Kraak, JG Van Pijlen 1992 Flow cytometric determination of nuclear replication stages in tomato seeds during priming and germination. Ann Bot 69:231–236.

Burt A, DA Carter, GL Koenig, TL White, JW Taylor 1996 Molecular markers reveal cryptic sex in the human pathogen *Coccidioides immitis*. Proc Natl Acad Sci USA 93:770–773.

Ceplitis A 2001 The importance of sexual and asexual reproduction in the recent evolution of *Allium vineale*. Evolution 58:1581–1591.

Chapman H, B Robson, ML Pearson 2004 Population genetic structure of a colonizing weed, *Hieracium lepidulum*. Heredity 92:182–188.

Dart S, P Kron, BK Mable 2004 Characterizing polyploidy in *Arabidopsis lyrata* using chromosome counts and flow cytometry. Can J Bot 82:185–197.

Dorken ME, KJ Neville, CG Eckert 2004 Evolutionary vestigialization of sex in a clonal plant: selection versus neutral mutation in geographically peripheral populations. Proc R Soc B 271:2375–2380.

Douglas GW, GW Argus, HL Dickson, DF Brunton 1981 The rare vascular plants of the Yukon. National Museum of Natural Sciences, Ottawa. 63 pp.

Doyle JJ, JD Doyle 1987 A rapid DNA isolation procedure for small quantities of fresh leaf tissue. Phytochem Bull 19:11–15.

Eckert CG 2002 The loss of sex in clonal plants. Evol Ecol 15:501–520.

Eckert CG, ME Dorken, SA Mitchell 1999 Loss of sex in clonal populations of a flowering plant, *Decodon verticillatus* (Lythraceae). Evolution 53:1079–1092.

Eckert CG, K Lui, K Bronson, P Corradini, A Bruneau 2003 Population genetic consequences of extreme variation in sexual and clonal reproduction in an aquatic plant. Mol Ecol 12:331–344.

Ellstrand NC, ML Roose 1987 Patterns of genotypic diversity in clonal plant species. Am J Bot 74:123–131.

Faith DP, PS Cranston 1991 Could a cladogram this short have arisen by chance alone?: on permutation tests for cladistic structure. Cladistics 7:1–28.

Gaiotto FA, M Bramucci, D Grattapaglia 1997 Estimation of outcrossing rate in a breeding population of *Eucalyptus urophylla*

with dominant RAPD and AFLP markers. Theor Appl Genet 95:842–849.

Gandolfi A, IR Sanders, V Rossi, P Menozzi 2003 Evidence of recombination in putative ancient asexuals. Mol Biol Evol 20:754–761.

Goddard MR, HCJ Godfray, A Burt 2005 Sex increases the efficacy of natural selection in experimental yeast populations. Nature 434:636–640.

Hamrick JL, MJW Godt 1989 Allozyme diversity in plant species. Pages 43–63 *in* AHD Brown, MT Clegg, AL Kahler, BS Weir, eds. Plant population genetics, breeding, and genetic resources. Sinauer, Sunderland, MA.

Harper JL 1977 Population biology of plants. Academic Press, London.

Hill WG, A Robertson 1968 Linkage disequilibrium in finite populations. Theor Appl Genet 38:226–231.

Husband BC, B Ozimec, SL Martin, L Pollock 2008 Mating consequences of polyploid evolution in flowering plants: current trends and insights from synthetic polyploids. Int J Plant Sci 169:195–206.

Johnson JS, MD Bennett, AL Rayburn, DW Galbraith, HJ Price 1999 Reference standards for determination of DNA content of plant nuclei. Am J Bot 86:609–613.

Kjølner S, SM Såstad, C Brochmann 2006 Clonality and recombination in the arctic plant *Saxifraga cernua*. Bot J Linn Soc 152:209–217.

Legendre P, L Legendre 1998 Numerical ecology. Elsevier, Amsterdam.

Levin DA 1975 Minority cytotype exclusion in local plant populations. Taxon 24:35–43.

Liu N, Y Shan, F Wang, C Xu, K Peng, X Li, Q Zhang 2001 Identification of an 85-kb DNA fragment containing *pms*1, a locus for photoperiod-sensitive genic male sterility in rice. Mol Genet Genomics 266:271–275.

Löve A 1968 IOPB chromosome number reports. XV. Taxon 17:91–104.

Loxdale HD, G Lushai 2003 Rapid changes in clonal lines: the death of a "sacred cow." Biol J Linn Soc 79:3–16.

Mark Welch DB, M Meselson 2000 Evidence for the evolution of bdelloid rotifers without sexual reproduction or genetic exchange. Science 288:1211–1215.

Mark Welch JL, DB Mark Welch, M Meselson 2004 Cytogenetic evidence for asexual evolution of bdelloid rotifers. Proc Natl Acad Sci USA 101:1618–1621.

Mayer SS 1991 Artificial hybridization in Hawaiian *Wikstroemia* (Thymelaeaceae). Am J Bot 78:122–130.

Maynard Smith J, NH Smith 1998 Detecting recombination from gene trees. Mol Biol Evol 15:590–599.

Meirmans PG, CM den Nijs, PH Van Tienderen 2006 Male sterility in triploid dandelions: asexual females vs. asexual hermaphrodites. Heredity 96:45–52.

Mes THM 1998 Character compatibility of molecular markers to distinguish asexual and sexual reproduction. Mol Ecol 7:1719–1727.

Muller HJ 1964 The relation of recombination to mutational advance. Mutat Res 1:2–9.

Noyes RD, Rieseberg LH 2000 Two independent loci control agamospermy (apomixis) in the triploid flowering plant *Erigeron annuus*. Genetics 155:379–390.

Parker ED 1979 Ecological implications of clonal diversity in parthenogenetic morphospecies. Am Zool 19:753–762.

Peck JR, JM Yearsley, D Waxman 1998 Explaining the geographic distributions of sexual and asexual populations. Nature 391: 889–892.

Ramsey J, DW Schemske 1998 Pathways, mechanisms, and rates of polyploid formation in polyploid plants. Annu Rev Ecol Syst 29: 467–501.

Reddy OUK, EA Siddiq, NP Sarma, J Ali, AJ Hussain, P Nimmakayala, P Ramasamy, S Pammi, AS Reddy 2000 Genetic analysis of temperature-sensitive male sterility in rice. Theor Appl Genet 100:794–801.

Richards AJ 1997 Plant breeding systems. 2nd ed. Chapman & Hall, London. 544 pp.

Rozenblum E, S Maldonado, CE Waisman 1988 Apomixis in *Eupatorium tanacetifolium* (Compositae). Am J Bot 75:311–322.

Schaefer I, K Domes, M Heethoff, K Schneider, I Schön, RA Norton, S Scheu, M Maraun 2006 No evidence for the "Meselson effect" in parthenogenetic oribatid mites (Oribatida, Acari). J Evol Biol 19: 184–193.

Silander JA 1985 Microevolution in clonal plants. Pages 107–152 *in* JBC Jackson, LW Buss, and RE Cook, eds. Population biology and evolution of clonal organisms. Yale University Press, New Haven, CT.

Silvertown J 2008 The evolutionary maintenance of sexual reproduction: evidence from the ecological distribution of asexual reproduction in clonal plants. Int J Plant Sci 169:157–168.

Smith G 1963 Studies in *Potentilla* L. I. Embryological investigations into the mechanism of agamospermy in British *P. tebernaemontani* Aschers. New Phytol 62:264–282.

Stenberg P, M Lundmark, A Saura 2003 MLGsim: a program for detecting clones using a simulation approach. Mol Ecol Notes 3: 329–331.

Swofford DL 2000 PAUP*: phylogenetic analysis using parsimony (*and other methods). Sinauer, Sunderland, MA.

Thompson SL 2006 Genetic structure and mating patterns of diploid and polyploid Easter daisies (*Townsendia hookeri*, Asteraceae). PhD diss. Unversity of British Columbia, Vancouver.

——— 2007 A simple procedure for joint estimation of the long-term rates of sexuality and mutation in predominantly clonal populations, for use with dominant molecular markers. Mol Ecol Notes 7:567–569.

Thompson SL, K Ritland 2006 A novel mating system analysis for modes of self-oriented mating applied to diploid and polyploid arctic Easter daisies (*Townsendia hookeri*). Heredity 97:119–126.

Thompson SL, J Whitton 2006 Patterns of recurrent evolution and geographic parthenogenesis within apomictic polyploid Easter daisies (*Townsendia hookeri*). Mol Ecol 15:3389–3400.

Tibayrenc M, F Kjellberg, J Arnaud, B Oury, SF Brenière, ML Dardé, FJ Ayala 1991 Are eukaryotic microorganisms clonal or sexual? a population genetics vantage. Proc Natl Acad Sci USA 88: 5129–5133.

van der Hulst RGM, THM Mes, JCM den Nijs, K Bachmann 2000 Amplified fragment length polymorphism (AFLP) markers reveal that population structure of triploid dandelions (*Taraxacum officinale*) exhibits both clonality and recombination. Mol Ecol 9:1–8.

van der Hulst RGM, THM Mes, M Falque, P Stam, JCM den Nijs, K Bachmann 2003 Genetic structure of a population sample of apomictic dandelions. Heredity 90:326–335.

Vos P, R Hogers, M Bleeker, M Reijans, T van de Lee 1995 AFLP: a new technique for DNA fingerprinting. Nucleic Acids Res 23: 4407–4414.

Whitton J, CJ Sears, EJ Baack, SP Otto 2008 The dynamic nature of apomixis in the angiosperms. Int J Plant Sci 169:169–182.

Widén B, N Cronberg, M Widén 1994 Genotypic diversity, molecular markers and spatial distribution of genets in clonal plants: a literature survey. Folia Geobot Phytotaxon 29:245–263.

Wilkinson M 2001 PICA 4.0: software and documentation. Department of Zoology, Natural History Museum, London.

Xing QH, ZG Ru, CJ Zhou, X Xue, CY Liang, DE Yang, DM Jin, B Wang 2003 Genetic analysis, molecular tagging and mapping of the thermo-sensitive genic male-sterile gene (*wtms1*) in wheat. Theor Appl Genet 107:1500–1504.

Int. J. Plant Sci. 169(1):195–206. 2008.
1058-5893/2008/16901-0016$15.00   DOI: 10.1086/523367

MATING CONSEQUENCES OF POLYPLOID EVOLUTION IN FLOWERING PLANTS: CURRENT TRENDS AND INSIGHTS FROM SYNTHETIC POLYPLOIDS

Brian C. Husband, Barbara Ozimec, Sara L. Martin, and Lisa Pollock

Department of Integrative Biology, University of Guelph, Ontario N1G 2W1, Canada

The evolutionary transition from diploidy to polyploidy is prevalent in flowering plants and may result in correlated changes in mating system (outcrossing rate). Most theory predicts a shift toward self-fertilization (decrease in outcrossing) in polyploids, but empirical evidence for this pattern and its underlying mechanisms is inconclusive or restricted to a few cases. In an analysis of variation in outcrossing rates among diploid-polyploid species pairs from the literature, polyploids had lower outcrossing rates (higher selfing; $t = 0.23$, $SE = 0.09$) than diploids ($t = 0.52$, $SE = 0.12$). Among polyploids, however, allopolyploids were predominantly selfing ($t = 0.20$, $SE = 0.099$), whereas autopolyploids had significantly higher outcrossing rates ($t = 0.64$, $SE = 0.087$), raising the question of what limits the evolution of selfing in autopolyploids. To address this, we examined the magnitude of inbreeding depression in synthetic polyploids of the plant *Chamerion angustifolium*. The intrinsic cost of selfing in newly formed polyploids was negligible compared with extant polyploids, thus promoting the spread of selfing. However, there was weak evidence that inbreeding depression increases with history of inbreeding, suggesting that the rise in selfing may be ephemeral and that selection ultimately favors mixed or outcrossed mating systems in autopolyploids. Such constraints on selfing have some theoretical support, but additional research on patterns of variation and genetic mechanisms governing polyploid mating systems are needed.

Keywords: allopolyploid, autopolyploid, *Chamerion angustifolium*, inbreeding depression, outcrossing rate, polyploidy.

Introduction

The evolutionary transition from diploidy to polyploidy by genome duplication is widespread among flowering plants and is associated with significant phenotypic effects (Levin 1983; Soltis and Soltis 1999; Otto and Whitton 2000; Levin 2002; Vamosi and Dickinson 2006). Based on chromosome number, between 57% and 70% of angiosperm species have been estimated to be polyploid (Grant 1963; Masterson 1994); this value may be closer to 100% if one considers the incidence of paleopolyploids detected through comparative genomics (Wolfe 2001; Cui et al. 2006). Phylogenetic evidence (Soltis and Soltis 1999) and studies of polyploid formation (Ramsey and Schemske 1998; Burton and Husband 2001; Husband 2004) suggest that polyploids have established repeatedly within species and that such duplications may arise within populations at rates comparable to or exceeding the genic mutation rate. These shifts in ploidy can have significant effects on morphological, physiological, and life-history traits, owing to direct effects on cell size, cell cycle, and development rate and indirect effects on heterozygosity and adaptive potential (Levin 1983, 2002; Otto and Whitton 2000). Because polyploidy affects the entire genome, it can influence many aspects of the phenotype simultaneously, perhaps more so than any other single genetic event.

The consequences of genome duplication for the evolution of reproductive systems in flowering plants have long been of interest to biologists and may serve as a general model for understanding the role of polyploidy in phenotypic diversification. An association between mating system, defined as the rate of cross-fertilization (t), and polyploidy was first described by Grant (1956), who identified a higher incidence of polyploidy in selfing lineages than outcrossing lineages of plants. Today, theoretical models are consistent in arguing for some effect of genome multiplication on mating systems, mostly toward increased self-fertilization, although predictions differ with respect to direction of effect and underlying mechanisms. The state of empirical evidence is similarly unresolved, with relatively few studies and no general consensus regarding the effect of genome multiplication on mating systems.

Here, we summarize the state of research on the mating consequences of genome duplication and present new results regarding the patterns of variation and underlying selective causes. We begin by reviewing the main theoretical predictions and summarizing the empirical patterns for the evolution of mating systems in polyploids. From these results, we identify incongruence between theoretical predictions and empirical results as it relates to autopolyploids, and we argue that the most pertinent question is not whether selfing evolves in autopolyploids but rather what limits the evolution of selfing in autopolyploids. Finally, we address this question with an experimental analysis of inbreeding depression in synthesized polyploids (hereafter, "neopolyploids") of the perennial plant *Chamerion angustifolium* (Onagraceae). Our results suggest that the costs of selfing are dynamic in polyploids and may change as a function of the age of the polyploid and its history of inbreeding.

Theoretical Arguments

Historically, hypotheses regarding the relationship between polyploidy and mating system have been developed from two perspectives: (1) the effect of self-fertilization on the likelihood of polyploid formation and establishment, and (2) the effect of polyploidy on the costs of selfing, specifically inbreeding depression. These perspectives are not mutually exclusive; in fact, a full understanding of this problem will likely come from considering both perspectives.

In the early twentieth century, plant biologists interpreted mating patterns in polyploids in terms of their effects on the likelihood of polyploid formation and establishment. The most common process by which polyploids formed was thought to be through the union of unreduced gametes in interspecific hybrids (Clausen et al. 1945). As a result, early attention focused more on allopolyploids (duplication of a hybrid genome) than autopolyploids (duplication of a single species' genome). In this context, Gustafsson (1947, pp. 183–371, and 1948, as cited in Grant 1956) suggested that polyploidy would be more likely to arise in cross-fertilizing species, presumably because of the higher incidence of hybridization. Based on observations of selfing in several polyploid taxa, Grant (1956) and Stebbins (1950, 1957) suggested the opposite of Gustafsson. They argued that chromosome doubling of the hybrid, through union of unreduced gametes, would be more likely to occur through selfing, especially in short-lived species. These two perspectives are not directly competing but rather argue for the importance of different steps in the pathway of polyploid formation.

Ramsey and Schemske (1998) examined this issue quantitatively by calculating the average rates of allopolyploid and autotetraploid formation from published estimates of key parameters in the formation pathway. They found that selfing increased the likelihood of allopolyploid formation by 34 times compared with outcrossing. In contrast, the benefits of selfing to autotetraploid formation were only 1.7 times higher. While this result is consistent with the hypothesis by Grant (1956) and Stebbins (1957), the calculation for allopolyploid formation does not consider the likelihood of initial hybrid formation and thus is unable to evaluate the relative importance of hybrid formation versus polyploid formation phases in polyploid evolution. It is conceivable that Gufstaffson and Stebbins are both correct, in which case one might expect that lineages with mixed mating, or those in which the diploid progenitor is outcrossing and the hybrid and polyploid derivatives are selfing, will maximize allopolyploid formation.

Predictions about mating system evolution have also focused on the factors affecting establishment of polyploids. For example, Stebbins (1957, 1971) argued that autotetraploids are more likely to be outcrossing, as any fitness advantages accrued by them would arise from their heterozygosity and adaptive potential. In contrast, allopolyploids would "tolerate" selfing, as they often exhibit fixed heterozygosity formed from the union of divergent genomes. In these hypotheses, mating systems are evaluated in terms of their long-term or group benefits to polyploid establishment, but these selective forces are not well explored (but see Schoen and Busch 2008) and may have relatively minor effects unless similar benefits are incurred by individuals in the short term.

More recent theoretical analyses, which consider the immediate fitness effects of mating system, are uniform in their prediction that selfing will favor establishment in autopolyploids. Levin (1975) showed that newly formed autopolyploids that randomly outcross should experience a strong mating disadvantage (called "minority cytotype exclusion"), as most of their ovules would be fertilized by diploids, yielding triploids of low fitness. Selfing would reduce the frequency of hybrid matings and thereby enhance polyploid fecundity and establishment within diploid populations. Additional mathematical models by Rodriguez (1996) and Rausch and Morgan (2005) and an experimental study (Husband 2000) have corroborated this result for autopolyploids. The assertion is also likely to apply to allopolyploids, given that hybrid matings would impose a similar fitness cost; however, this has not been explored mathematically.

The second theoretical avenue of research on polyploidy and mating systems focuses on how genome duplication affects the fitness costs of selfing, specifically inbreeding depression, defined as the reduced fitness of offspring derived from selfing compared with outcrossing (Charlesworth and Charlesworth 1987; Goodwillie et al. 2005). Theoretical models predict different effects of genome duplication depending on the genetic basis of inbreeding depression. Busbice and Wilsie (1966) proposed that fitness may be enhanced by allelic diversity or allelic overdominance. If inbreeding depression is caused by a reduction in allelic combinations, then autopolyploids should have greater inbreeding depression than their diploid progenitors. This occurs because autopolyploids can retain >2 alleles per homologous locus and because potential combinations of alleles will decrease more rapidly upon selfing than in diploids. Tests of allelic overdominance as a primary cause of inbreeding depression have not been conducted for natural populations of tetraploids, and evidence is limited in diploids (reviewed by Goodwillie et al. 2005). Nevertheless, circumstantial support for this hypothesis exists in the crop literature, where the negative effects of selfing on seed production and yield are larger in polyploids than in diploid relatives (Tysdal et al. 1942; Busbice and Wilsie 1966; Dewey 1966; Hecker 1972; reviewed by Bingham 1980).

Lande and Schemske (1985) modeled the evolution of inbreeding depression in tetraploids under the assumption that it is caused by partially or completely recessive deleterious mutations (i.e., partial dominance model). Under these conditions, and when mutations are recessive and lethal in homozygote form, equilibrium inbreeding depression in tetraploids will be half that of diploids. This effect seems counterintuitive, as the frequency of deleterious mutations in polyploids is expected to exceed that in diploids at equilibrium. However, this result is balanced by the fact that full homozygosity (aa or $aaaa$) increases at a faster rate (2.9–3.8 times, depending on segregation type—chromosomal or chromatid) upon selfing in diploids compared with autopolyploids (Parsons 1959; Bever and Felber 1992). Hedrick (1987) showed a similar theoretical result when applied to homosporous, allopolyploid ferns, based on the fact that offspring will be strongly homozygous due to intragametophytic selfing, and thus the genetic load will not segregate on selfing. Evidence to support Lande and Schemske's prediction is limited, in part due to the lack of research on inbreeding depression in polyploids and their diploid progenitors. In the most comprehensive study to date, Husband and Schemske (1997) reported that inbreeding depression in tetraploid populations of *Chamerion* (formerly *Epilobium*) *angustifolium* was 29% lower than in diploid populations. A similar pattern was observed by Rosquist (2001) in diploid and tetraploid species of *Anthericum*. Johnston and Schoen (1996) observed the reverse trend in *Amsinckia*; however, their study species were highly selfing and may already have purged most recessive lethals.

Arguably the most comprehensive theoretical model was developed by Ronfort (1999), who examined the evolution of inbreeding depression in autotetraploids with different mating systems, and dominance coefficients for the three heterozygote genotypes (*Aaaa*, *AAaa*, *AAAa*). Contrary to Lande and Schemske (1985), the Ronfort model predicts that tetraploids and diploids will have similar inbreeding depression at equilibrium when mutations are completely recessive; however, inbreeding depression should be lower in polyploids when mutations are partially dominant. Moreover, the equilibrium relationship between inbreeding depression and the selfing rate is expected to differ depending on the dominance effect. In the case of duplex mutation dominance (a deleterious allele requires two copies to affect fitness), inbreeding depression should decrease as a function of selfing rate, as expected in diploids (Charlesworth et al. 1990). With dosage-like dominance (expression of mutations is proportional to allele copy number), inbreeding depression should increase monotonically or temporarily with increased selfing. The rise in inbreeding depression occurs because the equilibrium frequency of those heterozygotes that are most likely to segregate full homozygotes actually increases with selfing. This study makes the novel conclusion that while selfing may be favored initially in polyploids, the cost of selfing increases with inbreeding history, and hence, mixed mating systems may be favored under some circumstances. Unfortunately, little empirical data are available regarding the dominance relationships in polyploids or the correlated evolution of selfing rate and inbreeding depression.

The two main theoretical approaches used to examine relationships between polyploidy and outcrossing are from the perspective of either mating system evolution or polyploid evolution. A more complete understanding of the correlated evolution of mating systems and polyploidy will require models that focus on both simultaneously (i.e., joint evolution of polyploidy and selfing), and this has not been attempted. Rausch and Morgan (2005) have taken the first step in this direction by incorporating the costs of selfing into their model of polyploid evolution. Inbreeding depression is not allowed to evolve, but it can vary between ploidy levels. They show that polyploids will be more likely to evolve when polyploids have higher selfing or lower inbreeding depression than their diploid progenitors.

Empirical Evidence

Empirical tests of the association between ploidy and mating system in flowering plants can only be described as incomplete and often case specific. Grant (1956) first reported that of several genera of short-lived species (*Layia*, *Madia*, *Gilia*, *Amsinckia*), those containing self-fertilizing species sometimes contained allopolyploid derivatives, while genera that contained outcrossing species were strictly diploid. Further, Stebbins's (1957) analysis of annual and perennial polyploids showed that all selfing polyploids were allopolyploids and that autopolyploids were strictly outcrossing. These conclusions are compelling but are based on morphological indicators of the reproductive system rather than estimates of fertilization patterns, and the generality of such conclusions has not yet been fully established.

The most direct method for evaluating patterns of polyploid mating is to quantify their outcrossing rates (t, selfing rate $s = 1 - t$) or inbreeding coefficients from natural populations by comparing the genotypic diversity in offspring to the genotypes of their parents (Ritland 1990). Unfortunately, scoring genotypes of polyploids, particularly in autopolyploids with greater than two copies of each homologous chromosome, can be challenging because of the difficulty of assessing dosage. This obstacle may have limited the number of outcrossing rate estimates published for polyploids. Furthermore, published values are more likely to be based on ancient allopolyploids with restructured genomes or autopolyploids with bivalent chromosome pairing at meiosis. Their value is questionable for this analysis because they likely obey the genetic constraints of diploids (disomic inheritance), and as a result, predictions of mating system evolution in polyploids based on models with tetrasomic inheritance or multiple alleles may not apply.

As a first step toward understanding the patterns of variation in mating system, we examined the relationship between ploidy and outcrossing rate, using data from the literature. Using an existing database of outcrossing rates (S. C. H. Barrett and C. G. Eckert, personal communication) and more recent values from the literature, we sought estimates for as many diploid and polyploid congeners (or conspecifics) as possible. By restricting our sampling to closely related species pairs, we controlled for the influence of phylogeny on mating systems and ploidy variation. However, these paired taxa could differ in other respects, such as life history, making it difficult to completely isolate the effects of genome duplication on mating system. We identified 10 congeneric pairs from seven flowering plant families (table 1). Each species was categorized as diploid or polyploid by comparing their chromosome number to the published base number for the genus. Polyploids were identified as allopolyploids in six pairs, whereas the others were either autopolyploid or of unknown origin. Outcrossing rates were averaged across populations and species when more than one estimate was available for a single cytotype. The mean outcrossing rate for diploids ($\bar{x} = 0.512$, SE = 0.119) was significantly higher than for polyploids ($\bar{x} = 0.227$, SE = 0.091) in a paired t-test ($t = 2.49$, $n = 10$, $P = 0.034$). Polyploids had lower outcrossing rates than their diploid relatives in eight of 10 pairs, the two exceptions being the genera *Tragopogon* and *Avena* (table 1). The difference in mating system between ploidies was not confounded by associated changes in life history (perennial vs. annual: $\chi^2 = 0.81$, $P = 0.370$) or mode of pollination (insect vs. wind: $\chi^2 = 0.0$, $P = 1.00$). Our results are consistent with much of the existing theory and with a recent study by Barringer (2007), who conducted phylogenetic comparisons of selfing rates and ploidal level for angiosperms. Based on 235 taxa and 32

Table 1

Outcrossing Rate Estimates for 10 Diploid-Polyploid Pairs of Angiosperms

Family, genus, and species	2n	Ploidy	Outcrossing rate (t)	Reference
Amaryllidaceae:				
Narcissus longispathus	14	Diploid	.635	Barrett et al. 2004
Narcissus assoanus	14	Diploid	.990	A. M. Baker and S. C. H. Barrett, unpublished data
Narcissus papyraceeus	22	Diploid	.810	Arroyo et al. 2002
Narcissus dubius	54	Polyploid (allo)	.690	A. M. Baker and S. C. H. Barrett, unpublished data
Asteraceae:				
Senecio squalidus	20	Diploid	.930	Abbott and Forbes 1993
Senecio vulgaris	40	Polyploid (allo)	.058	Marshall and Abbott 1982
Townsendia hookeri	18	Diploid	.230	Thompson and Ritland 2006
T. hookeri	36	Polyploid (auto)	.030	Thompson and Ritland 2006
Tragopogon dubius	12	Diploid	.155	Cook and Soltis 1999
Tragopogon mirus	24	Polyploid (allo)	.418	Cook and Soltis 1999
Boraginaceae:				
Amsinckia spectabilis	10	Diploid	.307	Ganders et al. 1985
Amsinckia gloriosa	24	Polyploid (allo)	.020	Johnston and Schoen 1996
Caryophyllaceae:				
Spergularia media	18	Diploid	.120	Sterk and Dijkhuizen 1972
Spergularia marina	36	Polyploid	.015	Sterk and Dijkhuizen 1972
Lythraceae:				
Cuphea lanceolata	12	Diploid	.812	Knapp et al. 1991
Cuphea lutea	28	Polyploid	.293	Krueger and Knapp 1991
Onagraceae:				
Chamerion angustifolium	36	Diploid	.900	Husband and Schemske 1995; Ozimec 2006
C. angustifolium	72	Polyploid (auto)	.715	Parker et al. 1995; Husband and Schemske 1997; Ozimec 2006
Poaceae:				
Triticum speltoides	14	Diploid	.850	Zohary and Imber 1963
Triticum dicoccoides	28	Polyploid (allo)	.000	Golenberg 1988
Avena barbata	14	Diploid	.013	Hamrick and Allard 1972
Avena fatua	42	Polyploid (allo)	.025	Imam and Allard 1965

Note. For each species, we report the somatic chromosome number (2n), ploidy (relative to the base number for the genus), and mean outcrossing rate. When multiple population estimates are available, or when a population consists of a floral polymorphism, we present the mean. All estimates are based on natural populations except *Cuphea*, which was included because diploid and polyploid estimates were estimated using the same method.

phylogenetically independent comparisons, polyploids had lower outcrossing rates than diploids. Interestingly, while polyploids were more selfing, the actual rate of outcrossing was 0.48 across all polyploid species, indicating that they are not predominantly selfing, as some models might predict.

To reexamine Stebbins's (1957) assertion that autopolyploids have different mating patterns than allopolyploids, we assembled outcrossing rate estimates for 18 species of polyploids that have been classified in the literature based on cytogenetic, genetic, and morphological criteria as either allopolyploid or autopolyploid (table 2). We consider these to be putative designations, because the criteria used to classify polyploids vary among taxa, and the power to distinguish these classes can be obscured by divergence between progenitor taxa and by genomic evolution subsequent to the duplication (Soltis et al. 2003). Autopolyploids had a significantly higher outcrossing rate ($t = 0.64$, SE $= 0.087$) than allopolyploids ($t = 0.20$, SE $= 0.099$; $F_{1,14} = 10.88$, $P = 0.0045$). In fact, nine of 11 autopolyploids had greater than 50% outcrossing, whereas six of seven allopolyploids were less than 50% outcrossing. The fact that autopolyploids often have

mixed or outcrossed mating systems runs counter to theoretical predictions based on arguments related to selection for polyploid establishment or costs of selfing. While sample sizes are relatively small, this pattern raises the question of what limits the evolution of selfing in many autopolyploids and may reveal our lack of understanding of the underlying mechanisms governing mating system evolution in polyploids.

What Limits Selfing in Autopolyploids?

The discrepancy between theoretical expectations of selfing in autopolyploids and observations of mixed mating to outcrossing is currently not addressed in the literature. One explanation raised by existing models is that inbreeding depression at the time of formation differs from well-established polyploids and may actually be greater in neoautopolyploids than diploids (Busbice and Wilsie 1966; Bingham 1980). Alternatively, inbreeding depression may be reduced upon polyploid formation but actually increases with selfing due to dosage-like dominance (Ronfort 1999), thus preventing the fixation of selfing variants. Current estimates of inbreeding

Table 2

Summary of Outcrossing Rates (*t*) for Autopolyploid and Allopolyploid Plant Species

Species	Polyploid origin	Outcrossing rate	Reference
Acacia nilotica ssp. *leiocarpa*	Auto	.384	Mandal et al. 1994
Amsinckia douglasiana	Auto	.750	Schoen et al. 1997
Aster kantoensis	Auto	.883	Inoue et al. 1998
Campanula americana	Auto	.938	Galloway et al. 2003
Chamerion angustifolium	Auto	.715	Husband and Schemske 1997; Parker et al. 1995; Ozimec 2006
Pachycereus pringlei	Auto	.625	Murawski et al. 1994
Rutidosis leptorrhynchoides	Auto	.913	Young and Brown 1999
Silene virginica	Auto	.890	Dudash and Fenster 2001
Silene douglasii var. *oraria*	Auto	.620	Kephart et al. 1999
Townsendia hookeri	Auto	.030	Thompson and Ritland 2006
Turnera ulmifolia var. *elegans*	Auto	1.000	Barrett and Shore 1987
Amsinckia gloriosa	Allo	.020	Johnston and Schoen 1996
Avena fatua	Allo	.025	Imam and Allard 1965
Narcissus dubius	Allo	.690	Barrett, unpublished
Senecio vulgaris	Allo	.058	Marshall and Abbott 1982
Tragopogon mirus	Allo	.418	Cook and Soltis 1999
Triticum dicoccoides	Allo	.000	Golenberg 1988
Turnera ulmifolia var. *angustifolia*	Allo	.190	Barrett and Shore 1987

Note. Putative origin of polyploid (auto vs. allo) was reported as given by authors, based on a combination of morphological, cytogenetic, and genetic criteria. All estimates are based on segregation of genetic markers in progeny arrays, with the exception of *Silene douglasii*, which is based on seed production in self- and outcross-pollination. We report means when multiple population estimates were available. Estimates for *Pachycereus* are the mean of estimates of hermaphrodites and females. The estimate for *Turner ulmifolia* var. *intermedia* is the mean of long- and short-styled morphs, truncated to the maximum value of 1.

depression for natural tetraploid and diploid populations (Husband and Schemske 1997; Rosquist 2001) indicate that polyploids have less inbreeding depression than diploids; however, this pattern is limited to few studies and is based on extant polyploids, which may not reflect the effects of selfing early in polyploid formation. Synthesized polyploids, or neopolyploids, may be particularly valuable for studying the selective forces acting on mating systems near the time of genome duplication; however, this nonequilibrium perspective has not been explored previously for natural populations.

Here, we examine the dynamics of inbreeding depression by focusing on the costs of selfing in newly synthesized polyploids of the perennial herbaceous plant *Chamerion angustifolium* (Onagraceae). This insect-pollinated species occurs throughout the Northern Hemisphere in open or disturbed habitats. Within North America it comprises diploid and tetraploid forms. The tetraploid has been classified as autotetraploid based on its morphological resemblance to diploids (Mosquin 1966), its multivalent formation during meiosis (Mosquin 1966), its multisomic inheritance of allozymes (Husband and Schemske 1997), and its interfertility with induced tetraploids (H. A. Sabara and B. C. Husband, unpublished data). A previous study (Husband and Schemske 1997) found that extant tetraploids had consistently less inbreeding depression than diploids, corroborating the theoretical expectation that selfing should be favored in this cytotype (Lande and Schemske 1985). However, outcrossing rate estimates for tetraploid *C. angustifolium* ($t = 0.715$, $N = 3$) are only marginally lower than diploids ($t = 0.900$, $N = 3$; table 1; Husband and Schemske 1995, 1997; Ozimec 2006) and can best be described as

mixed mating to outcrossing. In this study, neopolyploids were generated using colchicine from diploid stock and were used to address two questions: (1) what is the magnitude of inbreeding depression in neopolyploids compared with diploid and extant polyploids? and (2) what is the effect of inbreeding history on the magnitude of inbreeding depression?

Synthesizing Polyploids

We synthesized tetraploids using plants from two diploid populations, Fortress Mountain (lat. 50°49.380′N, long. 115°11.670′W) and Mount Norquay (lat. 51°12.240′N, long. 115°35.782′W), located approximately 50 km apart in western Alberta, Canada. Based on previous estimates of mating system, both populations are expected to be outcrossing and highly heterozygous. We sowed and germinated 50 seeds from each of 20 maternal plants per population on moist filter paper in petri dishes. We bathed 10-d old seedlings in 5 mL of 0.2% colchicine solution for 12 h, rinsed them in distilled water, and transplanted them into soil. Because colchicine-induced doubling is often incomplete and can result in ploidy chimeras, we screened all plants for ploidy using flow cytometry (Dart et al. 2004) at the 4–6-leaf stage and then again at flowering (using leaves subtending the flowers). We identified at least one completely neopolyploid plant from 11 maternal families from Fortress Mountain and 14 families from Mount Norquay. To avoid any secondary effects of colchicine exposure on the plants (Ramsey and Schemske 2002), we cross-pollinated each neopolyploid with an unrelated neopolyploid within the same population to produce the first seed generation of neopolyploids.

Estimating Inbreeding Depression in Diploid, Neotetraploid, and Tetraploid C. angustifolium

We examined the fitness of selfed and outcrossed offspring for two diploid populations (Fortress Mountain, Mount Norquay), two synthesized (neotetraploid) populations, and two naturally occurring tetraploid populations (Barrier Station: lat. 51°01.634′N, long. 115°02.187′W; Moose Meadows: lat. 51°15.383′N, long. 115°52.035′W). The naturally occurring tetraploid populations were in close proximity to the diploid populations. We sowed and germinated seeds from 15 open-pollinated families from each diploid and tetraploid population and 24 neotetraploid families, and seedlings were then transplanted into plug trays. We screened seedlings from neopolyploid families for ploidy level, and all families with any triploid or diploid offspring were eliminated, yielding 10 and nine neopolyploid families from Fortress Mountain and Mount Norquay, respectively. All seedlings were grown to flowering in half-gallon pots. At flowering, we emasculated six flowers per plant: two flowers were then cross-pollinated with a randomly chosen donor, and four flowers were self-pollinated. In each case, we saturated the stigmas with pollen. We pollinated plants over a 4-wk period, with no more than two flowers per plant pollinated on a given day.

We estimated the fitnesses of selfed and outcrossed offspring in terms of (1) likelihood of producing a fruit (having at least one filled seed), (2) seeds per fruit, (3) germination rate, (4) survival to 10 wk, (5) flower production, and (6) ovule and pollen production. Seed number was expressed as the proportion of fertilized ovules maturing to plump seeds. Fertilized ovules are generally larger than unfertilized ovules, although unfertilized ovules may have included some early aborted seeds. Germination was estimated as the rate of seedling emergence from 28 outcrossed seeds and 22 selfed seeds per plant. We measured the subsequent fitness components on up to five selfed and five outcrossed seedlings per family, grown for 10 wk in a randomized design in a glasshouse with edge plants around the periphery. Plant height was used as a surrogate measure of reproductive effort because it was correlated with flower number ($R^2 = 0.81$, $P < 0.001$, $N = 28$). We calculated gamete production as the mean of pollen and ovule number per flower. Pollen number was calculated as the average of three anthers per plant (one from each of three flowers) and was estimated using a particle counter (Multisizer III, Beckman Coulter). Ovule production was estimated from the number of seeds produced in a randomly cross-pollinated flower on each plant (assuming that few zygotes are aborted with outcrossing). We estimated total fitness over the 10-wk period (referred to as cumulative fitness) as the product of mean fruit set, seed set per fruit, germination, survival, plant height, and gamete number for each maternal family for which data were available. We transformed plant height and gamete number per flower to relative measures by dividing over the maximum value to ensure that each fitness measure ranged from 0 to 1 and thus was weighted equally in the cumulative fitness calculation.

We evaluated the effects of ploidy, maternal family (random effect), and pollination treatment (self vs. outcross) for each fitness component using ANOVA (JMP, SAS Institute

2002). Because family size (i.e., number of offspring) varied widely, we based our analysis on the mean values for selfed and outcrossed offspring for each family. Since no effect of population (within ploidy) was detected, it was removed from the model. All fitness components were log transformed to meet assumptions of ANOVA and to evaluate relative differences between selfing and outcrossing rather than absolute differences (Johnston and Schoen 1994). We also conducted a priori comparisons between self and outcross means for each separate ploidy using Student's t test with an experiment-wise error rate of 0.05. Inbreeding depression for each ploidy was calculated as $\delta = 1 - [w_s/w_o]$ (or $\delta = [w_o/w_s] - 1$ when $w_s > w_o$), where w_o and w_s are the mean cumulative fitnesses of outcrossed and selfed progeny, respectively.

Estimating Inbreeding Depression as a Function of Inbreeding History

We also compared the magnitude of inbreeding depression between outbred and inbred neopolyploid plants. Using the protocol above ("Synthesizing Polyploids"), we generated 25 seed families (11 from Fortress Mountain, 14 from Mount Norquay) with at least one neopolyploid plant. Neopolyploids were grown to flowering and subject to two pollination treatments: flowers were pollinated with an unrelated plant from the same population to yield outbred neopolyploid offspring, and two flowers were self-pollinated to produce inbred offspring. Assuming that the initial diploids used to create the neopolyploids were outbred, the expected inbreeding coefficients for the outbred and inbred neopolyploid plants were $F = 0.11$ and $F = 0.44$, respectively (following calculations by Jones and Bingham [1957] and Kempthorne [1957]). We then grew the inbred and outbred offspring to flowering and subjected each to self- and outcross-pollinations to estimate inbreeding depression. Initially, we sowed 10 seeds per neotetraploid seed family (inbred and outbred), germinated them, and grew all seedlings to flowering. After eliminating any families with ploidy reversions, the outbred group consisted of 19 families (10 from Fortress Mountain, nine from Mount Norquay) and the inbred group comprised 19 families (13 from Fortress Mountain, six from Mount Norquay). We grew one plant from each family to reproductive maturity, emasculated six flowers, and then self- ($N = 4$ flowers) and cross-pollinated ($N = 2$ flowers) them. We estimated fitnesses of selfed and outcrossed progeny for inbred and outbred neopolyploids in a glasshouse using the following fitness components: (1) fruit set (at least one filled seed), (2) seeds set per fruit, (3) germination rate, (4) survival to 10 wk, and (5) total height (base of stem to top of inflorescence). Germination percentage was based on a mean of 19 outcrossed seeds and 17 selfed seeds per plant. When numbers permitted, we grew five selfed and five outcrossed progeny per maternal plant in a randomized design in the glasshouse. The effects of inbreeding, maternal parent (random effect), and pollination treatment on each fitness component were analyzed with ANOVA. The population term was removed from the final model because no effect was detected. All fitness components were log transformed to meet assumptions of ANOVA and to evaluate relative differences between selfing and outcrossing rather than absolute differences (Johnston and Schoen 1994). We also conducted a priori comparisons between self and outcross means for each inbreeding treatment using Student's t-test, with an experimentwise error rate of 0.05. Total effect of selfing and outcrossing was estimated as the product of family means for fruit set, seed set, germination, survival, and plant height. Inbreeding depression was estimated using $\delta = 1 - [w_s/w_o]$ or $\delta = [w_o/w_s] - 1$ when $w_s > w_o$, where w_s and w_o are the mean overall fitness values for selfed and outcrossed families.

Inbreeding Depression in Neopolyploids, Diploids, and Natural Tetraploids

Of the 78 individuals crossed, nine (six diploid, three neopolyploid) were pollen sterile and five (including four neopolyploids) were ovule sterile, as indicated by the lack of fertilized ovules in self- or cross-pollinations. Pollen-sterile plants were used only as pollen recipients. All ovule-sterile plants were excluded from the analysis. Sterility also varied among flowers within plants. In total, 24% of flowers did not yield any fertilized ovules and could not be used in the analysis. The specific cause of pollen and ovule sterility is not known. Pollen sterility has been previously observed in the diploid populations. The neopolyploids may be expressing the genes associated with their diploid parents; in addition, pollen and ovule sterility are often observed in neopolyploids as a result of meiotic irregularities (Ramsey and Schemske 2002).

Pollination treatment (self or outcross) had a significant effect on fruit set, seed set, germination, and cumulative fitness but not survival rate, plant height, or gamete production (table 3). For all significant fitness measures, values for selfed offspring were lower than for outcrossed offspring (fig. 1; fruit set not shown). The effect of pollination differed among ploidy categories for seed set, germination, and cumulative fitness. Relative to outcrossing, selfing had a smaller effect on neopolyploids than on tetraploids and diploids at all three stages (fig. 1). Specifically, the effect of selfing on seed set was significant for diploids ($w_s/w_o = 0.197$) and tetraploids ($w_s/w_o = 0.500$) but not neopolyploids ($w_s/w_o = 0.566$). At germination, only diploids were significantly affected ($w_s/w_o = 0.429$), while tetraploids ($w_s/w_o = 0.754$) and neotetraploid ($w_s/w_o = 0.960$) were not. With respect to cumulative fitness, there was a significant effect of selfing on diploids ($w_s/w_o = 0.079$), a smaller effect on tetraploids ($w_s/w_o = 0.384$), and no detectable effect on neopolyploids ($w_s/w_o = 1.15$). Weak overall effects of selfing on neopolyploids likely reflect the neutral to positive influences of selfing on plant height and gamete production. Total inbreeding depression, measured as the combined effects of fruit set, seed set, germination, survival, total height, and gamete production (mean of pollen and ovule number per flower), was $\delta = 0.92$ in diploids, $\delta = 0.62$ in natural tetraploids, and $\delta = -0.13$ in neopolyploids.

Inbreeding Depression in Inbred versus Outbred Neopolyploids

Ovule sterility was common in both inbred and outbred neotetraploid populations, as observed in the previous experiment, with 51% of all pollinated flowers producing no fertilized (aborted or filled) ovules. These flowers were not included in the analysis. Compared with data from naturally

Table 3

Analysis of the Effects of Selfing versus Outcrossing on Neopolyploid, Tetraploid, and Diploid *Chamerion angustifolium*

Source of variation	df	SS	F	P
Fruit set:				
Ploidy	2	1.02	3.68	.029
Maternal family (ploidy)	73	11.24	1.96	.010
Pollination treatment	1	.56	7.13	.011
Ploidy × pollination interaction	2	.09	.60	.556
Residual	42	3.30		
Seed set:				
Ploidy	2	6.10	1.45	.241
Maternal family (ploidy)	72	184.07	5.55	<.001
Pollination treatment	1	17.45	37.89	<.001
Ploidy × pollination interaction	2	5.90	6.40	.004
Residual	39	17.97		
Germination:				
Ploidy	2	1.13	1.06	.352
Maternal family (ploidy)	56	34.16	1.81	.045
Pollination treatment	1	1.98	5.86	.022
Ploidy × pollination interaction	2	2.81	4.16	.026
Residual	28	9.45		
Survival:				
Ploidy	2	.07	.72	.489
Maternal family (ploidy)	54	3.24	1.93	.035
Pollination treatment	1	.037	1.19	.286
Ploidy × pollination interaction	2	.11	1.78	.188
Residual	26	.81		
Total height:				
Ploidy	2	1.01	4.49	.014
Maternal family (ploidy)	54	6.69	1.40	.175
Pollination treatment	1	.17	1.88	.182
Ploidy × pollination interaction	2	.30	1.68	.205
Residual	26	2.30		
Gamete production:				
Ploidy	2	.07	.19	.824
Maternal family (ploidy)	44	9.74	1.31	.230
Pollination treatment	1	.0004	.003	.961
Ploidy × pollination interaction	2	.58	1.70	.218
Residual	14	2.37		
Cumulative fitness:				
Ploidy	2	2.82	.68	.508
Maternal family (ploidy)	43	147.15	3.83	.004
Pollination treatment	1	4.18	4.68	.048
Ploidy × pollination	2	16.80	9.40	.003
Residual	14	12.51		

Note. Results for ANOVA are given for six fitness components and cumulative fitness, which is a multiplicative function of all components. Total height is height to top of inflorescence at 10 wk expressed relative to maximum height (×100). Gamete production represents the mean of ovule and pollen production per flower expressed relative to maximum gamete production (×100). All fitness components and cumulative fitness were log transformed.

occurring polyploids in previous studies, fruit set, seed set, germination, and survival were generally low, causing a significant reduction in family number in this experiment, particularly at later stages. As a result, analyses of most stages suffered from low power. To provide insight into the effects of selfing, we therefore restricted our focus to three stages: fruit set, seed set, and germination.

Pollination treatment had a strong effect on fruit set, seed set, and cumulative fitness but not on germination (table 4).

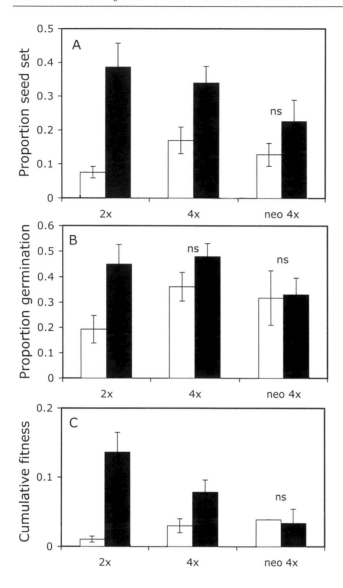

Fig. 1 Mean (SE) fitness of self-fertilized (open bar) and cross-fertilized (solid bar) offspring of diploid, tetraploid, and neotetraploid *Chamerion angustifolium*. Means are shown for (*A*) seed maturation, (*B*) germination, and (*C*) cumulative fitness, which is the product of six fitness components. Values for fruit set, survival, and reproductive effort (measured as growth) are not shown because there were no variations in effects of pollination among ploidies. Each bar represents the mean of all maternal families of a given ploidy. Results from a priori comparisons (Student's *t*-test, experimentwise error = 0.05) between self and outcrossed means for each ploidy are significant (*P* < 0.05) unless indicated (ns = not significant).

In all cases, means of selfed progeny were less than outcrossed progeny. Inbreeding history had a significant effect only on fruit set. The interaction between inbreeding history and pollination treatment was not significant at 0.05 for any fitness component, but planned comparisons between selfed and outcrossed means for inbred and outbred lines showed a significant effect of selfing on fitness of inbred lines but not outbred lines. This effect was strong in fruit set, seed set, and cumulative fitness (fig. 2). The pattern was the same at germi-

nation but not significant. Combining fruit set and seed set into a single cumulative measure, the mean relative fitness of self-fertilized ovules was 0.45 in outbred families and 0.01 in inbred families. Inbreeding depression in these two groups was therefore $\delta = 0.55$ and 0.99, respectively. When cumulative fitness was calculated based on the first three fitness components, inbreeding depression was -0.23 and 0.80 for outbred and inbred groups, respectively.

Conclusions and Future Directions

On theoretical grounds, the transition from diploidy to polyploidy should be associated with correlated evolution of the mating system, and in most models, predominant self-fertilization. However, empirical evidence for this pattern or its causes is limited and has largely been derived from specific case studies using reproductive traits such as gender and self-incompatibility rather than quantitative estimates of cross- and self-fertilization. Our review of outcrossing rates in diploid-polyploid species pairs corroborates the selfing hypothesis and is congruent with mating system studies on ferns (Masuyama and Watano 1990). On closer inspection, however, strong selfing seems to be associated mostly with allopolyploids, while autopolyploids have more mixed or outcrossed mating systems.

Table 4

Analysis of the Effects of Selfing and Outcrossing on Inbred ($F = \sim 0.44$) and Outbred ($F = \sim 0.11$) Neopolyploid *Chamerion angustifolium*

Source of variation	df	SS	F	P
Fruit set:				
Inbreeding history	1	16.95	4.68	.036
Maternal family (inbreeding)	34	125.53	1.15	.375
Pollination treatment	1	23.59	7.37	.013
Inbreeding × pollination interaction	1	1.09	.34	.565
Residual	20	64.01		
Seed set:				
Inbreeding history	1	.56	.18	.674
Maternal family (inbreeding)	27	96.41	5.26	.003
Pollination treatment	1	3.18	4.68	.053
Inbreeding × pollination interaction	1	1.31	1.93	.192
Residual	11	7.47		
Germination:				
Inbreeding history	1	.45	1.18	.290
Maternal family (inbreeding)	15	5.59	.91	.595
Pollination treatment	1	.00	.00	.986
Inbreeding × pollination interaction	1	.23	.57	.478
Cumulative fitness:				
Inbreeding history	1	.73	.18	.672
Maternal family (inbreeding)	26	121.08	4.56	.008
Pollination treatment	1	7.77	7.61	.020
Inbreeding × pollination interaction	1	3.22	3.15	.106

Note. Results for ANOVA are given for three fitness components and cumulative fitness. Cumulative fitness is a multiplicative function of the first two or three components, due to low sample sizes in later stages. Total height is height to top of inflorescence at 10 wk expressed relative to maximum height (×100). Gamete production represents the mean of ovule and pollen production per flower expressed relative to maximum gamete production (×100). All fitness components and cumulative fitness were log transformed.

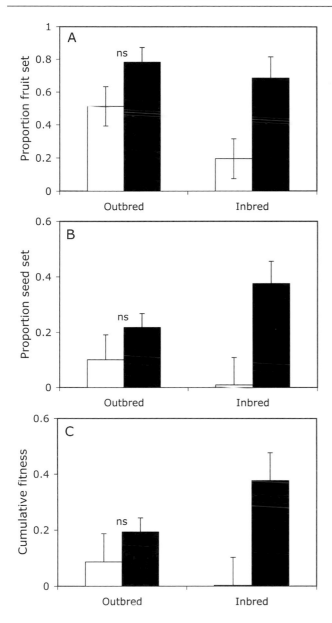

Fig. 2 Mean (SE) fitness of self- and cross-fertilized offspring of outbred ($F = 0.11$) and inbred ($F = 0.44$) neotetraploids of *Chamerion angustifolium*. Fitness values are shown for fruit set (A), seed set (B), and cumulative fitness (C), which is the product of all three fitness components. Each bar represents the mean of all maternal families for a given treatment. Results from a priori comparisons (Student's t-test, experimentwise error = 0.05) between self and outcrossed means for each treatment are significant ($P < 0.05$) unless indicated (ns = not significant).

This result is consistent with Stebbins's (1957) qualitative observation that autopolyploids are frequently outcrossing, but it runs counter to predictions from most theoretical models (although see Ronfort 1999) based on autopolyploids.

The observed difference in outcrossing between polyploids and diploids is somewhat consistent with analyses of the relationship between ploidy and other reproductive traits in flowering plants. Many investigators have reported an association

between polyploidy and self-fertility or self-compatibility in particular taxonomic groups (Lawrence 1930; Lewis 1943; Ross 1981; Miller and Venable 2000), suggesting that the potential for selfing is greater in polyploids. However, in a recent review on this topic, Mable (2004) found no association between ploidy and self-compatibility when viewed across all self-incompatible taxa, but the pattern was stronger in species with gametophytic self-incompatibility than sporophytic or heteromorphic systems. Miller and Venable (2000) did report a strong association between ploidy and self-compatibility in 12 genera with SI, but they also showed a higher incidence of gender dimorphism in polyploids compared with related diploids. They hypothesize that genome duplication leads to self-fertility but that costs of selfing (inbreeding depression) are still sufficiently high to cause evolution of inbreeding avoidance through gender dimorphism. In the future, it may be valuable to consider the association between ploidy and other reproductive attributes such as autogamy, floral form, flower size, and inflorescence architecture, which may facilitate mating-system shifts.

The difference in outcrossing between allopolyploids and autopolyploids observed is striking but must be interpreted cautiously due to limited sample sizes. Nevertheless, selfing might be expected to evolve in allopolyploids based on both genetic and ecological grounds. The formation of allopolyploids may benefit more from selfing than autopolyploids because of its effect on the likelihood that two unreduced gametes would unite. In addition, allopolyploids may have reduced inbreeding depression because of the low production of full homozygotes (Lande and Schemske 1985) and the ability to retain fixed heterozygosity at divergent homoeologous loci in the face of selfing. Alternatively, allopolyploids may effectively behave as diploids if homoeologous chromosomes have low homology, or if duplicate genes are lost or silenced during genomic restructuring (Soltis et al. 2003; Adams and Wendel 2005), or if allelic variation becomes fixed at one homoeologous locus (Lande and Schemske 1985). In these cases, where phenotypic variation is the product of expression of one of the homoeologous genomes, allopolyploids will have the same capacity to mask deleterious mutations as a diploid, and thus selective forces on their mating systems may be unlikely to differ (Soltis and Soltis 2000; Comai 2005; Galloway and Etterson 2007). Clearly, more estimates of selfing rates for allopolyploid and autopolyploids are necessary to corroborate the differences observed here. In addition, we need further research on inbreeding depression in both allopolyploids and autopolyploids, particularly comparing them with their diploid progenitors and with other polyploids that differ in age (neopolyploids vs. paleopolyploids).

The relatively high rate of outcrossing in autopolyploids is surprising given the benefits of selfing to the formation of autopolyploids and the reduced costs of selfing due to sheltering of deleterious alleles (Lande and Schemske 1985). This pattern is either an artifact of relatively small sample sizes or suggests that evolutionary dynamics of selfing in polyploids have not been adequately captured in most models. The number of outcrossing-rate estimates available for polyploids and polyploid-diploid pairs are indeed limited, and thus sampling error may have exaggerated the differences between autopolyploids and allopolyploids. Our analysis is also subject to error associated with classifying plants as either autopolyploid or allopolyploid, which is based on a variety of different criteria. Given these potential sources of variation, and the observed heterogeneity in mating system among flowering plant species generally, the uniformity of outcrossing rates among autopolyploid species is noteworthy.

One potential deficiency of current models is that they do not depict the relationship between ploidy and inbreeding depression in the initial stages after polyploid formation. This masks a critical stage in the evolution of mating systems, and it is likely that inbreeding depression in neopolyploids will differ from established polyploids. Our studies of inbreeding depression in neotetraploid *Chamerion angustifolium* provide a novel perspective on the dynamics of inbreeding depression. Neopolyploids have negligible inbreeding depression and substantially less than their diploid progenitors and established tetraploids. This is opposite to the predictions of the overdominance model of Busbice and Wilsie (1966) and to many studies involving crop species (Bingham 1980). The discrepancy between results for crop and natural plant populations remains unresolved but may be related to the reduced genetic variability and history of strong viability selection associated with crop breeding. The breeding history may effectively diminish differences in genetic load between diploids and tetraploids, a hypothesis supported by Johnston and Schoen's (1996) study of *Amsinkia* species, which are largely self-fertilizing. In this case, polyploids have as much if not more inbreeding depression than diploids.

Low inbreeding depression in neopolyploids is, however, consistent with the partial dominance model. Neopolyploids have less inbreeding depression than diploids because they have the same frequency of deleterious mutations as diploids but greater sheltering due to low homozygosity. They differ from established tetraploids because there has been insufficient opportunity for the frequency of deleterious mutations to reach equilibrium. Regardless of the mechanisms, there appears to be virtually no cost to selfing in neopolyploids. This further supports previous arguments that selfing should be favored in autopolyploids.

What, then, favors the evolution of outcrossing in established autotetraploids? Our results provide weak but consistent evidence that while inbreeding depression is low upon initial polyploid formation, its magnitude rises quickly upon repeated inbreeding. Based on three fitness components, inbreeding depression in neopolyploids after one generation of selfing rose from effectively 0 to 0.80. The precise mechanism for this increase is not known, although Ronfort's model (1999) may offer some insight. Despite being an equilibrium model, it shows that inbreeding depression will increase with selfing rate in polyploids under dosage-like dominance in which expression of a deleterious mutation at a locus is proportional to its copy number. Inbreeding depression increases primarily because the dominant form of heterozygote changes with history of inbreeding. As selfing increases, heterozygotes are increasingly of the *AAaa* or *Aaaa* form, which are most likely to yield full homozygotes after selfing. As a result, inbreeding depression increases. Therefore, our results are consistent with the dosage-dominance model for polyploids and present a working hypothesis for the genetic constraints opposing the evolution of selfing in autopolyploids.

Research on the mating consequences of genome duplication has been ongoing for nearly 60 yr, and there is much room for growth. Theoretical models have largely outpaced empirical research. This is probably related to the complexities of estimating traits such as polyploidy and mating systems on a large scale. As a result, data are available for specific taxonomic groups, but few generalities have emerged. Hopefully, techniques such as flow cytometry in combination with the proliferation of the use of genetic markers will facilitate research on a broader array of polyploid taxa and their diploid progenitors. Our understanding of the relationship between polyploidy and mating system must also advance through additional theoretical studies, such as modeling the joint evolution of selfing and polyploidy. Collectively, these avenues of research will advance our understanding of the evolutionary dynamics of mating systems in polyploids. Given the prevalence of genome and gene duplication, this approach may offer insights into the mechanisms of phenotypic diversification and adaptation of angiosperms in general.

Acknowledgments

We thank S. C. H. Barrett and C. G. Eckert for access to the outcrossing-rate database, Brian Barringer for providing an unpublished manuscript, and the Natural Sciences and Engineering Research Council of Canada, the Canadian Foundation for Innovation, and the Canada Research Chair program for funding to B. C. Husband.

Literature Cited

Abbott RJ, DG Forbes 1993 Outcrossing rate and self-incompatibility in the colonizing species *Senecio squalidus*. Heredity 71:155–159.

Adams KL, JF Wendel 2005 Polyploidy and genome evolution in plants. Curr Opin Plant Biol 8:135–141.

Arroyo J, SCH Barrett, R Hidalgo, WW Cole 2002 Evolutionary maintenance of stigma-height dimorphism in *Narcissus papyraceus* (Amaryllidaceae). Am J Bot 89:1242–1249.

Barrett SCH, WW Cole, CM Herrera 2004 Mating patterns and genetic diversity in the wild daffodil *Narcissus longispathus* (Amaryllidaceae). Heredity 92:459–465.

Barrett SCH, JS Shore 1987 Variation and evolution of breeding systems in the *Turnera ulmifolia* L. complex (Turneraceae). Evolution 41:340–354.

Barringer BC 2007 Polyploidy and self-fertilization in flowering plants. Am J Bot 94:1527–1533.

Bever JD, F Felber 1992 The theoretical population genetics of autopolyploidy. Pages 185–217 *in* J Antonovics, D Futuyma, eds. Oxford surveys in evolutionary biology. Oxford University Press, New York.

Bingham ET 1980 Maximizing heterozygosity in autopolyploids. Pages 471–490 *in* WH Lewis, ed. Polyploidy, biological relevance. Plenum, New York.

Burton TL, BC Husband 2001 Fecundity and offspring ploidy in matings among diploid, triploid and tetraploid *Chamerion angustifoium* (Onagraceae): consequences for tetraploid establishment. Heredity 87:573–582.

Busbice TH, CP Wilsie 1966 Inbreeding depression and heterosis application to *Medicago sativa* L. Euphytica 15:52–67.

Charlesworth D, B Charlesworth 1987 Inbreeding depression and its evolutionary consequences. Annu Rev Ecol Syst 18:237–268.

Charlesworth D, MT Morgan, B Charlesworth 1990 Inbreeding depression, genetic load, and the evolution of outcrossing rate in a multilocus system with no linkage. Evolution 44:1469–1489.

Clausen J, DD Keck, WM Heisey 1945 Experimental studies on the nature of species. II. Plant evolution through amphiploidy and autoploidy, with examples from the Madiinae. Carnegie Inst Wash Publ 564.

Comai L 2005 The advantages and disadvantages of being polyploid. Nat Rev Genet 6:836–846.

Cook LM, PS Soltis 1999 Mating systems of diploid and allotetraploid populations of *Tragopogon* (Asteraceae). I. Natural populations. Heredity 82:237–244.

Cui L, PK Wall, JH Leebens-Mack, BG Lindsay, DE Soltis, JJ Doyle, PS Soltis, et al 2006 Widespread genome duplications throughout the history of flowering plants. Genome Res 16:738–749.

Dart S, P Kron, BK Mable 2004 Characterizing polyploidy in *Arabidopsis lyrata* using chromosome counts and flow cytometry. Can J Bot 82:185–197.

Dewey DR 1966 Inbreeding depression in diploid, tetraploid and hexaploid crested wheatgrass. Crop Sci 6:144–147.

Dudash MR, CB Fenster 2001 The role of breeding system and inbreeding depression in the maintenance of an outcrossing mating strategy in *Silene virginica* (Caryophyllaceae). Am J Bot 88:1953–1959.

Galloway LF, JR Etterson 2007 Inbreeding depression in an autotetraploid herb: a three cohort field study. New Phytol 173:383–392.

Galloway LF, JR Etterson, JL Hamrick 2003 Outcrossing rate and inbreeding depression in the herbaceous autotetraploid, *Campanula americana*. Heredity 90:308–315.

Ganders FR, SK Denny, D Tsai 1985 Breeding systems and genetic variation in *Amsinckia spectabilis* (Boraginaceae). Can J Bot 63:533–538.

Golenberg EM 1988 Outcrossing rates and their relationship to phenology in *Triticum dicoccoides*. Theor Appl Genet 75:937–944.

Goodwillie C, S Kalisz, CG Eckert 2005 The evolution enigma of mixed mating systems in plants: occurrence, theoretical explanations, and empirical evidence. Annu Rev Ecol Evol Syst 36:47–79.

Grant V 1956 The influence of breeding habit on the outcome of natural hybridization in plants. Am Nat 90:319–322.

——— 1963 The origin of adaptations. Columbia University Press, New York.

Gustafsson A 1947 Apomixis in higher plants. Pt III. Biotype and species formation. Gleerup, Lund.

——— 1948 Polyploidy, life-form and vegetative reproduction. Hereditas 34:1–22.

Hamrick JL, RW Allard 1972 Microgeographical variation in allozyme frequencies in *Avena barbata*. Proc Natl Acad Sci USA 69:2100–2104.

Hecker RJ 1972 Inbreeding depression in diploid and autotetraploid sugarbeet, *Beta vulgaris* L. Euphytica 21:106–111.

Hedrick PW 1987 Genetic load and the mating system in homosporous ferns. Evolution 41:1282–1289.

Husband BC 2000 Constraints on polyploid evolution: a test of the minority cytotype exclusion principle. Proc R Soc B 267:217–223.

——— 2004 The role of triploid hybrids in the evolutionary dynamics of mixed-ploidy populations. Biol J Linn Soc 82:537–546.

Husband BC, DW Schemske 1995 Magnitude and timing of inbreeding depression in a diploid population of *Epilobium angustifolium* (Onagraceae). Heredity 75:206–215.

——— 1997 The effect of inbreeding in diploid and tetraploid populations of *Epilobium angustifolium* (Onagraceae): implications

for the genetic basis of inbreeding depression. Evolution 51:737–746.

Imam AG, RW Allard 1965 Population studies in predominantly self-pollinated species. VI. Genetic variability between and within natural populations of wild oats from differing habitats in California. Genetics 51:49–62.

Inoue K, M Masuda, M Maki 1998 Inbreeding depression and outcrossing rate in the endangered autotetraploid plant *Aster kantoensis* (Asteraceae). J Hered 89:559–562.

Johnston MO, DJ Schoen 1994 On the measurement of inbreeding depression. Evolution 48:1735–1741.

——— 1996 Correlated evolution of self-fertilization and inbreeding depression: an experimental study of nine populations of *Amsinckia* (Boraginaceae). Evolution 50:1478–1491.

Jones JS, ET Bingham 1957 Inbreeding depression in alfalfa and cross-pollinated crops. Plant Breed Rev 13:209–233.

Kempthorne O 1957 An introduction to genetic statistics. Wiley, New York.

Kephart SR, E Brown, J Hall 1999 Inbreeding depression and partial selfing: evolutionary implications of mixed-mating in a coastal endemic, *Silene douglasii* var. *oraria* (Caryophyllaceae). Heredity 82:543–554.

Knapp SJ, LA Tagliani, BH Liu 1991 Outcrossing rates of experimental populations of *Cuphea lanceolata*. Plant Breed 106:334–337.

Krueger SK, SJ Knapp 1991 Mating systems of *Cuphea laminuligera* and *Cuphea lutea*. Theor Appl Genet 82:221–226.

Lande R, DW Schemske 1985 The evolution of self-fertilization and inbreeding depression in plants. I. Genetic models. Evolution 39:24–40.

Lawrence WJ 1930 Incompatibility in polyploids. Genetica 12:269–296.

Levin DA 1975 Minority cytotype exclusion in local plant populations. Taxon 24:35–43.

——— 1983 Polyploidy and novelty in flowering plants. Am Nat 122:1–25.

——— 2002 The role of chromosomal change in plant evolution. Oxford University Press, New York.

Lewis D 1943 The incompatibility sieve for producing polyploids. J Genet 45:261–264.

Mable BK 2004 Polyploidy and self-compatibility: is there an association? New Phytol 162:803–811.

Mandal AK, RA Ennos, CW Fagg 1994 Mating system analysis in a natural population of *Acacia nilotica* subspecies *leiocarpa*. Theor Appl Genet 89:931–935.

Marshall DF, RJ Abbott 1982 Polymorphism for outcrossing frequency at the ray floret locus in *Senecio vulgaris* L. I. Evidence. Heredity 48:227–235.

Masterson J 1994 Stomatal size in fossil plants: evidence for polyploidy in majority of angiosperms. Science 264:421–423.

Masuyama S, Y Watano 1990 Trends for inbreeding in polyploid pteridophytes. Plant Species Biol 5:13–17.

Miller JS, DL Venable 2000 Polyploidy and the evolution of gender dimorphism in plants. Science 289:2335–2338.

Mosquin T 1966 A new taxonomy for *Epilobium angustifolium* L. (Onagraceae). Brittonia 18:167–188.

Murawski DA, TH Fleming, K Ritland, JL Hamrick 1994 Mating system of *Pachycereus pringlei*: an autotetraploid cactus. Heredity 72:86–94.

Otto SP, J Whitton 2000 Polyploid incidence and evolution. Annu Rev Genet 34:401–437.

Ozimec B 2006 Inbreeding depression and mating system evolution in the autotetraploid *Chamerion angustifolium* (Onagraceae). MS thesis. University of Guelph.

Parker IM, RR Nakamura, DW Schemske 1995 Reproductive allocation and the fitness consequences of selfing in two sympatric species of *Epilobium* (Onagraceae) with contrasting mating systems. Am J Bot 82:1007–1016.

Parsons PA 1959 Some problems in inbreeding and random mating in tetrasomics. Agron J 51:465–467.

Ramsey J, DW Schemske 1998 Pathways, mechanisms and rates of polyploid formation in flowering plants. Annu Rev Ecol Syst 29: 467–501.

——— 2002 Neopolyploidy in flowering plants. Annu Rev Ecol Syst 33:589–639.

Rausch JH, MT Morgan 2005 The effect of self-fertilization, inbreeding depression and population size on autopolyploid establishment. Evolution 59:1867–1875.

Ritland K 1990 A series of FORTRAN computer programs for estimating plant mating systems. J Hered 81:235–237.

Rodriguez DJ 1996 A model for the establishment of polyploidy in plants. Am Nat 147:33–46.

Ronfort J 1999 The mutation load under tetrasomic inheritance and its consequences for the evolution of the selfing rate in autotetraploid species. Genet Res 74:31–42.

Rosquist R 2001 Reproductive biology in diploid *Anthericum ramosum* and tetraploid *A. liliago* (Anthericaceae). Oikos 92:143–152.

Ross R 1981 Chromosome counts, cytology and reproduction in Cactaceae. Am J Bot 68:463–470.

SAS Institute 2002 Statistical discovery software, version 5.1. SAS Institute, Cary, NC.

Schoen DJ, JW Busch 2008 On the evolution of self-fertilization in a metapopulation. Int J Plant Sci 169:119–127.

Schoen DJ, MO Johnston, AM L'Heureux, JV Marsolais 1997 Evolutionary history of mating system in *Amsinckia* (Boraginaceae). Evolution 51:1090–1099.

Soltis DE, PS Soltis 1999 Polyploidy: recurrent formation and genome evolution. Trends Ecol Evol 14:348–352.

Soltis DE, PS Soltis, JA Tate 2003 Advances in the study of polyploidy since plant speciation. New Phytol 161:173–191.

Soltis PS, DE Soltis 2000 The role of genetic and genomic attributes in the success of polyploids. Proc Natl Acad Sci USA 97: 7051–7057.

Stebbins GL 1950 Variation and evolution in plants. Columbia University Press, New York.

——— 1957 Self fertilization and population variability in the higher plants. Am Nat 91:337–354.

——— 1971 Chromosomal evolution in higher plants. Arnold, London.

Sterk AA, L Dijkhuizen 1972 The relation between the genetic determination and the ecological significance of the seed wing in *Spergularia media* and *S. marina*. Acta Bot Neerlandica 21:481–490.

Thompson SL, K Ritland 2006 A novel mating system analysis for models of self-oriented mating applied to diploid and polyploid arctic Easter daisies (*Townsendia hookeri*). Heredity 97:119–126.

Tysdal HM, TA Kiesselbach, HL Westover 1942 Alfalfa breeding. University of Nebraska, College of Agriculture, Agriculture Experiment Station, Lincoln. 46 pp.

Vamosi JC, TA Dickinson 2006 Polyploidy and diversification: a phylogenetic investigation in Rosaceae. Int J Plant Sci 167:349–358.

Wolfe K 2001 Yesterday's polyploids and the mystery of diploidization. Nat Rev Genet 2:333–341.

Young AG, AHD Brown 1999 Paternal bottlenecks in fragmented populations of the grassland daisy *Rutidosis leptorrhynchoides*. Genet Res 73:111–117.

Zohary D, D Imber 1963 Genetic dimorphism in fruit types in *Aegilops speltoides*. Heredity 18:223–231.

Index

Entries in bold refer to pages containing a figure relating to the entry.